Extending Structures
Fundamentals and Applications

Monographs and Research Notes in Mathematics

Series Editors:

John A. Burns, Thomas J. Tucker, Miklos Bona, Michael Ruzhansky

About the Series

This series is designed to capture new developments and summarize what is known over the entire field of mathematics, both pure and applied. It will include a broad range of monographs and research notes on current and developing topics that will appeal to academics, graduate students, and practitioners. Interdisciplinary books appealing not only to the mathematical community, but also to engineers, physicists, and computer scientists are encouraged.

This series will maintain the highest editorial standards, publishing well-developed monographs as well as research notes on new topics that are final, but not yet refined into a formal monograph. The notes are meant to be a rapid means of publication for current material where the style of exposition reflects a developing topic.

Sturm-Liouville Problems
Theory and Numerical Implementation
Ronald B. Guenther, John W Lee

Analysis on Function Spaces of Musielak-Orlicz Type
Jan Lang, Osvaldo Mendez

Analytic Methods for Coagulation-Fragmentation Models, Volume I
Jacek Banasiak, Wilson Lamb, Philippe Laurencot

Analytic Methods for Coagulation-Fragmentation Models, Volume II
Jacek Banasiak, Wilson Lamb, Philippe Laurencot

Neutrices and External Numbers
A Flexible Number System
Dinis Bruno, Imme van den Berg

Extending Structures
Fundamentals and Applications
Ana Agore, Gigel Militaru

For more information about this series please visit: https://www.crcpress.com/Chapman--HallCRC-Monographs-and-Research-Notes-in-Mathematics/book-series/CRCMONRESNOT

Extending Structures

Fundamentals and Applications

Ana Agore
Institute of Mathematics of the Romanian Academy
Vrije Universiteit Brussel

Gigel Militaru
University of Bucharest

CRC Press
Taylor & Francis Group
Boca Raton London New York

CRC Press is an imprint of the
Taylor & Francis Group, an **informa** business
A CHAPMAN & HALL BOOK

CRC Press
Taylor & Francis Group
6000 Broken Sound Parkway NW, Suite 300
Boca Raton, FL 33487-2742

Printed on acid-free paper

International Standard Book Number-13: 978-0-8153-4784-2 (Hardback)

Visit the Taylor & Francis Web site at
http://www.taylorandfrancis.com

and the CRC Press Web site at
http://www.crcpress.com

Contents

Introduction

The classification up to an isomorphism of all *finite objects* of a certain type (e.g., all groups of a given finite order, all associative/Lie algebras of a given finite-dimension over a field k, etc.) is one of the famous and very difficult problems which are still open although it has attracted the attention of many mathematicians.

At the level of groups, the problem was first considered in 1854 by Cayley ([73]) who classified all groups whose order is at most 6. Later on, in 1881, Peirce ([200]) formulated the analogue problem in the context of associative algebras and initiated the classification of all associative algebras of dimension at most 3 over the field of complex numbers; this classification was completed in 1890 by Study ([215]).

In the same time, the problem was considered for Lie algebras by S. Lie himself ([162]) who classified all Lie algebras of dimension 3 over the field of complex numbers. The first important distinction between associative and Lie algebras was observed: even in dimension 3, there exists infinitely many Lie algebras and their bracket can be explicitly described (see, for instance [99, 141]). Although the classification problem was intensively studied ever since using various methods, including computer software, very few general results are known. For instance, in the case of groups we have complete classifications only up to order less than 2048; we refer to [82] for the number of types of groups of order less than 2048 and to [49] for a detailed account on the fascinating problem of classifying finite groups. For associative (resp. Lie) algebras there is even less information on the classification problem. Over an algebraically closed field of characteristic zero, associative algebras have been classified only up to dimension 5 by Mazzola [182] while, according to [201], Lie algebras have been classified completely up to dimension 7.

The study of the classification problem led to several subsequent problems, among which we mention the *extension problem* of Hölder ([133]) and the *factorization problem* of Ore ([192]). We recall them briefly as they constitute the starting point of what we have called the *extending structures problem*, the subject of this monograph. Let H and G be two given groups. Loosely speaking, the extension problem consists of describing and classifying all groups E containing H as a normal subgroup such that the quotient E/H is isomorphic to G. The precise statement of the problem is given in Chapter 1 while a comprehensive treatment of the problem is contained in [2, 210]. The factorization problem of Ore is, in a certain sense, the "dual" of the extension problem: it consists of describing and classifying up to isomorphism all groups E that *factorize* through two given groups H and G, i.e., E contains H and G as subgroups such that $E = HG$ and $H \cap G = \{1\}$. Surprisingly, although its statement is quite natural, the factorization problem turned out to be more difficult than the extension problem. For further details and background information on the two problems we refer to Chapter 1.

Both problems were first introduced in group theory and subsequently they were considered and intensively studied for many other mathematical objects such as associative/Lie algebras, quantum groups, Poisson algebras, etc. Although the two problems are completely different and they were studied independently, there is however a common point between them: namely, in both cases the group E contains H as a subgroup. Therefore the question of whether we can formulate a general problem which will unify the extension problem and the factorization problem arises naturally. The answer to this question is affirmative and the present monograph is concerned with the study of this problem which we called the *extending structures problem* or the ES problem for short. Formulated in the most general context which will allow us to approach it for various fields of research, the ES problem comes down to the following:

Extending structures problem. *Let H be a fixed object in a certain category $\mathcal{C}$ (for instance, the category of groups, associative algebras, Lie algebras etc.) and c a given cardinal. Describe and classify, up to an isomorphism that stabilizes H, all objects $C \in \mathcal{C}$ which contain and stabilize H as a subobject of codimension c.*

Obviously, when we formulate the problem in the concrete case of groups by codimension we mean the index of H in C while in the case of associative or Lie algebras the codimension is precisely the ordinary concept used in linear algebra. An isomorphism that stabilizes H is nothing but an isomorphism which acts as the identity on H.

The ES problem can be formulated in many equivalent ways: for instance, inspired by Lagrange's theorem, at the level of groups, the ES problem can be rephrased as follows (note that we replaced the fixed cardinal c by another fixed set E):

Let H be a group and E a set such that $H \subseteq E$. Describe and classify, up to an isomorphism that stabilizes H, the set of all group structures $\cdot$ that can be defined on E such that H is a subgroup of $(E, \cdot)$.

This monograph is dedicated to the study of the ES problem and its applications for five different categories, namely: groups, Leibniz algebras, Lie algebras, associative algebras and Jacobi algebras. Each of these mathematical objects has its own dedicated chapter which can be read independently of the others.

The monograph is structured as follows. The first section, called Generalities, sets the notation and recalls some basic concepts which will be used throughout the five chapters. Chapter 1 is dedicated to the study of the ES problem at the level of groups. We start by recalling both the extension and the factorization problem and the objects used to approach them, namely: the crossed and the bicrossed product of groups. In particular, we revisit two classical results concerning the aforementioned problems: the theorems of Schreier and Takeuchi as they served as a model for our approach of the ES problem. The first important results of this chapter are given in Section 1.2: first we construct a new product called the *unified product*, which unifies and generalizes both the crossed and the bicrossed product of groups. We establish the necessary and sufficient axioms for the construction of the unified product in Theorem 1.2.1. This new product is responsible for the classification part of the ES problem: any group structure on the set E which contains H as a subgroup is proved to be isomorphic to a certain unified product. The answer to the classification part of the ES problem is given in Theorem 1.2.16: the set of all group structures $\cdot$ that can be defined on E such that H becomes a subgroup of $(E, \cdot)$ are classified up to an isomorphism of groups that stabilizes H by a non-abelian cohomological type object which is explicitly constructed. Moreover, as an application, a general Schreier type theorem for unified products is proven. The most important application is contained in Section 1.3 where we consider a question related to the ES problem called the *classifying complements problem* or the CCP problem for short:

Classifying complements problem. *Let A be a subgroup of G. If an A-complement of G (that is, a subgroup $H \leq G$ such that $G = AH$ and $A \cap H = \{1\}$) exists, describe explicitly, classify all A-complements of G and compute the factorization index $[G : A]^f$ (that is, the number of isomorphism types of all A-complements of G).*

In the case of abelian groups the classifying complements problem is trivial: if a subgroup A of an abelian group G has a complement then it can be easily seen that this complement is unique up to isomorphism. The situation changes radically in the case of non-abelian groups, where the problem is far from being trivial and has very important consequences. As an argument in this direction we mention that the factorization index of the canonical inclusion $S_{n-1} \subseteq S_n$ (where S_n is the group of permutations on n letters) is equal to $g(n)$, the number of types of isomorphisms of groups of order n. We provide the full answer to the classifying complements problem in three steps called deformation of complements, description of complements and classification of complements. The main result of this section is Theorem 1.3.11 which shows that there exists a bijection between the set of isomorphism types of all A-complements of G and a new cohomological type object which is explicitly constructed. In particular, an explicit formula is provided for computing the factorization index $[G : A]^f$. Section 1.4 contains the main application of the theory previously developed. More precisely, by applying our results to the factorization $S_n = S_{n-1}C_n$, where C_n is the cyclic group of order n, we obtain a combinatorial formula for computing the number of isomorphism types of all groups of order n which arises from a minimal set of data: the factorization $S_n = S_{n-1}C_n$.

Chapter 2 is dedicated to the study of both the ES problem and the CCP problem for Leibniz algebras. The theory we developed in the first chapter can be transposed, by "linearizing" the constructions, into the new context of Leibniz algebras with the necessary changes. In this context we introduce a new general product called the unified product for Leibniz algebras, which will play a key role in solving the ES problem. Let $\mathfrak{g}$ be a Leibniz algebra and E a vector space containing $\mathfrak{g}$ as a subspace. It is proved that any Leibniz algebra structure on E which contains $\mathfrak{g}$ as a subalgebra is isomorphic to a unified product $\mathfrak{g} \ltimes V$, where V is a fixed vector space of dimension equal to the codimension of $\mathfrak{g}$ in E. The main result of this chapter, which provides the theoretical answer to the ES problem, is given in Theorem 2.1.9: we construct explicitly a cohomological type object which will be the classifying object of all extending structures of a given Leibniz algebra $\mathfrak{g}$ to a given vector space E—the classification is given up to an isomorphism of Leibniz algebras which stabilizes $\mathfrak{g}$.

Two special cases of the unified product are presented in detail, namely the crossed and the bicrossed product (we kept the terminology used in the case of groups) of Leibniz algebras which, exactly as in group theory, play a key role in solving the extension problem and respectively the factorization problem for Leibniz algebras. Moreover, we introduced a new class of Leibniz algebras called *flag Leibniz algebras*: these are finite-dimensional Leibniz algebras E which admit a sequence of Leibniz subalgebras of the form $E_0 := 0 \subset E_1 \subset \cdots \subset E_m := E$, such that E_i has codimension 1 in E_{i+1}, for all $i = 0, \cdots, m-1$. Section 2.2 shows the efficiency of our theoretical results in classifying flag Leibniz algebra. Section 2.4 deals with the classifying complements problem, this time at the level of Leibniz algebras. The strategy developed for groups can be transferred to this context providing a complete answer to the CCP problem in Theorem 2.4.5. The last section of the chapter is devoted to the proof of Ito's theorem in the context of Leibniz algebras.

Chapter 3 is dedicated to the study of the same two problems in the context of Lie algebras. Since the latter are special cases of Leibniz algebras, the results of the first three sections of this chapter can be deduced from the corresponding results for Leibniz algebras proved in Chapter 2. We point out that given the antisymmetry of the bracket of a Lie algebra, the input data for the construction of unified products is significantly reduced, two of the actions of an extending datum being trivial in this case. These considerations allow us to provide several explicit examples in Section 3.4 and Section 3.5 respectively. Among the important results of these sections we mention the explicit description of all Lie algebras $\mathfrak{L}$ containing a given Lie algebra $\mathfrak{h}$ as a subalgebra of codimension 1 which are proved to be parameterized by the space $\mathrm{TwDer}(\mathfrak{h})$ of what we have called "twisted derivations" of $\mathfrak{h}$. As an application, the automorphism group of this type of Lie algebras is explicitly described. Several other examples are presented in detail. Finally, in Section 3.6 we take the first steps towards developing a Galois type theory for Lie algebra extensions, analogous to the classical Galois theory for fields. If $\mathfrak{g} \subseteq \mathfrak{h}$ is an extension of Lie algebras, we give an explicit description of the Galois group $\mathrm{Gal}\,(\mathfrak{h}/\mathfrak{g})$ as a subgroup of the canonical semidirect product $\mathrm{GL}_k(V) \rtimes \mathrm{Hom}_k(V, \mathfrak{g})$ of groups, where V is a vector space that measures the codimension of $\mathfrak{g}$ in $\mathfrak{h}$, i.e., the 'degree' of the extension $\mathfrak{h}/\mathfrak{g}$. The counterpart of Artin's Theorem for Lie algebras is also proved: if G is a finite group of invertible order in the field k acting on a Lie algebra $\mathfrak{h}$, then the Lie algebra $\mathfrak{h}$ is reconstructed as a *"skew crossed product"* $\mathfrak{h} \cong \mathfrak{h}^G \#^{\bullet} V$ between the Lie subalgebra of invariants $\mathfrak{h}^G$ and the kernel V of the Reynolds operator $t : \mathfrak{h} \to \mathfrak{h}^G$. The Galois group $\mathrm{Gal}\,(\mathfrak{h}/\mathfrak{h}^G)$ is also described and we show that the group $\mathrm{Gal}\,(\mathfrak{h}/\mathfrak{h}^G)$ is different from G, as opposed to the classical Galois theory of fields where the two groups coincide. The next step proves a version of Hilbert's 90 Theorem for Lie algebras: if G is a cyclic group then the kernel of the Reynolds operator $t : \mathfrak{h} \to \mathfrak{h}^G$ is determined. As an application we show that if $\mathfrak{g} \subseteq \mathfrak{h}$ is a Lie subalgebra of codimension 1 in $\mathfrak{h}$, then the Galois group $\mathrm{Gal}\,(\mathfrak{h}/\mathfrak{g})$ is metabelian (in particular, solvable). Based on this, the Lie algebra counterpart of the concept of a radical extension of fields is introduced. As in the classical Galois theory, we prove that the Galois group $\mathrm{Gal}\,(\mathfrak{h}/\mathfrak{g})$ of a radical extension $\mathfrak{g} \subseteq \mathfrak{h}$ of finite-dimensional Lie algebras is a solvable group. Several other applications and concrete examples of Galois groups are presented.

Chapter 4 is dedicated to the same two problems, namely the ES problem and the classification of complements, this time for unital associative algebras. Mutatis mutandis, all results proved in Chapter 2 for Leibniz algebras can be transposed at the level of associative algebras, with the obvious changes imposed by the new context. Therefore, instead of listing here all results contained in this chapter, we only point out that these results are used in the next chapter in approaching the same two problems in the more general setting of Jacobi and Poisson algebras.

Chapter 5 deals with *Jacobi* (in particular *Poisson*) *algebras*. A Jacobi algebra is a commutative associative algebra A with a Lie algebra bracket such that the following compatibility is fulfilled: $[ab,\, c] = a\,[b,\, c] + [a,\, c]\,b - ab\,[1_A,\, c]$, for all $a, b, c \in A$. A Poisson algebra is a Jacobi algebra A such that $[1_A,\, a] = 0$, for all $a \in A$. Jacobi (resp. Poisson) algebras are algebraic counterparts of Jacobi (resp. Poisson) manifolds: a smooth manifold M is a Jacobi (resp. Poisson) manifold if and only if the algebra $A := C^{\infty}(M)$ of real smooth functions on M is a Jacobi (resp. Poisson) algebra. Poisson algebras are very important and interesting objects since they appear in various research areas situated at the border between mathematics and physics. The first steps towards the classification of low dimensional Poisson manifolds, which is a very difficult task, are taken in [122, 159] using mainly differential geometry tools. The classification of finite-dimensional Poisson algebras is equally difficult being the algebraic counterpart of the classification problem in differential geometry. To illustrate this, Chapter 5 begins with the classification of all complex Jacobi algebras of dimension 2 or 3. Frobenius Jacobi algebras are introduced and characterized by using what we have called, inspired by Hopf algebra theory, an *integral* of a Jacobi algebra. The main sections of this chapter deal with the two problems subsequent to the classification problem for finite-dimensional Poisson/Jacobi algebras, namely the ES problem and the classification of complements. Using the constructions previously introduced in Chapter 3 (resp. 4) for Lie algebras (resp. associative algebras) we are able to construct the unified product $A \ltimes V$ associated to a Jacobi algebra A and a vector space V which are connected through four *actions* and two *cocycles.* As expected, the unified product is the object used in approaching the ES problem. Theorem 5.2.7 provides the theoretical answer to the ES problem: for a given Jacobi algebra A and a vector space V, a non-abelian cohomological type object $\mathcal{JH}^2\,(V,\,A)$ is explicitly constructed; it parameterizes and classifies all Jacobi algebras containing A as a subalgebra of codimension equal to $\dim(V)$. Several examples for computing $\mathcal{JH}^2\,(V,\,A)$ are provided in the case of flag extensions. The last part of this chapter is concerned with the classifying complements problem for arbitrary extensions $P \subset R$ of Poisson algebras. The main result is Theorem 5.4.9 which proves that there exists a bijection between the isomorphism classes of all P-complements of R and a new cohomological object whose construction is made explicit. A formula for computing the factorization index $[R : P]^f$ is provided as well. Several examples are also included: in particular, an extension $P \subseteq R$ of Poisson algebras such that P has infinitely many non-isomorphic complements in R is constructed.

Finally, we mention that the approach used in this monograph for solving the ES problem (resp. the classifying complements problem) can constitute a source of inspiration for considering the problem in other categories. The first steps toward this goal were taken for Lie conformal algebras in [136] and for associative conformal algebras (resp. left-symmetric algebras) in [134, 135]. Furthermore, a special case of the ES problem for Hopf algebras (more precisely, corresponding to the forgetful functor from the category of Hopf algebras to the category of coalgebras) was considered in [15].

Acknowledgement The first author was supported by a grant of Romanian Ministery of Research and Innovation, CNCS - UEFISCDI, project number PN-III-P1-1.1-TE-2016-0124, within PNCDI III and is a fellow of FWO (Fonds voor Wetenschappelijk Onderzoek—Flanders).

Generalities: Basic notions and notation

For a family of sets $(X_i)_{i\in I}$ we shall denote by $\sqcup_{i\in I} X_i$ their coproduct in the category of sets, i.e., $\sqcup_{i\in I} X_i$ is the disjoint union of the X_i's.

Throughout, k will be a field whose group of units is denoted by k^*. All vector spaces, associative/Lie/Leibniz/Poisson/Jacobi algebras, linear or bilinear maps, tensor products and so on are over k. A linear homomorphism $f : V \to W$ between two vector spaces is called the *trivial map* if $f(v) = 0$, for all $v \in V$. $V^* = \mathrm{Hom}_k(V, k)$ and $\mathrm{End}_k(V)$ denote the dual, respectively the endomorphisms ring of a vector space V. If $\mathfrak{g} \leq E$ is a subspace in a vector space E, another subspace V of E such that $E = \mathfrak{g} + V$ and $V \cap \mathfrak{g} = 0$ is called a *complement of* $\mathfrak{g}$ *in* E. Such a complement is unique up to an isomorphism and its dimension is called the *codimension of* $\mathfrak{g}$ *in* E. For two vector spaces V and W we denote by $\mathrm{Hom}_k(V, W)$ the abelian group of all linear maps from V to W and by $\mathrm{GL}_k(V) := \mathrm{Aut}_k(V)$ the group of all linear automorphisms of V; if V has dimension m over k then $\mathrm{GL}_k(V)$ is identified with the general linear group $\mathrm{GL}(m, k)$ of all $m \times m$ invertible matrices over k. As usual, $\mathrm{SL}(m, k)$ stands for the special linear group of degree m over k which is the normal subgroup of $\mathrm{GL}(m, k)$ consisting of all $m \times m$ matrices of determinant 1.

We briefly recall below the construction of the semidirect product of groups using the right-hand side convention. Let G and H be two groups and $\triangleleft : H \times G \to H$ a right action as automorphisms of the group G on the group H, i.e., the following compatibility conditions hold for all g, $g' \in G$ and h, $h' \in H$:

$$h \triangleleft 1 = h, \qquad h \triangleleft (gg') = (h \triangleleft g) \triangleleft g' \qquad (hh') \triangleleft g = (h \triangleleft g)(h' \triangleleft g)$$

The associated semidirect product $G \ltimes H$ is the group structure on $G \times H$ with multiplication given for any g, $g' \in G$ and h, $h' \in H$ by:

$$(g, h) \cdot (g', h') := \big(gg', (h \triangleleft g')h'\big) \tag{0.1}$$

Let V and W be two vector spaces. Then there exists a canonical right action as automorphisms of the group $\mathrm{GL}_k(V)$ on the abelian group $(\mathrm{Hom}_k(V, W), +)$ given for any $r \in \mathrm{Hom}_k(V, W)$ and $\sigma \in \mathrm{GL}_k(V)$ by:

$$\triangleleft : \mathrm{Hom}_k(V, W) \times \mathrm{GL}_k(V) \to \mathrm{Hom}_k(V, W), \qquad r \triangleleft \sigma := r \circ \sigma$$

We shall denote by $\mathbb{G}_W^V := \mathrm{GL}_k(V) \ltimes \mathrm{Hom}_k(V, W)$ the corresponding semidirect product, i.e., $\mathbb{G}_W^V := \mathrm{GL}_k(V) \times \mathrm{Hom}_k(V, W)$, with the multiplication given for any σ, $\sigma' \in \mathrm{GL}_k(V)$ and r, $r' \in \mathrm{Hom}_k(V, W)$ by:

$$(\sigma, r) \cdot (\sigma', r') := (\sigma \circ \sigma', r \circ \sigma' + r') \tag{0.2}$$

The unit of the group $\mathbb{G}_W^V$ is $(\mathrm{Id}_V, 0)$. Moreover, $\mathrm{GL}_k(V) \cong \mathrm{GL}_k(V) \times \{0\}$ is a subgroup of $\mathbb{G}_W^V$ and the abelian group $\mathrm{Hom}_k(V, W) \cong \{\mathrm{Id}_V\} \times \mathrm{Hom}_k(V, W)$ is a normal subgroup of $\mathbb{G}_W^V$. The relation $(\sigma, r) = (\sigma, 0) \cdot (\mathrm{Id}_V, r)$ gives an exact factorization $\mathbb{G}_W^V = \mathrm{GL}_k(V) \cdot \mathrm{Hom}_k(V, W)$ of the group $\mathbb{G}_W^V$ through the subgroup $\mathrm{GL}_k(V)$ and the abelian normal subgroup $\mathrm{Hom}_k(V, W)$. Being a semidirect product, the group $\mathbb{G}_W^V$ is a split extension of $\mathrm{GL}_k(V)$ by the abelian group $\mathrm{Hom}_k(V, W)$; that is, it fits into an exact sequence of groups $0 \to \mathrm{Hom}_k(V, W) \to \mathbb{G}_W^V \to \mathrm{GL}_k(V) \to 1$ and the canonical projection $\mathbb{G}_W^V \to \mathrm{GL}_k(V) \to 1$ has a section that is a morphism of groups. The group $\mathbb{G}_W^V$ constructed above will play the key role in describing the Galois group of an arbitrary extension of Lie algebras. If $V \cong k$ is a 1-dimensional vector space, then the group $\mathbb{G}_W^k$ identifies with the semidirect product $k^* \rtimes W$ of the multiplicative group of units $(k^*, \cdot)$ with the abelian group $(W, +)$ and will be denoted simply by $\mathbb{G}_W$. The multiplication on $\mathbb{G}_W = k^* \rtimes W$ is given for any u, $u' \in k^*$ and x, $x' \in W$ by:

$$(u, x) \cdot (u', x') := (uu', u'x + x') \tag{0.3}$$

The non-abelian group $\mathbb{G}_W$ is an extension of the abelian group k^* by the abelian group $W = (W, +)$; hence, $\mathbb{G}_W$ is a metabelian group (i.e., the derived subgroup $[\mathbb{G}_W, \mathbb{G}_W]$ is abelian). In particular, $\mathbb{G}_W$ is a 2-step solvable group. On the other hand, if $W \cong k$ is a 1-dimensional vector space then $\mathbb{G}_k^V = \mathrm{GL}_k(V) \rtimes V^*$, and for finite-dimensional vector spaces V the group can be identified with the general affine group $\mathrm{Aff}\,(V) = \mathrm{GL}_k(V) \rtimes V$.

Unless otherwise stated, by an algebra A we mean an associative and unitary algebra; the unit of A will be denoted by 1_A while the multiplication m_A is denoted by juxtaposition $m_A(a, b) = ab$. However, whenever the algebras are not unitary it will be explicitly mentioned. For an algebra A, we shall denote by ${}_A\mathcal{M}_A$ the category of all A-bimodules, i.e., triples $(V, \rightharpoonup, \leftharpoonup)$ consisting of a vector space V and two bilinear maps $\rightharpoonup\colon A \times V \to V$, $\leftharpoonup\colon V \times A \to V$ such that $(V, \rightharpoonup)$ is a left A-module, $(V, \leftharpoonup)$ is a right A-module and $a \rightharpoonup (x \leftharpoonup b) = (a \rightharpoonup x) \leftharpoonup b$, for all a, $b \in A$ and $x \in V$. All algebra maps preserve units and any left/right A-module is unital. $\mathrm{Alg}\,(A, k)$ denotes the space of all algebra maps $A \to k$ and $\mathrm{Aut}_{\mathrm{Alg}}(A)$ the group of algebra automorphisms of A. If $(V, \rightharpoonup, \leftharpoonup) \in {}_A\mathcal{M}_A$, then the *trivial extension* of A by V is the algebra $A \times V$, with the multiplication given for any a, $b \in A$, x, $y \in V$ by:

$$(a, x) \cdot (b, y) := (ab,\ a \rightharpoonup y + x \leftharpoonup b)$$

A *Lie algebra* is a vector space $\mathfrak{g}$, together with a bilinear map $[-, -] : \mathfrak{g} \times \mathfrak{g} \to \mathfrak{g}$ called a bracket, satisfying the following two properties:

$$[g, g] = 0, \qquad [g, [h, l]] + [h, [l, g]] + [l, [g, h]] = 0$$

for all $g, h, l \in \mathfrak{g}$. The second condition is called the *Jacobi identity*. For two given Lie algebras $\mathfrak{g}$ and $\mathfrak{h}$ we denote by $\mathrm{Aut}_{\mathrm{Lie}}(\mathfrak{g})$ the group of automorphisms of $\mathfrak{g}$ and by $\mathrm{Hom}_{\mathrm{Lie}}(\mathfrak{g}, \mathfrak{h})$ the space of all Lie algebra homomorphisms between $\mathfrak{g}$ and $\mathfrak{h}$. Let $\mathfrak{g}$ be a Lie algebra and $\mathfrak{g}' := [\mathfrak{g}, \mathfrak{g}]$ be the *derived algebra* of $\mathfrak{g}$; $\mathfrak{g}$ is called *perfect* if $\mathfrak{g}' = \mathfrak{g}$ and *abelian* if $\mathfrak{g}' = 0$. The abelian Lie algebra of dimension n will be denoted by k_0^n. Furthermore, $\mathfrak{gl}(m, k)$ (resp. $\mathfrak{sl}(m, k)$) stands for the general (resp. special) linear Lie algebra of all $m \times m$ matrices (resp. all $m \times m$ matrices of trace 0) having the bracket $[A, B] := AB - BA$.

Representations of a Lie algebra $\mathfrak{g}$ will be viewed as modules over $\mathfrak{g}$; moreover, we shall work with both concepts of right and left $\mathfrak{g}$-modules. Explicitly, a *right* $\mathfrak{g}$*-module* is a vector space V together with a bilinear map $\triangleleft : V \times \mathfrak{g} \to V$, called a right action of $\mathfrak{g}$ on V, satisfying the following compatibility

$$x \triangleleft [g, h] = (x \triangleleft g) \triangleleft h - (x \triangleleft h) \triangleleft g \tag{0.4}$$

for all $x \in V$ and $g, h \in \mathfrak{g}$. A *left* $\mathfrak{g}$*-module* is a vector space V together with a bilinear map $\triangleright : \mathfrak{g} \times V \to V$, called a left action of $\mathfrak{g}$ on V such that:

$$[g, h] \triangleright x = g \triangleright (h \triangleright x) - h \triangleright (g \triangleright x) \tag{0.5}$$

for all $g, h \in \mathfrak{g}$ and $x \in V$. Any right $\mathfrak{g}$-module is a left $\mathfrak{g}$-module via $g \triangleright x := -x \triangleleft g$ and vice versa, that is the category of right $\mathfrak{g}$-modules is isomorphic to the category of left $\mathfrak{g}$-modules and both of them are isomorphic to the category of representations of $\mathfrak{g}$. The category of right Lie $\mathfrak{g}$-modules will be denoted by $\mathcal{LM}^{\mathfrak{g}}$.

A Lie algebra $\mathfrak{h}$ is called *self-dual* (or *metric*)([183]) if there exists a non-degenerate invariant bilinear form $B : \mathfrak{h} \times \mathfrak{h} \to k$, i.e., $B([a, b], c) = B(a, [b, c])$, for all $a, b, c \in \mathfrak{h}$. Self-dual Lie algebras generalize finite-dimensional complex semisimple Lie algebras (the second Cartan's criterion shows that any finite-dimensional complex semisimple Lie algebra is self-dual since its Killing form is non-degenerate and invariant). Besides the mathematical interest in studying self-dual Lie algebras, they are also important and have been intensively studied in physics [107, 197].

$\mathrm{Der}(\mathfrak{g})$ denotes the *Lie algebra of all derivations of* $\mathfrak{g}$, that is all linear maps $D : \mathfrak{g} \to \mathfrak{g}$ such that

$$D([g, h]) = [D(g), h] + [g, D(h)] \tag{0.6}$$

for all $g, h \in \mathfrak{g}$. $\mathrm{Der}(\mathfrak{g})$ is a Lie algebra with the bracket $[D_1, D_2] := D_1 \circ D_2 - D_2 \circ D_1$ and the map

$$\mathrm{ad} : \mathfrak{g} \to \mathrm{Der}(\mathfrak{g}), \quad \mathrm{ad}(g) := [g, -] : \mathfrak{g} \to \mathfrak{g}, \quad h \mapsto [g, h]$$

is called the *adjoint representation* of $\mathfrak{g}$. Then, $\mathrm{Ker}(\mathrm{ad}) = Z(\mathfrak{g})$, the center of $\mathfrak{g}$, and $\mathrm{Im}(\mathrm{ad})$ is called the *space of inner derivations* of $\mathfrak{g}$ and will be denoted by $\mathrm{Inn}(\mathfrak{g})$. $\mathrm{Inn}(\mathfrak{g})$ is a Lie ideal in $\mathrm{Der}(\mathfrak{g})$ and

$$\mathrm{Out}(\mathfrak{g}) := \mathrm{Der}(\mathfrak{g})/\mathrm{Inn}(\mathfrak{g})$$

is the Lie algebra of *outer derivations* of $\mathfrak{g}$. If $\mathfrak{g}$ is semisimple, then $\mathfrak{g}$ is perfect, $\mathrm{Inn}(\mathfrak{g}) = \mathrm{Der}(\mathfrak{g})$ and $Z(\mathfrak{g}) = 0$ ([138]). A Lie algebra $\mathfrak{h}$ is called *complete* (see [143, 216] for examples and structural results on this class of Lie algebras) if $\mathfrak{h}$ has a trivial center and any derivation is inner. A complete and perfect Lie algebra is called *sympathetic* [48]: semisimple Lie algebras over a field of characteristic zero are sympathetic and there exists a sympathetic non-semisimple Lie algebra in dimension 25.

For a given Lie algebra $\mathfrak{g}$ we denote by $\mathfrak{H}(\mathfrak{g})$ the *holomorph* Lie algebra [216] of $\mathfrak{g}$, i.e., $\mathfrak{H}(\mathfrak{g}) = \mathfrak{g} \times \mathrm{Der}(\mathfrak{g})$ endowed with the Lie bracket given by: $[(g, \varphi), (h, \psi)] = \{[g, h] + \varphi(h) - \psi(g), [\varphi, \psi]\}$, for all $g, h \in \mathfrak{g}$ and $\varphi, \psi \in \mathrm{Der}(\mathfrak{g})$.

For unexplained concepts pertaining to Lie algebra theory we refer to [138].

Leibniz algebras were introduced by Bloh [51] under the name of D-algebras and rediscovered later on by Loday [166] as non-commutative generalizations of Lie algebras. A Leibniz algebra is a vector space $\mathfrak{g}$, together with a bilinear map $[-, -] : \mathfrak{g} \times \mathfrak{g} \to \mathfrak{g}$ satisfying the *Leibniz identity*, that is:

$$[g, [h, l]] = [[g, h], l] - [[g, l], h] \tag{0.7}$$

for all $g, h, l \in \mathfrak{g}$. Any Lie algebra is a Leibniz algebra, and a Leibniz algebra satisfying $[g, g] = 0$, for all $g \in \mathfrak{g}$ is a Lie algebra. The typical example of a Leibniz algebra is the following [166]: let $\mathfrak{g}$ be a Lie algebra, $(M, \triangleleft)$ a right $\mathfrak{g}$-module and $\mu : M \to \mathfrak{g}$ a $\mathfrak{g}$-equivariant map, i.e., $\mu(m \triangleleft g) = [\mu(m), g]$, for all $m \in M$ and $g \in \mathfrak{g}$. Then M is a Leibniz algebra with the bracket $[m, n]_{(\triangleleft, \mu)} := m \triangleleft \mu(n)$, for all $m, n \in M$. Another important example was constructed in [156]: if $\mathfrak{g}$ is a Lie algebra, then $\mathfrak{g} \otimes \mathfrak{g}$ is a Leibniz algebra with the bracket given by $[x \otimes y, a \otimes b] := [x, [a, b]] \otimes y + x \otimes [y, [a, b]]$, for all $x, y, a, b \in \mathfrak{g}$. For other interesting examples of Leibniz algebras we refer to [167].

A subspace $I \leq \mathfrak{g}$ of a Leibniz algebra $\mathfrak{g}$ is called a *two-sided ideal* of $\mathfrak{g}$ if $[x, g] \in I$ and $[g, x] \in I$, for all $x \in I$ and $g \in \mathfrak{g}$. $\mathfrak{g}$ is called *perfect* if $[\mathfrak{g}, \mathfrak{g}] = \mathfrak{g}$ and *abelian* if $[\mathfrak{g}, \mathfrak{g}] = 0$. By $Z(\mathfrak{g})$ we shall denote the center of $\mathfrak{g}$, that is the two-sided ideal consisting of all $g \in \mathfrak{g}$ such that $[g, x] = [x, g] = 0$, for all $x \in \mathfrak{g}$. As in the case of Lie algebras, we denote by $\mathrm{Der}(\mathfrak{g})$ the *space of all derivations* of $\mathfrak{g}$, that is, all linear maps $\Delta : \mathfrak{g} \to \mathfrak{g}$ satisfying (0.6). For two subspaces A and B of a Leibniz algebra $\mathfrak{g}$ we denote by $[A, B]$ the vector space generated by all brackets $[a, b]$, for any $a \in A$ and $b \in B$. In particular, $\mathfrak{g}' := [\mathfrak{g}, \mathfrak{g}]$ is called the *derived subalgebra* of $\mathfrak{g}$. A Leibniz or a Lie algebra $\mathfrak{g}$ is called *metabelian* if $\mathfrak{g}'$ is an abelian subalgebra of $\mathfrak{g}$, i.e., $[[\mathfrak{g}, \mathfrak{g}], [\mathfrak{g}, \mathfrak{g}]] = 0$.

An algebra A is called a *Frobenius* algebra if $A \cong A^*$ as right A-modules, where A^* is viewed as a right A-module via $(a^* \cdot a)(b) := a^*(ab)$, for all $a^* \in A^*$ and $a, b \in A$. For the basic theory of Frobenius algebras we refer to [145]. The Lie algebra counterpart of Frobenius algebras was studied under different names such as *self-dual*, *metric* or Lie algebras having a non-degenerate invariant bilinear form. Throughout we will call them Frobenius

Lie algebras: a *Frobenius Lie* algebra is a Lie algebra $\mathfrak{g}$ such that $\mathfrak{g} \cong \mathfrak{g}^*$ as right Lie $\mathfrak{g}$-modules, where $\mathfrak{g}$ and $\mathfrak{g}^*$ are right Lie $\mathfrak{g}$-modules via the canonical actions: $b \leftharpoonup a := [b, a]$ and $(a^* \curvearrowleft a)(b) := a^*\big([a, b]\big)$, for all $a, b \in \mathfrak{g}$ and $a^* \in \mathfrak{g}^*$. We can easily see that a Lie algebra $\mathfrak{g}$ is Frobenius if and only if there exists a non-degenerate invariant bilinear form $B : \mathfrak{g} \times \mathfrak{g} \to k$, i.e., $B([a, b], c) = B(a, [b, c])$, for all $a, b, c \in \mathfrak{g}$. In light of this reformulation, the second Cartan's criterion shows that any finite-dimensional complex semisimple Lie algebra is Frobenius since its Killing form is non-degenerate and invariant. Besides the mathematical interest in studying Frobenius Lie algebras [147, 183], they are also important and have been intensively studied in physics [107, 197], in particular for the construction of Wess-Zumino-Novikov-Witten models. The property of being Frobenius reflects a certain natural symmetry: for instance, a functor $F : \mathcal{C} \to \mathcal{D}$ is called Frobenius [65] if F has the same left and right adjoint functor. This idea will be used in Section 5.1 where we introduce the notion of Frobenius Jacobi algebras.

A *Poisson* algebra is a triple $A = (A, m_A, [-, -])$, where (A, m_A) is a (not necessarily unital) commutative algebra, $(A, [-, -])$ is a Lie algebra such that the Leibniz law

$$[ab, c] = a\,[b, c] + [a, c]\,b$$

holds for any $a, b, c \in A$. For further details concerning the study of Poisson algebras arising from differential geometry see [159] and the references therein. If a Poisson algebra A has a unit 1_A, then by taking $a = b = 1_A$ in the Leibniz law we obtain that $[1_A, c] = [c, 1_A] = 0$, for all $c \in A$. Any non-unital Poisson algebra embeds into a unital Poisson algebra.

A (right) *Poisson A-module* [169, 232] is a vector space V equipped with two bilinear maps $\triangleleft : V \times A \to V$ and $\leftharpoonup : V \times A \to V$ such that $(V, \triangleleft)$ is a right A-module, $(V, \leftharpoonup)$ is a right Lie A-module satisfying the following two compatibility conditions for any $a, b \in A$ and $x \in V$:

$$x \leftharpoonup (ab) = (x \leftharpoonup a) \triangleleft b + (x \leftharpoonup b) \triangleleft a, \quad x \triangleleft [a, b] = (x \triangleleft a) \leftharpoonup b - (x \leftharpoonup b) \triangleleft a \tag{0.8}$$

We denote by $\mathcal{PM}_A^A$ the category of right Poisson A-modules having as morphisms all linear maps which are compatible with both actions.

A *Jacobi algebra* is a quadruple $A = (A, m_A, 1_A, [-, -])$, where $(A, m_A, 1_A)$ is a unital algebra, $(A, [-, -])$ is a Lie algebra such that for any $a, b, c \in A$:

$$[ab, c] = a\,[b, c] + [a, c]\,b - ab\,[1_A, c] \tag{0.9}$$

Any unital Poisson algebra is a Jacobi algebra. Any algebra A is a Jacobi algebra with the trivial bracket $[a, b] = 0$, for all $a, b \in A$; such a Jacobi algebra will be called *abelian* and will be denoted by A_0. A morphism between two Jacobi (resp. Poisson) algebras A and B is a linear map $\varphi : A \to B$ which is both a morphism of algebras as well as a morphism of Lie algebras. We denote by ${}_k\mathrm{Jac}$ (resp. ${}_k\mathrm{Poss}$) the category of Jacobi (resp. unitary Poisson) algebras over a field k. A *Jacobi ideal* of a Jacobi algebra A is a linear subspace I which is both an ideal with respect to the associative product as well as a Lie ideal of A. If I is a Jacobi ideal of A then A/I inherits a Jacobi algebra structure in the obvious way.

Chapter 1

Extending structures: The group case

The present chapter is the starting point for a study concerning what we have called the *extending structures problem* or the ES problem for short. At the level of groups the ES problem has a very natural statement:

Let H be a group and E a set such that $H \subseteq E$. Describe and classify up to an isomorphism that stabilizes H the set of all group structures $\cdot$ that can be defined on E such that H is a subgroup of $(E, \cdot)$.

In other words, the ES problem is trying to provide an answer to the very natural question: to what extent a group structure on H can be extended beyond H to a bigger set which contains H as a subset in such a way that H would become a subgroup within the new structure. The ES problem generalizes and unifies two famous problems in the theory of groups which served as model for our approach: the *extension problem* of Hölder [133] and the *factorization problem* of Ore [192]. Let us explain this briefly. Consider two groups H and G. The extension problem of Hölder consists of describing and classifying all groups E containing H as a normal subgroup such that $E/H \cong G$. An important step related to the extension problem was made by Schreier: any extension E of G by H is equivalent to a crossed product extension (see Theorem 1.1.6 below). For more details and references on the extension problem we refer to the monographs [2], [210]. The factorization problem is a "dual" of the extension problem and it was formulated by Ore [192]. It consists of describing and classifying up to an isomorphism all groups E that factorize through two given groups H and G: i.e., E contains H and G as subgroups such that $E = HG$ and $H \cap G = \{1\}$. The dual version of Schreier's theorem was proven by Takeuchi [224]: the bicrossed product associated to a matched pair of groups $(H, G, \triangleleft, \triangleright)$ was constructed and it was proven that a group E factorizes through H and G if and only if E is isomorphic to a bicrossed product $H \bowtie G$ (see Proposition 1.1.9). The factorization problem is even more difficult than the more popular extension problem and little progress has been made since then. For instance, in the case of two cyclic groups H and G, not both finite, the problem was started by L. Rédei in [208] and finished by P.M. Cohn in [80]. If H and G are both finite cyclic groups the problem seems to be still open, even though J. Douglas [95] has devoted four papers to the subject. The case of two cyclic groups, one of them being of prime order, was solved in [11]. The bicrossed product, also known as the knit product or Zappa-Szep product in the theory of groups, appeared for the first time in a paper by Zappa [231] and it was rediscovered later on by Szep [221].

As the next section will highlight, in the construction of a crossed product we use a weak action $\alpha : G \to \mathrm{Aut}\,(H)$ together with an α-cocycle $f : G \times G \to H$, while the construction of a bicrossed product involves two compatible actions $\triangleright : G \times H \to H$ and $\triangleleft : G \times H \to G$. Even if the starting points are different, the two constructions have something in common: the crossed product structure as well as the bicrossed product structure are defined on the same set, namely $H \times G$ and $H \cong (H, 1)$ is a subgroup in both the crossed as well as the bicrossed product. Furthermore, $H \cong (H, 1)$ is a normal subgroup in the crossed product. Conversely, any group structure on a set E containing H as a normal subgroup can be reconstructed from H and the quotient E/H as a crossed product (see Theorem 1.1.6 for details). Now, if we drop the normality assumption on H the construction cannot be performed anymore and we have to come up with a new method of reconstructing a group E from a given subgroup and another set of data. This is what we do in Theorem 1.2.1, which is the first important result of this chapter. Let $H \leq E$ be a subgroup of a group E. Using the axiom of choice, we can pick a retraction $p : E \to H$ of the canonical inclusion $i : H \hookrightarrow E$ which is left H-'linear'. Having this application we consider the pointed set $S := p^{-1}(1)$, the fiber of p in 1. The group H and the pointed set S are connected by four maps arising from p: two actions $\triangleright = \triangleright_p : S \times H \to H$, $\triangleleft = \triangleleft_p : S \times H \to S$, a cocycle $f = f_p : S \times S \to H$ and a multiplication $* = *_p : S \times S \to S$ constructed in Theorem 1.2.1. Using these maps, we shall prove that there exists an isomorphism of groups $E \cong (H \times S, \cdot)$, where the multiplication $\cdot$ on the set $H \times S$ is given by

$$(h_1, s_1) \cdot (h_2, s_2) := \Big(h_1(s_1 \triangleright h_2) f(s_1 \triangleleft h_2, s_2),\ (s_1 \triangleleft h_2) * s_2 \Big) \tag{1.1}$$

for all $h_1,\ h_2 \in H$ and $s_1,\ s_2 \in S$. In other words, even if we drop the normality assumption, the group E can still be rebuilt from a subgroup H and the fiber S of an H-linear retraction. Moreover, any group structure $\cdot$ that can be defined on a set E such that a given group H will be contained as a subgroup has the form (1.1) for some system $(S, \triangleright, \triangleleft, f, *)$. This new type of product will be called the *unified product* and it is easily seen that both the crossed and the bicrossed product of groups are special cases of it. In the next step we will perform the abstract construction of the unified product $H \ltimes S$: it is associated to a group H and a system of data $\Omega(H) = \big((S, 1_S, *), \triangleleft,\ \triangleright,\ f\big)$ called the extending datum of H. Theorem 1.2.5 establishes the system of axioms that has to be satisfied by $\Omega(H)$ such that $H \times S$ with the multiplication defined by (1.1) becomes a group structure, i.e., it is a unified product. In this case $\Omega(H) = \big((S, 1_S, *), \triangleleft,\ \triangleright,\ f\big)$ will be called a *group extending structure* of H. Based on Theorem 1.2.1 and Theorem 1.2.5 we answer the description part of the ES problem in Corollary 1.2.7. The answer to the classification part of the ES problem is given in Theorem 1.2.16: the set of all group structures $\cdot$ that can be defined on E such that $H \leq (E, \cdot)$ are classified up to an isomorphism of groups $\psi : (E, \cdot) \to (E, \cdot')$ that stabilizes H by a cohomological type set $\mathcal{K}^2_{\ltimes}(H, (S, 1_S))$ which is explicitly constructed.

As a special case, a more restrictive version of the classification is given in Corollary 1.2.19, which is a general Schreier theorem for unified products. This time all unified products $H \ltimes S$ are classified up to an isomorphism of groups that stabilizes both H and S by a set $\mathcal{H}^2_{\ltimes}(H, (S, 1_S), \triangleleft)$ which plays for the ES problem the same role as the second cohomology group from Hölder's extension problem. An explicit example is given in Proposition 1.2.22 where all groups that contain H as a subgroup of index 2 are classified up to an isomorphism that stabilizes H.

1.1 Crossed product and bicrossed product of groups

We start by setting up the notation used throughout. Let $(S, 1_S)$ be a pointed set, i.e., S is a non-empty set and $1_S \in S$ is a fixed element in S. The group structures on a set H will be denoted using multiplicative notation and the unit element will be denoted by 1_H or only 1 when there is no danger of confusion. Aut (H) denotes the group of automorphisms of a group H and $|S|$ the cardinal of a set S. A map $r : S \to H$ is called unitary if $r(1_S) = 1_H$. S is called a right H-set if there exists a right action $\triangleleft : S \times H \to S$ of H on S, i.e.,

$$s \triangleleft (h_1 h_2) = (s \triangleleft h_1) \triangleleft h_2 \quad \text{and} \quad s \triangleleft 1_H = s \tag{1.2}$$

for all $s \in S$, h_1, $h_2 \in H$. Left H-sets are defined in an analogous manner. The action $\triangleleft : S \times H \to S$ is called the trivial action if $s \triangleleft h = s$, for all $s \in S$ and $h \in H$. Similarly the maps $\triangleright : S \times H \to H$ and $f : S \times S \to H$ are called trivial maps if $s \triangleright h = h$ and respectively $f(s_1, s_2) = 1_H$, for all s, s_1, $s_2 \in S$ and $h \in H$. If G is a group and $\alpha : G \to \text{Aut}\,(H)$ is a map we use the similar notation $\alpha(g)(h) = g \triangleright h$, for all $g \in G$ and $h \in H$.

Crossed product of groups

We recall briefly the construction of the crossed product of groups, the main tool for studying the extension problem.

Let H and G be two given groups. An *extension of G by H* is a triple (E, i, π), where E is a group, $i : H \to E$ and $\pi : E \to G$ are group homomorphisms which fit into a short exact sequence of groups

$$1 \longrightarrow H \xrightarrow{\ i\ } E \xrightarrow{\ \pi\ } G \longrightarrow 1$$

Two such extensions (E, i, π) and (E', i', π') of G by H are called equivalent if there exists a group (iso)morphism $\theta : E' \to E$ such that the following diagram is commutative

$$\begin{array}{ccccc} H & \xrightarrow{i'} & E' & \xrightarrow{\pi'} & G \\ {\scriptstyle Id}\downarrow & & \downarrow{\scriptstyle \theta} & & \downarrow{\scriptstyle Id} \\ H & \xrightarrow{i} & E & \xrightarrow{\pi} & G \end{array}$$

It can be easily proved that we have defined an equivalence relation on the class of all extensions of G by H. The extension problem asks for the explicit description of all equivalence classes of this equivalence.

To start with, we introduce the following concept:

Definition 1.1.1 *A crossed system of groups is a quadruple (H, G, α, f), where H and G are two groups, $\alpha : G \to \mathrm{Aut}\,(H)$ and $f : G \times G \to H$ are two maps such that the following compatibility conditions hold:*

$$g_1 \triangleright (g_2 \triangleright h) = f(g_1, g_2)\big((g_1 g_2) \triangleright h\big) f(g_1, g_2)^{-1} \tag{1.3}$$
$$f(g_1, g_2) f(g_1 g_2, g_3) = \big(g_1 \triangleright f(g_2, g_3)\big) f(g_1, g_2 g_3) \tag{1.4}$$

for all $g_1, g_2, g_3 \in G$ and $h \in H$. The crossed system $\Gamma = (H, G, \alpha, f)$ is called normalized if $f(1, 1) = 1$. The map $\alpha : G \to \mathrm{Aut}\,(H)$ is called a weak action and $f : G \times G \to H$ is called an α-cocycle.

We should point out that in the classical group theory terminology, a pair (α, f) which fulfills the conditions in Definition 1.1.1 is called a factor set ([2], [210]). The term "crossed system" is borrowed from Hopf algebra theory where we have a similar construction which generalizes the crossed product of groups.

Now we give some useful formulas for a crossed system.

Lemma 1.1.2 *Let (H, G, α, f) be a crossed system. Then*

$$f(g, 1) = g \triangleright f(1, 1) \tag{1.5}$$
$$1 \triangleright h = f(1, 1)\, h\, f(1, 1)^{-1} \tag{1.6}$$
$$f(1, g) = f(1, 1) \tag{1.7}$$

for any $g \in G$ and $h \in H$. In particular, if (H, G, α, f) is a normalized crossed system then

$$f(1, g) = f(g, 1) = 1 \quad \text{and} \quad 1 \triangleright h = h \tag{1.8}$$

for any $g \in G$ and $h \in H$.

Proof: The condition (1.4) for $g_2 = g_3 = 1$ and $g_1 = g$ gives (1.5). Now if we set $g_1 = g_2 = 1$ in (1.3) and take into account that $\alpha(1)$ is surjective we obtain (1.6). If we set $g_1 = g_2 = 1$ and $g_3 = g$ in (1.4) and take into account (1.6) we obtain (1.7). □

Let H and G be groups, $\alpha : G \to \mathrm{Aut}\,(H)$ and $f : G \times G \to H$ two maps. Let $H \#_{\alpha}^{f} G := H \times G$ as a set with a binary operation defined by the formula:

$$(h_1, g_1) \cdot (h_2, g_2) := \big(h_1 (g_1 \triangleright h_2) f(g_1, g_2),\, g_1 g_2\big) \tag{1.9}$$

for all $h_1, h_2 \in H$, $g_1, g_2 \in G$. The following theorem gives the construction of the crossed product of groups.

Theorem 1.1.3 *Let H and G be groups, $\alpha : G \to \mathrm{Aut}\,(H)$ and $f : G \times G \to H$ two maps. The following statements are equivalent:*

1. *The multiplication on $H \#_{\alpha}^{f} G$ given by (1.9) is associative.*
2. *(H, G, α, f) is a crossed system.*

In this case $\big(H \#_{\alpha}^{f} G, \cdot\big)$ is a group with the unit $1_{H \#_{\alpha}^{f} G} = \big(f(1,1)^{-1},\, 1\big)$ called the crossed product of H and G associated to the crossed system (H, G, α, f).

Proof: For $h_1, h_2, h_3 \in H$ and $g_1, g_2, g_3 \in G$ we have

$$[(h_1, g_1)\cdot(h_2, g_2)]\cdot(h_3, g_3) = (h_1 (g_1 \triangleright h_2) \underline{f(g_1, g_2)((g_1 g_2) \triangleright h_3) f(g_1 g_2, g_3)},\, g_1 g_2 g_3)$$

and

$$\begin{aligned}(h_1, g_1) \cdot [(h_2, g_2) \cdot (h_3, g_3)] = (&h_1 (g_1 \triangleright h_2)\\ &\underline{(g_1 \triangleright (g_2 \triangleright h_3))\big(g_1 \triangleright f(g_2, g_3)\big) f(g_1, g_2 g_3)},\, g_1 g_2 g_3)\end{aligned}$$

Hence, the multiplication given by (1.9) is associative if and only if

$$f(g_1, g_2)((g_1 g_2) \triangleright h_3) f(g_1 g_2, g_3) = \big(g_1 \triangleright (g_2 \triangleright h_3)\big)\big(g_1 \triangleright f(g_2, g_3)\big) f(g_1, g_2 g_3) \tag{1.10}$$

for all $g_1, g_2, g_3 \in G$ and $h_3 \in H$. We shall prove now that (1.10) holds if and only if (1.3) and (1.4) holds. Assume first that (1.3) and (1.4) holds. Then

$$\begin{aligned}\underline{f(g_1, g_2)\big((g_1 g_2) \triangleright h_3\big)} f(g_1 g_2, g_3) &\overset{(1.3)}{=} \big(g_1 \triangleright (g_2 \triangleright h_3)\big) f(g_1, g_2) f(g_1 g_2, g_3)\\ &\overset{(1.4)}{=} \big(g_1 \triangleright (g_2 \triangleright h_3)\big)\big(g_1 \triangleright f(g_2, g_3)\big) f(g_1, g_2 g_3)\end{aligned}$$

i.e., (1.10) holds. Conversely, assume that (1.10) holds. After we specialize $h_3 = 1$ in (1.10) we obtain (1.4). Now,

$$\begin{aligned}\Big(g_1 \triangleright (g_2 \triangleright h)\Big) f(g_1, g_2) &\overset{(1.10)}{=} f(g_1, g_2)\big((g_1 g_2) \triangleright h\big) f(g_1 g_2, g_3)\\ &\qquad [(g_1 \triangleright f(g_2, g_3)) f(g_1, g_2 g_3)]^{-1} f(g_1, g_2)\\ &\overset{(1.4)}{=} f(g_1, g_2)\big((g_1 g_2) \triangleright h\big) f(g_1 g_2, g_3) f(g_1 g_2, g_3)^{-1}\\ &\qquad f(g_1, g_2)^{-1} f(g_1, g_2)\\ &= f(g_1, g_2)\big((g_1 g_2) \triangleright h\big)\end{aligned}$$

i.e., (1.3) holds; hence the first part of the theorem is proved. We assume now that (H, G, α, f) is a crossed system and we prove that $(H\#_{\alpha}^{f} G, \cdot)$ is a group. For $h \in H$ and $g \in G$ we have

$$\begin{aligned}(h, g) \cdot (f(1,1)^{-1}, 1) &= \Big(h\big(g \triangleright (f(1,1)^{-1})\big) f(g,1),\, g\Big)\\ &\overset{(1.5)}{=} \Big(h\big(g \triangleright (f(1,1)^{-1})\,(g \triangleright f(1,1)\big),\, g\Big)\\ &= \Big(h\big(g \triangleright (f(1,1)^{-1} f(1,1))\big),\, g\Big)\\ &= (h(g \triangleright 1), g) = (h, g)\end{aligned}$$

and

$$\begin{aligned}(f(1,1)^{-1}, 1) \cdot (h, g) &= (f(1,1)^{-1}(1 \triangleright h) f(1,g),\, g)\\ &\overset{(1.6)}{=} (f(1,1)^{-1} f(1,1) h f(1,1)^{-1} f(1,g),\, g)\\ &\overset{(1.7)}{=} (h f(1,1)^{-1} f(1,1),\, g) = (h, g)\end{aligned}$$

i.e., $(f(1,1)^{-1}, 1)$ is the unit of $(H\#_{\alpha}^{f} G, \cdot)$. Let now $(h, g) \in H\#_{\alpha}^{f} G$. Then it is easy to see that

$$(h, g)^{-1} = \big(f(1,1)^{-1} f(g^{-1}, g)^{-1} (g^{-1} \triangleright h^{-1}),\, g^{-1}\big)$$

is a left inverse of (h, g). Thus $H\#_{\alpha}^{f} G$ is a monoid and any element of it has a left inverse. Then $H\#_{\alpha}^{f} G$ is a group and we are done. □

Note that a crossed product with f the trivial cocycle (that is $f(g_1, g_2) = 1_H$, for all $g_1, g_2 \in G$) is just the semidirect product $H \ltimes_{\alpha} G$ of H and G.

Example 1.1.4 *Let C_2 (resp. C_4) be the cyclic group of order 2 (resp. 4) generated by a (resp. b) and define*

$$\alpha : C_2 \to \mathrm{Aut}\,(C_4), \qquad \alpha(1) = \mathrm{Id}_{C_4}, \quad \alpha(a)(x) = x^{-1}$$

for all $x \in C_4$ and

$$f : C_2 \times C_2 \to C_4, \qquad f(1,1) = f(1,a) = f(a,1) = 1, \quad f(a,a) = b$$

Then we can easily prove that (C_4, C_2, α, f) is a normalized crossed system and $C_4\#_{\alpha}^{f} C_2 \cong Q$, the quaternion group Q of order 8.

Any crossed product $(H\#_{\alpha}^{f} G, i_H, \pi_G)$ of groups gives a canonical extension of G by H as the next result shows.

Corollary 1.1.5 *Let (H, G, α, f) be a crossed system. Then*

$$1 \longrightarrow H \xrightarrow{i_H} H\#_{\alpha}^{f} G \xrightarrow{\pi_G} G \longrightarrow 1 \tag{1.11}$$

where $i_H(h) := \big(h f(1,1)^{-1}, 1\big)$ and $\pi_G(h, g) := g$ for all $h \in H$ and $g \in G$ is an exact sequence of groups, i.e., $(H\#_{\alpha}^{f} G, i_H, \pi_G)$ is an extension of G by H.

Proof: Straightforward. □

The famous Schreier's theorem states the converse of Corollary 1.1.5: any extension E of a group G by a group H is equivalent to a crossed product extension.

Theorem 1.1.6 *Let (E, i, π) be an extension of a group G by a group H, i.e., there exists an exact sequence of groups $1 \longrightarrow H \xrightarrow{i} E \xrightarrow{\pi} G \longrightarrow 1$. Then there exists (H, G, α, f) a normalized crossed system and an isomorphism of groups $\theta : H \#_{\alpha}^{f} G \to E$ such that the following diagram*

$$\begin{array}{ccccc} H & \xrightarrow{i_H} & H \#_{\alpha}^{f} G & \xrightarrow{\pi_G} & G \\ \downarrow{\scriptstyle Id} & & \downarrow{\scriptstyle \theta} & & \downarrow{\scriptstyle Id} \\ H & \xrightarrow{i} & E & \xrightarrow{\pi} & G \end{array} \tag{1.12}$$

is commutative.

Proof: We shall identify $H \cong i(H) \trianglelefteq E$. The crossed system is constructed as follows: let $s : G \to E$ be a section of $\pi : E \to G$ such that $s(1) = 1$ and define α and f by the formulas:

$$\alpha : G \to \mathrm{Aut}\,(H), \qquad \alpha(g)(h) := s(g)hs(g)^{-1} \tag{1.13}$$

$$f : G \times G \to H, \qquad f(g_1, g_2) := s(g_1)s(g_2)s(g_1g_2)^{-1} \tag{1.14}$$

for all g, g_1, $g_2 \in G$ and $h \in H$. Then, by a long but straightforward computation we can show that (H, G, α, f) is a normalized crossed system and

$$\theta : H \#_{\alpha}^{f} G \to E, \qquad \theta(h, g) := i(h)s(g)$$

is an isomorphism of groups and the diagram is commutative: $\pi\big(\theta(h, g)\big) = \pi(i(h))\pi(s(g)) = g = Id_G\big(\pi_G(h, g)\big)$, for all $h \in H$ and $g \in G$. □

Theorem 1.1.6 allows for a computational reformulation of Hölder's extension problem as follows: *Let H and G be two fixed groups. Describe all normalized crossed systems (H, G, α, f) and classify up to isomorphism all crossed products $H \#_{\alpha}^{f} G$.*

The description of all extensions of a group by a group (or, equivalently, of all normalized crossed systems between two fixed groups) has been a central problem in group theory during the last century. For more detail we refer to [2], [210].

Bicrossed product of groups

We recall now, following [224], the construction of the bicrossed product of two groups, which is the main toll for the study of the factorization problem.

Definition 1.1.7 *A matched pair of groups is a quadruple* $(A, H, \triangleright, \triangleleft)$, *where* A *and* H *are groups,* $\triangleright : H \times A \to A$ *is a left action of the group* H *on the set* A, $\triangleleft : H \times A \to H$ *is a right action of the group* A *on the set* H *satisfying the following compatibilities for any* $a, b \in A$, $h, g \in H$:

$$h \triangleright (ab) = (h \triangleright a)((h \triangleleft a) \triangleright b) \tag{1.15}$$

$$(hg) \triangleleft a = (h \triangleleft (g \triangleright a))(g \triangleleft a) \tag{1.16}$$

If $(A, H, \triangleright, \triangleleft)$ is a matched pair then the following normalizing conditions hold for any $a \in A$ and $h \in H$:

$$1 \triangleright a = a, \quad h \triangleleft 1 = h, \quad h \triangleright 1 = 1, \quad 1 \triangleleft a = 1 \tag{1.17}$$

Let $\triangleright : H \times A \to A$, $\triangleleft : H \times A \to H$ be two maps and $A \bowtie H := A \times H$ with the binary operation defined by the formula:

$$(a, h) \cdot (b, g) := \big(a(h \triangleright b), (h \triangleleft b)g\big) \tag{1.18}$$

for all $a, b \in A$, $h, g \in H$.

Proposition 1.1.8 *Let* A *and* H *be groups and* $\triangleright : H \times A \to A$, $\triangleleft : H \times A \to H$ *two maps. Then* $A \bowtie H$ *is a group with unit* $(1, 1)$ *if and only if* $(A, H, \triangleright, \triangleleft)$ *is a matched pair of groups. In this case* $A \bowtie H$ *is called the bicrossed product of* A *and* H.

Proof: As the proof is similar to the one given for Theorem 1.1.3, we leave it to the reader; a detailed proof can be found in [224, Proposition 2.2.]. Furthermore, it can be obtained as a special case of Theorem 1.2.5. □

We point out that there are several other names used in the literature to designate the bicrossed product such as: *doublecross product, knit product* or *Zappa-Szep product* associated to the matched pair $(H, G, \triangleright, \triangleleft)$.

If $A \bowtie H$ is a bicrossed product then $i_A : A \to A \bowtie H$, $i_A(a) = (a, 1)$ and $i_H : H \to A \bowtie H$, $i_H(h) = (1, h)$ are morphisms of groups. A and H will be viewed as subgroups of $A \bowtie H$ via the identifications $A \cong A \times \{1\}$, $H \cong \{1\} \times H$. If the right action $\triangleleft$ of a matched pair $(A, H, \triangleright, \triangleleft)$ is the trivial action then the bicrossed product $A \bowtie H$ is just the semidirect product $A \ltimes H$ of A and H. Thus, the bicrossed product is another generalization of the semidirect product to the case when none of the factors is required to be normal. The bicrossed product $A \bowtie H$ factorizes through $A \cong A \times \{1\}$ and $H \cong \{1\} \times H$ as for any $a \in A$ and $h \in H$ we have that $(a, h) = (a, 1) \cdot (1, h)$. Conversely, the main motivation for defining the bicrossed product of groups is the following result which can be viewed as a dual to Schreier's theorem (Theorem 1.1.6).

Proposition 1.1.9 *A group* G *factorizes through two subgroups* A *and* H *(i.e.,* G *contains* A *and* H *as subgroups such that* $G = AH$ *and* $A \cap H = 1$*)*

if and only if there exists a matched pair of groups $(A, H, \triangleright, \triangleleft)$ such that the multiplication map

$$m_G : A \bowtie H \to G, \qquad m_G(a,\, h) = ah$$

for all $a \in A$ and $h \in H$ is an isomorphism of groups.

Proof: We indicate only the construction of the matched pair $(A, H, \triangleright, \triangleleft)$ associated to the factorization $G = AH$. Indeed, if G factorizes through A and H then for any $g \in G$ there exists a unique pair $(a, h) \in A \times H$ such that $g = ah$. This allows us to attach to any $(a, h) \in A \times H$ a unique pair of elements $(h \triangleright a,\, h \triangleleft a) \in A \times H$ such that

$$h\, a = (h \triangleright a)(h \triangleleft a) \in AH \tag{1.19}$$

Then we can easily prove that $(A, H, \triangleright, \triangleleft)$ is a matched pair of groups and $m_G : A \bowtie H \to G$ is an isomorphism of groups. For more detail see [224, Proposition 2.4]. □

1.2 Group extending structures and unified products

The abstract definition of the unified product of groups will arise from the following elementary question subsequent to the ES problem: *let $H \leq E$ be a subgroup in E. Can we reconstruct the group structure on E from the one of H and some extra set of datum?* First we note that Schreier's classical construction from Theorem 1.1.6 cannot be used anymore. Thus we should come up with a new method of reconstruction. The next theorem indicates the way we can perform this reconstruction.

Theorem 1.2.1 *Let $H \leq E$ be a subgroup of a group E. Then:*

1. *There exists a map $p : E \to H$ such that $p(1) = 1$ and for any $h \in H$, $x \in E$*
$$p(h\, x) = h\, p(x) \tag{1.20}$$

2. *For such a map $p : E \to H$ we define $S = S_p := p^{-1}(1) = \{x \in E \,|\, p(x) = 1\}$. Then the multiplication map*
$$\varphi : H \times S \to E, \quad \varphi(h, s) := hs \tag{1.21}$$
for all $h \in H$ and $s \in S$ is bijective with the inverse given for any $x \in E$ by
$$\varphi^{-1} : E \to H \times S, \quad \varphi^{-1}(x) = \big(p(x),\, p(x)^{-1}\, x\big)$$

3. For p and S as above there exist four maps $\triangleright = \triangleright_p : S \times H \to H$, $\triangleleft = \triangleleft_p : S \times H \to S$, $f = f_p : S \times S \to H$ *and* $* = *_p : S \times S \to S$ *given by the formulas*

$$\begin{aligned} s \triangleright h &:= p(sh), \qquad s \triangleleft h := p(sh)^{-1} sh \\ f(s_1, s_2) &:= p(s_1 s_2), \quad s_1 * s_2 := p(s_1 s_2)^{-1} s_1 s_2 \end{aligned}$$

for all s, s_1, $s_2 \in S$ *and* $h \in H$. *Using these maps, the unique group structure* '$\cdot$' *on the set* $H \times S$ *such that* $\varphi : (H \times S, \cdot) \to E$ *is an isomorphism of groups given by:*

$$(h_1, s_1) \cdot (h_2, s_2) := \Big(h_1 (s_1 \triangleright h_2) f(s_1 \triangleleft h_2, s_2),\ (s_1 \triangleleft h_2) * s_2 \Big) \tag{1.22}$$

for all h_1, $h_2 \in H$ *and* s_1, $s_2 \in S$.

Proof: (1) Using the axiom of choice we can fix $\Gamma = (x_i)_{i \in I} \subset E$ to be a system of representatives for the right congruence modulo H in E such that $1 \in \Gamma$. Then for any $x \in E$ there exists an unique $h_x \in H$ and an unique $x_{i_0} \in \Gamma$ such that $x = h_x x_{i_0}$. Thus, there exists a well-defined map $p : E \to H$ given by the formula $p(x) := h_x$, for all $x \in E$. As $1 \in \Gamma$ we have that $p(1) = 1$. Moreover, for any $h \in H$ and $x \in E$ we have that $hx = hh_x x_{i_0}$. Thus $p(hx) = hh_x = hp(x)$, as needed.

(2) We note that $p(x)^{-1} x \in S$ as $p\big(p(x)^{-1} x\big) = p(x)^{-1} p(x) = 1$, for all $x \in E$. The rest is straightforward.

(3) First we note that $\triangleleft$ and $*$ are well-defined maps. Next, we can easily prove that the following two formulas hold:

$$p(s_1 h_2 s_2) = (s_1 \triangleright h_2) f(s_1 \triangleleft h_2, s_2) \tag{1.23}$$

$$(s_1 \triangleleft h_2) * s_2 = p(s_1 h_2 s_2)^{-1} s_1 h_2 s_2 \tag{1.24}$$

for all s_1, $s_2 \in S$ and $h_2 \in H$. Indeed,

$$\begin{aligned} (s_1 \triangleright h_2) f(s_1 \triangleleft h_2, s_2) &= p(s_1 h_2)\, p\big((s_1 \triangleleft h_2) s_2\big) \\ &= p(s_1 h_2)\, p\big(p(s_1 h_2)^{-1} s_1 h_2 s_2\big) \\ &\overset{(1.20)}{=} p(s_1 h_2 s_2) \end{aligned}$$

Similarly, we can prove that (1.24) holds. Now, $\varphi : H \times S \to E$ is a bijection between the set $H \times S$ and the group E. Thus, there exists a unique group structure $\cdot$ on the set $H \times S$ such that φ is an isomorphism of groups. This group structure is obtained by transferring the group structure from E via the bijection φ, i.e., is given by:

$$\begin{aligned} (h_1, s_1) \cdot (h_2, s_2) &= \varphi^{-1}\big(\varphi(h_1, s_1) \varphi(h_2, s_2)\big) = \varphi^{-1}(h_1 s_1 h_2 s_2) \\ &= \big(p(h_1 s_1 h_2 s_2),\ p(h_1 s_1 h_2 s_2)^{-1} h_1 s_1 h_2 s_2\big) \\ &\overset{(1.20)}{=} \big(h_1 p(s_1 h_2 s_2),\ p(s_1 h_2 s_2)^{-1} s_1 h_2 s_2\big) \\ &\overset{(1.23),(1.24)}{=} \Big(h_1 (s_1 \triangleright h_2) f(s_1 \triangleleft h_2,\ s_2),\ (s_1 \triangleleft h_2) * s_2 \Big) \end{aligned}$$

for all $h_1,\ h_2 \in H$ and $s_1,\ s_2 \in S$ as needed. □

Remark 1.2.2 *As* $1 \in S$ *and* $p(s) = 1$, *for all* $s \in S$ *the maps* $\triangleright = \triangleright_p$, $\triangleleft = \triangleleft_p$, $f = f_p$ *and* $* = *_p$ *constructed in (3) of Theorem 1.2.1 satisfy the following normalizing conditions:*

$$s \triangleright 1 = 1, \quad 1 \triangleright h = h, \quad 1 \triangleleft h = 1, \quad s \triangleleft 1 = s \tag{1.25}$$

$$f(s, 1) = f(1, s) = 1, \quad s * 1 = 1 * s = s \tag{1.26}$$

for all $s \in S$ *and* $h \in H$. *Hence, the multiplication* $*$ *on* S *has a unit but is not necessarily associative. In fact, we can easily prove that it satisfies the following compatibility:*

$$(s_1 * s_2) * s_3 = \big(s_1 \triangleleft f(s_2, s_3)\big) * (s_2 * s_3) \tag{1.27}$$

for all $s_1,\ s_2,\ s_3 \in S$, *i.e.,* $*$ *is associative up to the pair* $(\triangleleft, f)$. *Moreover, any element* $s \in S$ *is left invertible in* $(S, *)$; *more precisely we can show that for any* $s \in S$ *there exists a unique element* $s' \in S$ *such that* $s' * s = 1$.

The abstract construction of the unified product

Let H be a group and E a set such that $H \subseteq E$. Theorem 1.2.1 describes the way any group structure $\cdot$ on the set E such that H is a subgroup of $(E, \cdot)$ should look like. We are left to find the abstract axioms that need to be fulfilled by the system of maps $(*, \triangleleft, \triangleright, f)$ such that (1.22) is indeed a group structure. This will be done below.

Definition 1.2.3 *An extending datum of a group* H *is a system* $\Omega(H) = \big((S, 1_S, *), \triangleleft,\ \triangleright,\ f\big)$ *where:*

(1) $(S, 1_S)$ *is a pointed set,* $* : S \times S \to S$ *is a binary operation such that for any* $s \in S$

$$s * 1_S = 1_S * s = s \tag{1.28}$$

(2) *The maps* $\triangleleft : S \times H \to S$, $\triangleright : S \times H \to H$ *and* $f : S \times S \to H$ *satisfy the following normalizing conditions for any* $s \in S$ *and* $h \in H$:

$$s \triangleleft 1_H = s,\ 1_S \triangleleft h = 1_S,\ 1_S \triangleright h = h,\ s \triangleright 1_H = 1_H,\ f(s, 1_S) = f(1_S, s) = 1_H \tag{1.29}$$

Let H be a group and $\Omega(H) = \big((S, 1_S, *), \triangleleft,\ \triangleright,\ f\big)$ an extending datum of H. We denote by $H \ltimes_{\Omega(H)} S := H \ltimes S$ the set $H \times S$ with the binary operation defined by the formula:

$$(h_1, s_1) \cdot (h_2, s_2) := \Big(h_1 (s_1 \triangleright h_2) f(s_1 \triangleleft h_2, s_2),\ (s_1 \triangleleft h_2) * s_2\Big) \tag{1.30}$$

for all $h_1,\ h_2 \in H$ and $s_1,\ s_2 \in S$.

Definition 1.2.4 *Let H be a group and $\Omega(H) = \big((S, 1_S, *), \triangleleft, \triangleright, f\big)$ an extending datum of H. The object $H \ltimes S$ introduced above is called the unified product of H and $\Omega(H)$ if $H \ltimes S$ is a group with the multiplication given by (1.30). In this case the extending datum $\Omega(H)$ is called a group extending structure of H. The maps $\triangleright$ and $\triangleleft$ are called the actions of $\Omega(H)$ and f is called the $(\triangleright, \triangleleft)$-cocycle of $\Omega(H)$.*

Using (1.28) and (1.29) it is straightforward to prove that $(1_H, 1_S)$ is a unit of the multiplication (1.30) and the following relations hold in $H \ltimes S$:

$$(h_1, 1_S) \cdot (h_2, s_2) = (h_1 h_2, s_2) \tag{1.31}$$
$$(h_1, s_1) \cdot (1_H, s_2) = (h_1 f(s_1, s_2), s_1 * s_2) \tag{1.32}$$
$$(h_1, s_1) \cdot (h_2, 1_S) = (h_1 (s_1 \triangleright h_2), s_1 \triangleleft h_2) \tag{1.33}$$

for all h_1, $h_2 \in H$ and s_1, $s_2 \in S$. Next, we indicate the abstract system of axioms that need to be satisfied by the maps $(*, \triangleleft, \triangleright, f)$ such that $H \ltimes S$ becomes a unified product.

Theorem 1.2.5 *Let H be a group and $\Omega(H) = \big((S, 1_S, *), \triangleleft, \triangleright, f\big)$ an extending datum of H. The following statements are equivalent:*

(1) $A \ltimes H$ is an unified product;

(2) The following compatibilities hold for any s, s_1, s_2, $s_3 \in S$ and h, h_1, $h_2 \in H$:

(ES1) The map $\triangleleft : S \times H \to S$ is a right action of the group H on the set S;

(ES2) $(s_1 * s_2) * s_3 = \big(s_1 \triangleleft f(s_2, s_3)\big) * (s_2 * s_3)$;

(ES3) $s \triangleright (h_1 h_2) = (s \triangleright h_1)\big((s \triangleleft h_1) \triangleright h_2\big)$;

(ES4) $(s_1 * s_2) \triangleleft h = \big(s_1 \triangleleft (s_2 \triangleright h)\big) * (s_2 \triangleleft h)$;

(ES5) $\big(s_1 \triangleright (s_2 \triangleright h)\big) f\big(s_1 \triangleleft (s_2 \triangleright h),\, s_2 \triangleleft h\big) = f(s_1, s_2)\big((s_1 * s_2) \triangleright h\big)$;

(ES6) $f(s_1, s_2) f(s_1 * s_2, s_3) = \big(s_1 \triangleright f(s_2, s_3)\big) f\big(s_1 \triangleleft f(s_2, s_3),\, s_2 * s_3\big)$;

(ES7) For any $s \in S$ there exists $s' \in S$ such that $s' * s = 1_S$.

Before going into the proof of the theorem we note that $(ES3)$ and $(ES4)$ are exactly, mutatis-mutandis, the compatibility conditions (1.15) and (1.16) from the definition of a matched pair of groups while $(ES5)$ and $(ES6)$ are deformations via the right action $\triangleleft$ of the compatibility conditions (1.3) and (1.4) from the definition of a crossed system of groups. The axiom $(ES2)$ is called the *twisted associativity condition* as it measures how far $*$ is from being associative, i.e., from being a group structure on S.

Proof: We know that $(1_H, 1_S)$ is a unit for the operation defined by (1.30). We prove now that $\cdot$ given by (1.30) is associative if and only if the compatibility conditions $(ES1) - (ES6)$ hold. Assume first that $\cdot$ is associative and let h, h_1, $h_2 \in H$ and s, s_1, $s_2 \in S$. The associativity condition

$$[(1_H, s) \cdot (h_1, 1_S)] \cdot (h_2, 1_S) = (1_H, s) \cdot [(h_1, 1_S) \cdot (h_2, 1_S)]$$

gives, after we use the cross relations (1.33) and (1.31), $(s \triangleright h_1,\ s \triangleleft h_1) \cdot (h_2, 1_S) = (1_H, s) \cdot (h_1 h_2,\ 1_S)$. Thus $\big((s \triangleright h_1)((s \triangleleft h_1) \triangleright h_2),\ (s \triangleleft h_1) \triangleleft h_2\big) = \big(s \triangleright (h_1 h_2),\ s \triangleleft (h_1 h_2)\big)$ and hence $(ES1)$ and $(ES3)$ hold. Now, by writing the associativity condition $[(1_H, s_1) \cdot (1_H, s_2)] \cdot (1_H, s_3) = (1_H, s_1) \cdot [(1_H, s_2) \cdot (1_H, s_3)]$ and computing this equality using the cross relation (1.32) it follows that the compatibility conditions $(ES2)$ and $(ES6)$ hold. Finally, if we write the associativity condition $[(1_H, s_1) \cdot (1_H, s_2)] \cdot (h, 1_S) = (1_H, s_1) \cdot [(1_H, s_2) \cdot (h, 1_S)]$ and use (1.32) and then (1.33) we obtain that $(ES4)$ and $(ES5)$ hold. Conversely, assume that the compatibility conditions $(ES1) - (ES6)$ hold. Then, by a rather long but straightforward computation we can prove that the operation $\cdot$ is associative, that is $(h_1, s_1) \cdot [(h_2, s_2) \cdot (h_3, s_3)] = [(h_1, s_1) \cdot (h_2, s_2)] \cdot (h_3, s_3)$, for all h_1, h_2, $h_3 \in H$ and s_1, s_2, $s_3 \in S$.

To conclude, we have proved that $(H \ltimes S, \cdot)$ is a monoid if and only if $(ES1) - (ES6)$ hold. Assume now that $(H \ltimes S, \cdot)$ is a monoid: it remains to be proved that the monoid is actually a group if and only if $(ES7)$ holds. Indeed, in the monoid $(H \ltimes S, \cdot)$ we have:

$$(h, 1_S) \cdot (1_H, s) = (h, s), \qquad (h_1, 1_S) \cdot (h_2, 1_S) = (h_1 h_2, 1_S)$$

for all h, h_1, $h_2 \in H$ and $s \in S$. In particular, any element of the form $(h, 1_S)$, for $h \in H$ is invertible in $(H \ltimes S, \cdot)$. Now, a monoid is a group if and only if each of his elements has a left inverse. As $\cdot$ is associative it follows from:

$$(h^{-1}, 1_S) \cdot (h, s) = (1_H, s)$$

that $(H \ltimes S, \cdot)$ is a group if and only if $(1_H, s)$ has a left inverse for all $s \in S$. Hence, for any $s \in S$ there exist elements $s' \in S$ and $h' \in H$ such that

$$(h', s') \cdot (1_H, s) = \big(h' f(s', s),\ s' * s\big) = (1_H, 1_S)$$

This is of course equivalent to the fact that $s' * s = 1_S$ for all $s \in S$ and $h' = f(s',\ s)^{-1}$. The proof is now finished. We note that the inverse of an element (h, s) in the group $(H \ltimes S, \cdot)$ is given by the formula

$$(h,\ s)^{-1} = \big(f(s',\ s)^{-1}(s' \triangleright h^{-1}),\ s' \triangleleft h^{-1}\big)$$

where $s' * s = 1_S$. □

Remark 1.2.6 *Let H be a group, $\Omega(H) = \big((S, 1_S, *), \triangleleft, \triangleright, f\big)$ a group extending datum of H and $H \ltimes_\Omega S$ the associated unified product. Then the canonical inclusion*

$$i_H : H \to H \ltimes_\Omega S, \qquad i_H(h) := (h, 1_S)$$

is a morphism of groups and the map

$$p_H : H \ltimes_\Omega S \to H, \qquad p_H(h,s) := h$$

satisfies condition (1.20) of Theorem 1.2.1. Moreover, if we identify $H \cong (H, 1_S) \leq H \ltimes_\Omega S$ *and* $S \cong (1_H, S) \subset H \ltimes_\Omega S$ *we can easily show that the maps* $*$, $\triangleleft$, $\triangleright$ *and* f *from the definition of* $\Omega(H)$ *are exactly the ones given in (3) of Theorem 1.2.1 associated to the splitting map* p_H.

We record these observations in the following result which gives the answer to the description part of the ES problem.

Corollary 1.2.7 *Let* H *be a group and* E *a set such that* $H \subseteq E$*. Then there exists a group structure* $\cdot$ *on* E *such that* H *is a subgroup of* $(E, \cdot)$ *if and only if there exists a group extending structure* $\Omega(H) = \big((S, 1_S, *), \triangleleft, \triangleright, f\big)$ *of* H *such that* $H \ltimes S \cong (E, \cdot)$.

Proof: It follows from Remark 1.2.6, Theorem 1.2.1 and Theorem 1.2.5. □

In what follows we provide some special cases of unified products. First of all we show that the unified product unifies both the crossed product as well as the bicrossed product of groups.

Example 1.2.8 *Let* $\Omega(H) = \big((S, 1_S, *), \triangleleft, \triangleright, f\big)$ *be an extending datum of* H *such that* $\triangleleft$ *is the trivial action, that is* $s \triangleleft h := s$*, for all* $s \in S$ *and* $h \in H$*. Then* $\Omega(H)$ *is a group extending structure of* H *if and only if* $(S, *)$ *is a group structure on the set* S *and* $(H, (S, *), \triangleright, f)$ *is a crossed system of groups. In this case, the associated unified product* $H \ltimes_\Omega S = H \#_\triangleright^f G$ *is the crossed product of groups.*

Example 1.2.9 *Let* $\Omega(H) = \big((S, 1_S, *), \triangleleft, \triangleright, f\big)$ *be an extending datum of* H *such that* f *is the trivial cocycle, that is* $f(s_1, s_2) = 1$*, for all* s_1*,* $s_2 \in S$*. Then* $\Omega(H)$ *is a group extending structure of* H *if and only if* $(S, *)$ *is a group structure on the set* S *and* $(H, (S, *), \triangleright, \triangleleft)$ *is a matched pair of groups. In this case, the associated unified product* $A \ltimes_\Omega H = H \bowtie G$ *is the bicrossed product of groups.*

Example 1.2.10 *There are examples of groups that cannot be written as a crossed product nor as a bicrossed product of two groups of smaller order. Such a group should be a simple group (otherwise it can be written as a crossed product). The simple group of smallest order that cannot be written as a bicrossed product is the alternating group* A_6 *([230]). The above results allows us to write* A_6*, and in fact any other simple group which is not a bicrossed product, as a unified product between one of its subgroups and an extending structure. For instance, we can write*

$$A_6 \cong A_4 \ltimes_\Omega S$$

for an extending structure $\Omega(A_4) = \big((S, 1_S, *), \triangleleft, \triangleright, f\big)$*, where* S *is a set with 30 elements.*

Finally, an example of a group extending structure was constructed in [17] as follows:

Example 1.2.11 *Let H be a group, $(S, 1_S, *)$ a pointed set with a binary operation $* : S \times S \to S$ having 1_S as a unit. Let $\triangleleft : S \times H \to S$ be a map such that $(S, \triangleleft)$ is a right H-set and $\gamma : S \to H$ be a map with $\gamma(1_S) = 1_H$ such that the following compatibilities hold*

$$\begin{aligned}(x * y) * z &= \Big(x \triangleleft \big(\gamma(y)\,\gamma(z)\,\gamma(y * z)^{-1}\big)\Big) * (y * z)\\ (x * y) \triangleleft g &= \Big(x \triangleleft \big(\gamma(y)\,g\,\gamma(y \triangleleft g)^{-1}\big)\Big) * (y \triangleleft g)\end{aligned}$$

for all $g \in H$, x, y, $z \in S$. Using the transition map γ we define a left action $\triangleright$ and a cocycle f via:

$$x \triangleright g := \gamma(x)\,g\,\gamma(x \triangleleft g)^{-1}, \quad f(x,\, y) := \gamma(x)\,\gamma(y)\,\gamma(x * y)^{-1}$$

for all x, $y \in S$ and $g \in H$. Then we can prove that $\Omega(H) = \big((S, 1_S, *), \triangleleft, \triangleright, f\big)$ is a group extending structure of H.

Universal properties of the unified product

In this subsection we prove the universality of the unified product. Let H be a group and $\Omega(H) = \big((S, 1_S, *), \triangleleft, \triangleright, f\big)$ a group extending structure of H. We associate to $\Omega(H)$ two categories ${}_{\Omega(H)}\mathcal{C}$ and $\mathcal{D}_{\Omega(H)}$ such that the unified product becomes an initial object in the first category and a final object in the second category. Define the category ${}_{\Omega(H)}\mathcal{C}$ as follows: the objects of ${}_{\Omega(H)}\mathcal{C}$ are pairs $(G, (u, v))$, where G is a group, $u : H \to G$ is a morphism of groups and $v : S \to G$ is a map such that:

$$\begin{aligned}v(s_1)v(s_2) &= u(f(s_1, s_2))v(s_1 * s_2) &(1.34)\\ v(s)u(h) &= u(s \triangleright h)v(s \triangleleft h) &(1.35)\end{aligned}$$

for all s, s_1, $s_2 \in S$ and $h \in H$. The morphisms of the category $f : (G_1, (u_1, v_1)) \to (G_2, (u_2, v_2))$ are morphisms of groups $f : G_1 \to G_2$ such that : $f \circ u_1 = u_2$ and $f \circ v_1 = v_2$.

Define the category $\mathcal{D}_{\Omega(H)}$ as follows: the objects of $\mathcal{D}_{\Omega(H)}$ are pairs $(G, (u, v))$, where G is a group, $u : G \to H$, $v : G \to S$ are maps such that:

$$\begin{aligned}u(xy) &= u(x)[v(x) \triangleright u(y)]f(v(x) \triangleleft u(y), v(y)) &(1.36)\\ v(xy) &= [v(x) \triangleleft u(y)] * v(y) &(1.37)\end{aligned}$$

for all $x, y \in G$ while the morphisms of this category $f : (G_1, (u_1, v_1)) \to (G_2, (u_2, v_2))$ are morphisms of groups $f : G_1 \to G_2$ such that: $u_2 \circ f = u_1$ and $v_2 \circ f = v_1$.

Theorem 1.2.12 *Let H be a group and $\Omega(H) = \big((S, 1_S, *), \triangleleft, \triangleright, f\big)$ a group extending structure of H. Then:*

(1) $(H \ltimes_\Omega S, (i_H, i_S))$ is an initial object of ${}_{\Omega(H)}\mathcal{C}$, where $i_H : H \to H \ltimes_\Omega S$ and $i_S : S \to H \ltimes_\Omega S$ are the canonical inclusions;

(2) $(H \ltimes_\Omega S, (\pi_H, \pi_S))$ is a final object of $\mathcal{D}_{\Omega(H)}$, where $\pi_H : H \ltimes_\Omega S \to H$ and $\pi_S : H \ltimes_\Omega S \to S$ are the canonical projections.

Proof: (1) It is easy to see that $(H \ltimes_\Omega S, (i_H, i_S))$ is an object in the category ${}_{\Omega(H)}\mathcal{C}$. Let $(G, (u, v))$ be an object in ${}_{\Omega(H)}\mathcal{C}$. We need to prove that there exists an unique morphism of groups $\psi : H \ltimes_\Omega S \to G$ such that the following diagram commutes:

$$\begin{array}{ccccc} H & \xrightarrow{i_H} & H \ltimes_\Omega S & \xleftarrow{i_S} & S \\ & \searrow^{u} & \downarrow{\scriptstyle \psi} & \swarrow_{v} & \\ & & G & & \end{array}$$

Assume first that ψ satisfies the above condition. We obtain: $\psi((h, s)) = \psi((h, 1_S) \cdot (1_H, s)) = \psi((h, 1_S))\psi((1_H, s)) = (\psi \circ i_H)(h)(\psi \circ i_S)(s)) = u(h)v(s)$, for all $h \in H$, $s \in S$ and we proved that ψ is uniquely determined by u and v. The existence of ψ can be proved as follows: we define $\psi : H \ltimes_\Omega S \to G$ by $\psi((h, s)) := u(h)v(s)$, for all $h \in H$ and $s \in S$. The fact that ψ is a morphism of groups is just a straightforward computation and the commutativity of the diagram is obvious.

(2) Similar to (1). □

The classification of unified products

In this subsection we provide the classification part of the ES problem. Using Corollary 1.2.7, the classification of all group structures on E that contain H as a subgroup, reduces to the classification of all unified products $H \ltimes_\Omega S$, associated to all group extending structures $\Omega(H) = \big((S, 1_S, *), \triangleleft, \triangleright, f\big)$, for a set S such that $|H||S| = |E|$.

From now on the group H and the pointed set $(S, 1_S)$ will be fixed. Let $\mathcal{GES}(H, (S, 1_S))$ be the set of all quadruples $(*, \triangleleft, \triangleright, f)$ such that $\big((S, 1_S, *), \triangleleft, \triangleright, f\big)$ is a group extending structure of H.

Definition 1.2.13 *Let $\Omega(H) = \big((S, 1_S, *), \triangleleft, \triangleright, f\big)$ and $\Omega'(H) = \big((S, 1_S, *'), \triangleleft', \triangleright', f'\big)$ be two group extending structures of H and $H \ltimes_\Omega S$, $H \ltimes_{\Omega'} S$ the associated unified products. For a morphism of groups $\psi : H \ltimes_\Omega S \to H \ltimes_{\Omega'} S$ we consider the following diagram*

$$\begin{array}{ccccc} H & \xrightarrow{i_H} & H \ltimes S & \xrightarrow{\pi_S} & S \\ {\scriptstyle Id_H}\downarrow & & \downarrow{\scriptstyle \psi} & & \downarrow{\scriptstyle Id_S} \\ H & \xrightarrow{i_H} & H \ltimes' S & \xrightarrow{\pi_S} & S \end{array} \qquad (1.38)$$

where $\pi_S : H \ltimes_\Omega S \to S$ is the canonical projection $\pi(h,s) := s$, for all $h \in H$ and $s \in S$. We say that $\psi : H \ltimes_\Omega S \to H \ltimes_{\Omega'} S$ stabilizes H *(resp.* stabilizes S*) if the left square (resp. the right square) of the diagram (1.38) is commutative.*

Proposition 1.2.14 *Let $\Omega(H) = \big((S, 1_S, *), \triangleleft, \triangleright, f\big)$ and $\Omega'(H) = \big((S', 1_{S'}, *'), \triangleleft', \triangleright', f'\big)$ be two group extending structures of a group H. Then there exists a bijective correspondence between the set of all morphisms of groups $\psi : H \ltimes_\Omega S \to H \ltimes_{\Omega'} S'$ that stabilize H and the set of all pairs (r, v), where $r : S \to H$, $v : S \to S'$ are two unitary maps such that:*

$$v(s \triangleleft h) = v(s) \triangleleft' h \tag{1.39}$$

$$(s \triangleright h)\, r(s \triangleleft h) = r(s)\, \big(v(s) \triangleright' h\big) \tag{1.40}$$

$$v(s_1 * s_2) = \big(v(s_1) \triangleleft' r(s_2)\big) *' v(s_2) \tag{1.41}$$

$$f(s_1, s_2)\, r(s_1 * s_2) = r(s_1)\, \big(v(s_1) \triangleright' r(s_2)\big)\, f'\big(v(s_1) \triangleleft' r(s_2),\, v(s_2)\big) \tag{1.42}$$

for all s, s_1, $s_2 \in S$ and $h \in H$. Through the above correspondence the morphism $\psi : H \ltimes_\Omega S \to H \ltimes_{\Omega'} S'$ corresponding to (r, v) is given by

$$\psi(h,\, s) = \big(h\, r(s),\, v(s)\big) \tag{1.43}$$

for all $h \in H$, $s \in S$. Furthermore, $\psi : H \ltimes_\Omega S \to H \ltimes_{\Omega'} S'$ is an isomorphism of groups if and only if $v : S \to S'$ is bijective.

Proof: A morphism of groups $\psi : H \ltimes_\Omega S \to H \ltimes_{\Omega'} S'$ that makes the left square of the diagram (1.38) commutative is uniquely determined by two maps $r = r_\psi : S \to H$, $v = v_\psi : S \to S'$ such that $\psi(1, s) = \big(r(s), v(s)\big)$ for all $s \in S$. In this case ψ is given by :

$$\psi(h, s) = \psi\big((h, 1_S) \cdot (1_H, s)\big) = (h, 1_S) \cdot \big(r(s), v(s)\big) = \big(hr(s), v(s)\big)$$

for all $h \in H$ and $s \in S$. Now, $\psi(1_H, 1_S) = (1_H, 1_{S'})$ if and only if r and v are unitary maps. Assuming this unitary condition, we can easily prove that ψ is a morphism of groups if and only if the compatibility conditions (1.39)-(1.42) hold for the pair (r, v). It remains to be proved that ψ given by (1.61) is an isomorphism if and only if $v : S \to S'$ is a bijective map. Assume first that ψ is an isomorphism. Then v is surjective and for s_1, $s_2 \in S$ such that $v(s_1) = v(s_2)$ we have:

$$\psi(1_H, s_2) = \big(r(s_2), v(s_2)\big) = \big(r(s_2), v(s_1)\big) = \psi\big(r(s_2) r(s_1)^{-1},\, s_1\big)$$

Hence $s_1 = s_2$ and v is injective. Conversely is straightforward. □

Definition 1.2.15 *Two elements $(*, \triangleleft, \triangleright, f)$ and $(*', \triangleleft', \triangleright', f')$ of $\mathcal{GES}$ $(H, (S, 1_S))$ are called* equivalent *and we denote this by $(*, \triangleleft, \triangleright, f) \sim (*', \triangleleft', \triangleright', f')$ if there exists a pair (r, v) of unitary maps $r : S \to H$, $v : S \to S$ such that v is a bijection on the set S and the compatibility conditions (1.39)–(1.42) are fulfilled.*

It follows from Proposition 1.2.14 that $(*, \triangleleft, \triangleright, f) \sim (*', \triangleleft', \triangleright', f')$ if and only if there exists $\psi : H \ltimes_{\Omega} S \to H \ltimes_{\Omega'} S$ an isomorphism of groups that stabilizes H. Thus, $\sim$ is an equivalence relation on the set $\mathcal{GES}(H, (S, 1_S))$. We denote by $\mathcal{K}^2_{\ltimes}(H, (S, 1_S))$ the quotient set $\mathcal{GES}(H, (S, 1_S))/\sim$.

Let $\mathcal{C}(H, (S, 1_S))$ be the category whose class of objects is the set $\mathcal{GES}(H, (S, 1_S))$. A morphism $\psi : (*, \triangleleft, \triangleright, f) \to (*', \triangleleft', \triangleright', f')$ in $\mathcal{C}(H, (S, 1_S))$ is a morphism of groups $\psi : H \ltimes_{\Omega} S \to H \ltimes_{\Omega'} S$ that stabilizes H.

The main result of this section which gives the full answer to the ES problem now follows as a direct application of Proposition 1.2.14: the classifying object for the ES problem is $\mathcal{K}^2_{\ltimes}(H, (S, 1_S))$ constructed above.

Theorem 1.2.16 *Let H be a group and $(S, 1_S)$ a pointed set. Then there exists a bijection between the set of objects of the skeleton of the category $\mathcal{C}(H, (S, 1_S))$ and $\mathcal{K}^2_{\ltimes}(H, (S, 1_S))$.*

Using Proposition 1.2.14 we can also prove a general Schreier classification theorem for unified products.

Proposition 1.2.17 *Let $\Omega(H) = \big((S, 1_S, *), \triangleleft, \triangleright, f\big)$, $\Omega'(H) = \big((S, 1_S, *'), \triangleleft', \triangleright', f'\big)$ be two group extending structures of a group H. Then there exists a morphism $\psi : H \ltimes_{\Omega} S \to H \ltimes_{\Omega'} S$ that stabilizes H and S if and only if $\triangleleft = \triangleleft'$ and there exists a unitary map $r : S \to H$ such that $\triangleright$, $*$ and f are implemented by $\triangleright'$, $*'$ and f' via r as follows:*

$$s \triangleright h = r(s)\,(s \triangleright' h)\,r(s \triangleleft h)^{-1} \tag{1.44}$$

$$s_1 * s_2 = \big(s_1 \triangleleft r(s_2)\big) *' s_2 \tag{1.45}$$

$$f(s_1, s_2) = r(s_1)\,\big(s_1 \triangleright' r(s_2)\big)\, f'(s_1 \triangleleft r(s_2),\, s_2)\, r(s_1 * s_2)^{-1} \tag{1.46}$$

for all s, s_1, $s_2 \in S$ and $h \in H$. Furthermore, any morphism of groups $\psi : H \ltimes S \to H \ltimes' S$ that stabilizes H and S is an isomorphism of groups and is given by

$$\psi(h, s) = \big(h\,r(s),\, s\big) \tag{1.47}$$

for all $h \in H$, $s \in S$.

Proof: Indeed, using Proposition 1.2.14, any morphism of groups $\psi : H \ltimes S \to H \ltimes' S$ that makes the left square of (1.38) commutative is given by (1.43) for some unique maps (u, v). Now, such a morphism $\psi = \psi_{u,v}$ makes the right square of (1.38) commutative if and only if v is the identity map on S. Now the proof follows from Proposition 1.2.14: (1.39) implies that the right actions $\triangleleft$ and $\triangleleft'$ should be equal while (1.44)–(1.46) are exactly (1.40)–(1.42) for $v = \mathrm{Id}_S$ and $\triangleleft = \triangleleft'$. □

Proposition 1.2.17 tells us that in order to obtain a Schreier type theorem for unified products we have to set the group H, the pointed set $(S, 1_S)$ and a right H-action $\triangleleft$ of the group H on the set S. Let $\mathcal{SES}(H, (S, 1_S), \triangleleft)$ be the set of all triples $(*, \triangleright, f)$ such that $\big((S, 1_S, *), \triangleleft, \triangleright, f\big)$ is a group extending structure of H.

Definition 1.2.18 *Let H be a group, $(S, 1_S)$ a pointed set and $\triangleleft : S \times H \to S$ a right action of H on S. Two elements $(*, \triangleright, f)$ and $(*', \triangleright', f')$ of $\mathcal{SES}(H, (S, 1_S), \triangleleft)$ are called* cohomologous *and we denote this by $(*, \triangleright, f) \approx (*', \triangleright', f')$ if there exists a unitary map $r : S \to H$ such that*

$$\begin{aligned} s_1 * s_2 &= \big(s_1 \triangleleft r(s_2)\big) *' s_2 \\ s \triangleright h &= r(s)\,\big(s \triangleright' h\big)\, r(s \triangleleft h)^{-1} \\ f(s_1, s_2) &= r(s_1)\,\big(s_1 \triangleright' r(s_2)\big)\, f'(s_1 \triangleleft r(s_2),\, s_2)\, r(s_1 * s_2)^{-1} \end{aligned}$$

for all s, s_1, $s_2 \in S$ and $h \in H$.

It follows from Proposition 1.2.17 that $(*, \triangleright, f) \approx (*', \triangleright', f')$ if and only if there exists $\psi : H \ltimes_\Omega S \to H \ltimes_{\Omega'} S$ a morphism of groups such that diagram (1.38) is commutative and moreover such a morphism is an isomorphism. Thus, $\approx$ is an equivalence relation on the set $\mathcal{SES}(H, (S, 1_S), \triangleleft)$. We denote by $\mathcal{H}^2_{\ltimes}(H, (S, 1_S), \triangleleft)$ the quotient set $\mathcal{SES}(H, (S, 1_S), \triangleleft)/ \approx$.

Let $\mathcal{D}(H, (S, 1_S), \triangleleft)$ be the category whose class of objects is the set $\mathcal{SES}(H, (S, 1_S), \triangleleft)$. A morphism $\psi : (*, \triangleright, f) \to (*', \triangleright', f')$ in $\mathcal{D}(H, (S, 1_S), \triangleleft)$ is a morphism of groups $\psi : H \ltimes_\Omega S \to H \ltimes_{\Omega'} S$ such that diagram (1.38) is commutative. The category $\mathcal{D}(H, (S, 1_S), \triangleleft)$ is a groupoid, that is, any morphism is an isomorphism. We obtain the Schreier type theorem for unified products:

Corollary 1.2.19 *Let H be a group, $(S, 1_S)$ a pointed set and $\triangleleft$ a right action of H on S. Then there exists a bijection between the set of objects of the skeleton of the category $\mathcal{D}(H, (S, 1_S), \triangleleft)$ and $\mathcal{H}^2_{\ltimes}(H, (S, 1_S), \triangleleft)$.*

Thus it follows from Corollary 1.2.19 that $\mathcal{H}^2_{\ltimes}(H, (S, 1_S), \triangleleft)$ is for the classification of unified products of groups the counterpart of the second cohomology group for the classification of extensions of an abelian group by a group [210, Theorem 7.34].

Corollary 1.2.20 *Let H be a group, $\Omega(H) = \big((S, 1_S, *), \triangleleft, \triangleright, f\big)$ a group extending structure of H and $(H, G, \triangleright', f')$ a crossed system of groups. Then there exists $\psi : H \ltimes_\Omega S \to H \#^{f'}_{\triangleright'} G$ an isomorphism of groups that stabilizes H if and only if the right action $\triangleleft$ of $\Omega(H)$ is the trivial one and there exists a pair (r, v), where $r : S \to H$ is a unitary map, $v : (S, *) \to G$ is an isomorphism of groups such that:*

$$s \triangleright h = r(s)\,\big(v(s) \triangleright' h\big)\, r(s)^{-1} \tag{1.48}$$

$$f(s_1, s_2) = r(s_1)\,\big(v(s_1) \triangleright' r(s_2)\big)\, f'\big(v(s_1),\, v(s_2)\big)\, r(s_1 * s_2)^{-1} \tag{1.49}$$

for all s, s_1, $s_2 \in S$ and $h \in H$.

Proof: We apply Proposition 1.2.14 in the case that $\triangleleft'$ is the trivial action in the group extending structure $\Omega'(H) = \big((G, 1_G, *'), \triangleleft', \triangleright', f'\big)$. □

Now we give necessary and sufficient conditions for a unified product to be isomorphic to a bicrossed product of groups such that H is stabilized.

Corollary 1.2.21 *Let H be a group, $\Omega(H) = \big((S, 1_S, *), \triangleleft, \triangleright, f\big)$ a group extending structure of H and $(H, G, \triangleleft', \triangleright')$ a matched pair of groups. Then there exists $\psi : H \ltimes_\Omega S \to H \bowtie G$ an isomorphism of group that stabilizes H if and only if there exists a pair (r, v), where $r : S \to H$ is a unitary map, $v : (S, \triangleleft) \to (G, \triangleleft')$ is a unitary map and an isomorphism of H-sets such that:*

$$(s \triangleright h)\, r(s \triangleleft h) = r(s)\, \big(v(s) \triangleright' h\big) \tag{1.50}$$

$$v(s_1 * s_2) = \big(v(s_1) \triangleleft' r(s_2)\big)\, v(s_2) \tag{1.51}$$

$$f(s_1, s_2) = r(s_1)\, \big(v(s_1) \triangleright' r(s_2)\big)\, r(s_1 * s_2)^{-1} \tag{1.52}$$

for all s, s_1, $s_2 \in S$ and $h \in H$.

Proof: We apply Proposition 1.2.14 in the case when f' is the trivial cocycle in the group extending structure $\Omega'(H) = \big((G, 1_G, *'), \triangleleft', \triangleright', f'\big)$. □

Theorem 1.2.16 offers the theoretical answer to the ES problem. The challenge that remains is a computational one: for a given group H and a pointed set $(S, 1_S)$ we have to compute explicitly the cohomological type object $\mathcal{K}^2_{\ltimes}(H, (S, 1_S))$: it classifies up to an isomorphism of groups that stabilizes H all groups that contain H as a subgroup of index $|S|$.

Now we provide an explicit example. For a group H we shall denote by $\mathcal{T}(H) \subseteq H \times \mathrm{Aut}(H)$ the set consisting of all pairs (h_0, D), where $h_0 \in H$, $D : H \to H$ is an automorphism of H such that for any $h \in H$:

$$D(h_0) = h_0, \qquad D^2(h) = h_0 h h_0^{-1} \tag{1.53}$$

Proposition 1.2.22 *Let H be a group and $S = \{1_S, c\}$ a set with two elements. Then:*

(1) There exists a bijection between the set $\mathcal{GES}(H, (S, 1_S))$ of all group extending structures of H and $\mathcal{T}(H)$. The bijection is given such that the group extending structure $\big((S, 1_S, *), \triangleleft, \triangleright, f\big)$ corresponding to $(h_0, D) \in \mathcal{T}(H)$ is defined as follows: $\triangleleft$ is the trivial action of H on S, $*$ is given by $c * c = 1_S$, the left action $\triangleright : S \times H \to H$ and the cocycle $f : S \times S \to H$ are given for any $h \in H$ by:

$$c \triangleright h := D(h), \qquad f(c, c) := h_0 \tag{1.54}$$

(2) There exists a bijection $\mathcal{K}^2_{\ltimes}(H, (S, 1_S)) \cong \mathcal{T}(H)/\sim$, where $\sim$ is the equivalence relation defined as follows: $(h_0, D) \sim (h_0', D')$ if and only if there exists $g \in H$ such that for any $h \in H$

$$h_0 = g D'(g)\, h_0', \qquad D(h) = g D'(h) g^{-1} \tag{1.55}$$

The bijection between $\mathcal{T}(H)/\sim$ and the isomorphism classes of all groups that contain and stabilize H as a subgroup of index 2 is given by

$$\overline{(h_0,\, D)} \mapsto H \ltimes_{(h_0,D)} S$$

where we denote by $H \ltimes_{(h_0,D)} S$ the unified product associated to the group extending structure constructed in (1.54) for a given $(h_0, D) \in \mathcal{T}(H)$ and $\overline{(h_0,\, D)}$ is the equivalence class of $(h_0,\, D)$ via the relation $\sim$. Explicitly, the multiplication on the group $H \ltimes_{(h_0,D)} S$ is given for any $h,\, h_1,\, h_2 \in H$ by:

$$\begin{aligned} (h_1, 1_S)\cdot(h_2, 1_S) &= (h_1h_2, 1_S), \qquad (1,c)\cdot(1,c) = (h_0, 1_S)\\ (h, 1_S)\cdot(1,c) &= (h,c), \qquad (1,c)\cdot(h,1_S) = (D(h),c) \end{aligned}$$

Proof: (1) We have to compute the set of all maps $(*, \triangleleft, \triangleright, f)$ satisfying the normalizing conditions (1.28), (1.29) as well as the compatibility conditions $(ES1)$-$(ES7)$. First we prove that $c \triangleleft h = c$, for all $h \in H$, i.e., $\triangleleft$ is the trivial action and $c * c = 1_S$, i.e., $S = C_2$, the cyclic group of order 2. Indeed, using (1.29) we already know that $1_S \triangleleft h = 1_S$, for all $h \in H$. Let $f : H \to S$ be a map such that $c \triangleleft h = f(h)$, for all $h \in H$. Then, f is a unit preserving map since (1.29) holds. We will prove that $f(h) = c$, for all $h \in H$, and hence $\triangleleft$ is the trivial action. Indeed, $(ES1)$ tells us that $\triangleleft$ is a right action: the condition $c \triangleleft (gh) = (c \triangleleft g) \triangleleft h$ takes the form $f(gh) = f(g) \triangleleft h$, for all g, $h \in H$. Assume, that there exists $g \in H$ such that $f(g) = 1_S$. Then we obtain that $f(gh) = 1_S \triangleleft h = 1_S$, i.e., $f(gh) = 1_S$, for all $h \in H$. Thus, f is the trivial map, that is $c \triangleleft h = 1_S$, for all $h \in H$. Then, axiom $(ES2)$ applied for $s_1 = s_2 = s_3 = c$ implies that $(c * c) * c = c * c$ and using $(ES7)$ we obtain $c = 1_S$, which is a contradiction.

Thus, $\triangleleft$ is the trivial action; it follows from $(ES1)$ that $*$ is a group structure on S i.e., $c * c = 1_S$. Now, any normalizing map $\triangleright : S \times H \to H$ is uniquely implemented by a map $D : H \to H$ such that $c \triangleright h = D(h)$, for all $h \in H$, and any normalized map $f : S \times S \to H$ is uniquely determined by an element $h_0 \in H$ such that $f(c, c) = h_0$. Now we can easily see that the compatibility conditions $(ES1)$, $(ES2)$, $(ES4)$ are trivially fulfilled, while $(ES3)$ is equivalent to the fact that D is an endomorphism of H, axiom $(ES5)$ takes the equivalent form given by the right-hand side of (1.53) while $(ES6)$ is the left-hand side of (1.53).

(2) Let $(*, \triangleleft, \triangleright, f)$ and $(*', \triangleleft', \triangleright', f')$ be two group extending structures associated to (h_0, D) and $(h_0', D') \in \mathcal{T}(H)$. Then, $(*, \triangleleft, \triangleright, f) \sim (*', \triangleleft', \triangleright', f')$ in the sense of Definition 1.2.15 if there exists a pair (r, v) of unitary maps $r : S \to H$, $v : S \to S$ such that v is a bijection on the set S and the compatibility conditions (1.39)–(1.42) are fulfilled. Since $|S| = 2$ and v is a unitary bijection, we obtain that v is the identity map on S. On the other hand, the unitary map $r : S \to H$ is given by an element $g \in H$ such that $r(c) = g$. Taking into account the construction of the group extending structures from

(1.54) we can easily show that the compatibility conditions (1.39)–(1.42) take the equivalent form (1.55). □

Examples 1.2.23 *1. Let H be a group which has no outer automorphisms. The typical example is S_n, for $n \neq 6$. Then, in this case $\mathcal{T}(H)$ identifies with the set of all pairs $(h_0, a) \in H \times H$ such that :*

$$a h_0 = h_0 a, \qquad a^2 h a^{-2} = h_0 h h_0^{-1} \tag{1.56}$$

for all $h \in H$. Moreover, $\mathcal{K}^2_{\ltimes}(H, (S, 1_S)) \simeq \mathcal{T}(H)/\sim$, where $(h_0, a) \sim (h_0', a')$ if and only if there exists $g \in H$ such that $h_0 = g a' g a'^{-1} h_0'$ and $a h a^{-1} = g a' h a'^{-1} g^{-1}$, for all $h \in H$.

2. Let H be an abelian group. Then $\mathcal{T}(H)$ is the set of all pairs $(h_0, D) \in H \times \mathrm{Aut}(H)$, such that $D^2 = \mathrm{Id}_H$ and $D(h_0) = h_0$. Two such pairs (h_0, D) and (h_0', D') are equivalent if and only if $D = D'$ and there exists $g \in H$ such that $h_0 = g D'(g) h_0'$.

In particular, if $H = \mathbb{Z}$ we obtain that $\mathcal{K}^2_{\ltimes}(\mathbb{Z}, (S, 1_S)) \cong \{\overline{(0, \mathrm{Id}_\mathbb{Z})}, \overline{(0, -\mathrm{Id}_\mathbb{Z})}, \overline{(1, \mathrm{Id}_\mathbb{Z})}\}$. Thus, up to an isomorphism of groups that stabilizes $\mathbb{Z}$ there are three groups that contain $\mathbb{Z}$ as a subgroup of index 2: the direct product $\mathbb{Z} \times C_2$ corresponding to $\overline{(0, \mathrm{Id}_\mathbb{Z})}$, the semidirect product $\mathbb{Z} \ltimes C_2$ corresponding to $\overline{(0, -\mathrm{Id}_\mathbb{Z})}$ and the crossed product $\mathbb{Z} \#^f C_2$ associated to the non-trivial cocycle corresponding to $\overline{(1, \mathrm{Id}_\mathbb{Z})}$, where C_2 is the cyclic group of order 2.

1.3 Classifying complements

In this section, as an application of the extending structures problem, we shall study what we have called the classifying complements problem for groups. First of all we need to introduce the following concept:

Definition 1.3.1 *Let $A \leq G$ be a subgroup of G. An A-*complement *of G is a subgroup $H \leq G$ such that G factorizes through A and H, that is $G = AH$ and $A \cap H = \{1\}$. We denote by $\mathcal{F}(A, G)$ the set of isomorphism types of all A-complements of G. We define the* factorization index *of A in G as the cardinal of $\mathcal{F}(A, G)$ and it will be denoted by $[G : A]^f := |\,\mathcal{F}(A, G)\,|$. We shall write $[G : A]^f = 0$, if the set $\mathcal{F}(A, G)$ is empty.*

The problem of existence of complements has to be treated "case by case" for every given subgroup A of G, a computational part of it cannot be avoided. It was studied in its global form: *find all factorizations of a given group G.* Particular attention was given to finding all factorizations of simple groups. Starting with the 1970s, a very rich literature on the subject was developed: see for instance [40, 54], [109, 110, 116, 117, 120], [164], [203], [228, 230]. For more details on this problem we refer to the two fundamental monographs

[163], [165] and the references therein. This section deals with the following question:

Classifying complements problem (CCP): *Let A be a subgroup of G. If an A-complement of G exists, describe explicitly, classify all A-complements of G and compute the factorization index $[G : A]^f$.*

We shall give the answer to the CCP in three steps called: deformation of complements, description of complements and classification of complements. Let H be a given A-complement of G and $(\triangleright, \triangleleft)$ the canonical left/right actions associated to the factorization $G = AH$, as described in Proposition 1.1.9, such that $(A, H, \triangleright, \triangleleft)$ is a matched pair of groups and $G = A \bowtie H$. Theorem 1.3.6 is called the *deformation of complements*: if $r : H \to A$ is a deformation map of the matched pair $(A, H, \triangleright, \triangleleft)$, then the group H is deformed to a new group H_r, called the r-deformation of H, such that H_r remains an A-complement of $G = A \bowtie H$. The key point is Theorem 1.3.7, called the *description of complements*: $\mathbb{H}$ is an A-complement of G if and only if $\mathbb{H}$ is isomorphic to H_r, for some deformation map $r : H \to A$ of the canonical matched pair $(A, H, \triangleright, \triangleleft)$. Finally, the *classification of complements* is proven in Theorem 1.3.11: there exists a bijection between the set of isomorphism types of all A-complements of G and a cohomological type object $\mathcal{D}(H, A \,|\, (\triangleright, \triangleleft))$ which is explicitly constructed. In particular, the factorization index is computed by the formula $[G : A]^f = |\, \mathcal{D}(H, A \,|\, (\triangleright, \triangleleft)) \,|$. Explicit examples will also be provided. Let S_n be the symmetric group and C_n the cyclic group of order n. By applying our results to the factorization $S_n = S_{n-1}C_n$ we obtain the following: (1) any group H of order n is isomorphic to $(C_n)_r$, the r-deformation of the cyclic group C_n for some deformation map $r : C_n \to S_{n-1}$ of the canonical matched pair $(S_{n-1}, C_n, \triangleright, \triangleleft)$ and (2) the number of isomorphism types of all groups of order n is equal to $|\, \mathcal{D}(C_n, S_{n-1} \,|\, (\triangleright, \triangleleft)) \,|$. Therefore, we obtain a combinatorial formula for computing the number of isomorphism types of all groups of order n which arises from a minimal set of data: the factorization $S_n = S_{n-1}C_n$.

Let $A \leq G$ be a given subgroup of G. We will see that a factorization $G = AH$ is not necessarily unique as there may exist other subgroups $H' \leq G$, not isomorphic to H, such that $G = AH'$. Such an example is presented below.

Example 1.3.2 *Let k be a positive integer. In what follows we view A_{4k-1} as a subgroup of A_{4k} by letting $4k$ be a fixed point in the alternating group A_{4k}. Then we have two factorizations: $A_{4k} = A_{4k-1}D_{4k} = A_{4k-1}(C_2 \times C_{2k})$, where D_{4k} is the dihedral group and C_m is the cyclic group of order m. Indeed, let σ, $\tau \in A_{4k}$ be the even permutations*

$$\begin{aligned}
\sigma &= (1, 3, 5, \cdots, 4k-1)(2, 4, 6, \cdots, 4k) \\
\tau &= (1, 2k+2)(2, 2k+1)(3, 2k+4)(4, 2k+3) \cdots (2k-1, 4k)(2k, 4k-1)
\end{aligned}$$

It is straightforward to check that σ and τ generate a subgroup of A_{4k} isomorphic to the dihedral group D_{4k} of order $4k$ and $A_{4k} = A_{4k-1}D_{4k}$. On the other hand, let σ', $\tau' \in A_{4k}$ given by

$$\sigma' = (1, 2, \cdots, 2k)(2k+1, 2k+2, \cdots, 4k), \quad \tau' = (1, 2k+1)(2, 2k+2) \cdots (2k, 4k)$$

Then $\sigma'\tau' = \tau'\sigma'$ and the subgroup of A_{4k} generated by σ and τ is $C_2 \times C_{2k}$. Moreover, we have $A_{4k} = A_{4k-1}(C_2 \times C_{2k})$.

More examples are presented below.

Examples 1.3.3 *1. Many group extensions $A \leq G$ have the factorization index $[G : A]^f$ equal to 0 (that is there exists no factorization $G = AH$) or 1. For instance, if G is an abelian group, then $[G : A]^f \in \{0, 1\}$, for any subgroup A of G ($[G : A]^f = 1$ if and only if A is a direct summand of G).*

Group extensions $A \leq G$ of factorization index 1 are exactly those for which the factorization is unique. A generic example of an extension of factorization index 1 is provided in Corollary 1.3.9 below: if $A \ltimes H$ is an arbitrary semidirect product of A and H, then $[A \ltimes H : A]^f = 1$.

2. Examples of extensions $A \leq G$ for which $[G : A]^f \geq 2$ are quite rare, makes them tempting to identify. Example 1.3.2 proves in fact that $[A_{4k} : A_{4k-1}]^f \geq 2$. We provide below an example of an extension of factorization index 2.

The extension $S_3 \leq S_4$ has factorization index 2. Indeed, let $C_4 =< (1234) >$ be the cyclic group of order 4 and $C_2 \times C_2$ the Klein's group viewed as a subgroup of S_4 being generated by (12)(34) and (13)(24). Then S_4 has two factorizations: $S_4 = S_3C_4 = S_3(C_2 \times C_2)$. Since there are no other groups of order four we obtain that $[S_4 : S_3]^f = 2$.

3. Example (2) above can be generalized as follows: the factorization index $[S_n : S_{n-1}]^f = g(n)$, the number of isomorphism types of groups of order n. Indeed, let H be a group of order n. We see H as a subgroup of S_n through the regular representation, i.e., $T : H \to S_n$ given by $T(h) = \sigma_h$, where $\sigma_h(x) = hx$, for all $h, x \in H$. It is now obvious that through this representation n is not fixed by any other element in H besides 1. Since we consider S_{n-1} as a subgroup in S_n by letting n be a fixed point, we have $H \cap S_{n-1} = 1$ and therefore $S_n = S_{n-1}H$.

From now on, the matched pair constructed in (1.19) will be called the *canonical matched pair* associated to the factorization $G = AH$. We use the above terminology in order to distinguish this matched pair among other possible matched pairs $(A, H, \triangleright', \triangleleft')$ such that $A \bowtie' H \cong G$ (isomorphism of groups that stabilizes A), where $A \bowtie' H$ is the bicrossed product associated to the matched pair $(A, H, \triangleright', \triangleleft')$. The following result provides more details: it is a special case of Proposition 1.2.14. However, we state the result below for the sake of completeness as it will be used in the sequel. We recall that if G, G' are two groups containing A as a subgroup, a morphism of groups $\psi : G \to G'$ *stabilizes* A if $\psi(a) = a$, for all $a \in A$.

Proposition 1.3.4 *Let $(A, H, \triangleright, \triangleleft)$ and $(A, H', \triangleright', \triangleleft')$ be two matched pairs of groups. There exists a bijection between the set of all morphisms of groups $\psi : A \bowtie' H' \to A \bowtie H$ that stabilize A and the set of all pairs (r, v), where*

$r : H' \to A$, $v : H' \to H$ are two unit preserving maps satisfying the following compatibilities for any h', $g' \in H'$, $a \in A$:

$$\begin{aligned} h' \triangleright' a &= r(h')\big(v(h') \triangleright a\big)\, r(h' \triangleleft' a)^{-1} && (1.57)\\ v(h' \triangleleft' a) &= v(h') \triangleleft a && (1.58)\\ r(h'g') &= r(h')\big(v(h') \triangleright r(g')\big) && (1.59)\\ v(h'g') &= \big(v(h') \triangleleft r(g')\big)\, v(g') && (1.60)\end{aligned}$$

Under the above correspondence, the morphism of groups $\psi : A \bowtie' H' \to A \bowtie H$ corresponding to (r, v) is given by:

$$\psi(a,\, h') = \big(a\, r(h'),\, v(h')\big) \tag{1.61}$$

for all $a \in A$, $h' \in H'$ and $\psi : A \bowtie' H' \to A \bowtie H$ is an isomorphism of groups if and only if the map $v : H' \to H$ is bijective.

Let H be a given A-complement of G and $(A, H, \triangleright, \triangleleft)$ the canonical matched pair associated to it as in (1.19) of Proposition 1.1.9. We shall describe all A-complements of G in terms of $(H, \triangleleft, \triangleright)$ and certain maps $r : H \to A$, called deformation maps defined below.

Definition 1.3.5 *Let $(A, H, \triangleright, \triangleleft)$ be a matched pair of groups. A* deformation map *of the matched pair $(A, H, \triangleright, \triangleleft)$ is a function $r : H \to A$ such that $r(1) = 1$ and for all g, $h \in H$ we have:*

$$r\big((h \triangleleft r(g))\, g\big) = r(h)\, \big(h \triangleright r(g)\big) \tag{1.62}$$

Let $\mathcal{DM}\,(H, A \,|\, (\triangleright, \triangleleft))$ be the set of all deformation maps of the matched pair $(A, H, \triangleright, \triangleleft)$. The trivial map $H \to A$, $h \mapsto 1$, for any $h \in H$ is a deformation map. If both actions $(\triangleright, \triangleleft)$ of the matched pair are trivial then a deformation map is just a morphism of groups $r : H \to A$. The following result is called the deformation of complements: it shows that any A-complement can be deformed to a new A-complement using a deformation map $r : H \to A$.

Theorem 1.3.6 *Let $(A, H, \triangleright, \triangleleft)$ be a matched pair of groups and $r : H \to A$ a deformation map. The following hold:*

(1) Let $H_r := H$, as a set, with the new multiplication $\bullet$ on H defined for any h, $g \in H$ as follows:

$$h \bullet g := \big(h \triangleleft r(g)\big)\, g \tag{1.63}$$

Then $(H_r, \bullet)$ is a group called the r-deformation of H.

(2) The map

$$\triangleright^r : H_r \times A \to A, \quad h \triangleright^r a := r(h)\, \big(h \triangleright a\big)\, r(h \triangleleft a)^{-1} \tag{1.64}$$

for all $h \in H_r$, $a \in A$ is a left action of the group H_r on the set A and $(A, H_r, \triangleright^r, \triangleleft)$ is a matched pair of groups. Furthermore, the map

$$\psi : A \bowtie^r H_r \to A \bowtie H, \quad \psi(a,\, h) = (a\, r(h),\, h) \tag{1.65}$$

for all $a \in A$ and $h \in H$ is an isomorphism of groups, where $A \bowtie^{r} H_r$ is the bicrossed product associated to the matched pair $(A, H_r, \triangleright^r, \triangleleft)$.

(3) *H_r is an A-complement of $A \bowtie H$.*

Proof: (1) Using the normalizing conditions (1.17) and the fact that $r : H \to A$ is a unitary map, 1 remains the unit for the new multiplication $\bullet$ given by (1.63). On the other hand for any $h, g, t \in H$ we have:

$$\begin{aligned}
(h \bullet g) \bullet t &= \big[(h \triangleleft r(g))g\big] \bullet t = \Big(\underline{((h \triangleleft r(g))g)\triangleleft r(t)}\Big)\, t \\
&\overset{(1.16)}{=} \Big(\underline{(h \triangleleft r(g))\triangleleft(g \triangleright r(t))}\Big)\, (g \triangleleft r(t))\, t \\
&= \Big(h \triangleleft \big(\underline{r(g)(g \triangleright r(t))}\big)\Big)\, (g \triangleleft r(t))\, t \\
&\overset{(1.62)}{=} \Big(h \triangleleft r\big((g \triangleleft r(t))\, t\big)\Big)\, (g \triangleleft r(t))\, t \\
&= h \bullet \big[(g \triangleleft r(t))\, t\big] = h \bullet (g \bullet t)
\end{aligned}$$

Thus, the multiplication $\bullet$ is associative and has 1 as a unit. We prove now that the inverse of an element $h \in H_r$ is given by $h^{-1} = h^{-1} \triangleleft r(h)^{-1}$, for all $h \in H$. Indeed, for any $h \in H$ we have:

$$\begin{aligned}
h^{-1} \bullet h &= (h^{-1} \triangleleft r(h)^{-1}) \bullet h = \Big((h^{-1} \triangleleft r(h)^{-1})\triangleleft r(h)\Big)\, h \\
&= \Big(h^{-1} \triangleleft \big(r(h)^{-1} r(h)\big)\Big)\, h = h^{-1}\, h = 1
\end{aligned}$$

Thus we proved that $(H_r, \bullet)$ is a monoid in which every element has a left inverse. Hence $(H_r, \bullet)$ is a group.

(2) Instead of using a rather long computation to prove that $(A, H_r, \triangleright^r, \triangleleft)$ satisfies the axioms (1.15)-(1.16) of a matched pair we proceed as follows: first, observe that the map $\psi : A \times H_r \to A \bowtie H$, $\psi(a, h) = (a\, r(h), h)$ is a bijection between the set $A \times H_r$ and the group $A \bowtie H$ with the inverse given by

$$\psi^{-1} : A \bowtie H \to A \times H_r, \quad \psi^{-1}(a, h) = (a\, r(h)^{-1}, h)$$

for all $a \in A$ and $h \in H$. Thus, there exists a unique group structure $\diamond$ on the set $A \times H_r$ such that ψ becomes an isomorphism of groups and this unique group structure $\diamond$ is obtained by transferring the group structure from the group $A \bowtie H$ via the bijection of sets ψ, i.e., is given by:

$$(a, h) \diamond (b, g) := \psi^{-1}\big(\psi(a, h) \cdot \psi(b, g)\big)$$

for all $a, b \in A$ and $h, g \in H_r = H$. If we prove that this group structure $\diamond$ on the direct product of sets $A \times H_r$ is exactly the one given by (1.18) associated to the pair of maps $(\triangleright^r, \triangleleft)$ the proof is finished by using Proposition 1.1.8.

Indeed, for any $a,\, b \in A$ and $g,\, h \in H$ we have:

$$\begin{aligned}
(a,\, h) \diamond (b,\, g) &= \psi^{-1}\big(\psi(a,\, h) \cdot \psi(b,\, g)\big) = \psi^{-1}\Big(\big(a\, r(h),\, h\big) \cdot \big(b\, r(g),\, g\big)\Big)\\
&= \psi^{-1}\Big(a\, r(h)\big(h \triangleright br(g)\big),\, \big(h \triangleleft br(g)\big)\, g\Big)\\
&= \Big(a\, r(h)\big(h \triangleright br(g)\big) r\Big(\big(h \triangleleft br(g)\big)\, g\Big)^{-1},\, \big(h \triangleleft br(g)\big)\, g\Big)\\
&= \Big(a\, r(h)\big(h \triangleright br(g)\big) \underline{r\Big(\big((h \triangleleft b) \triangleleft r(g)\big)\, g\Big)^{-1}},\, \big(h \triangleleft br(g)\big)\, g\Big)\\
&\overset{(1.62)}{=} \Big(a\, r(h)\underline{\big(h \triangleright br(g)\big)}\Big[r\big(h \triangleleft b\big)\Big(\big(h \triangleleft b\big) \triangleright r(g)\Big)\Big]^{-1},\\
&\qquad \big(h \triangleleft br(g)\big)\, g\Big)\\
&\overset{(1.15)}{=} \Big(a\, r(h)\big(h \triangleright b\big)\underline{\Big(\big(h \triangleleft b\big) \triangleright r(g)\Big)\Big(\big(h \triangleleft b\big) \triangleright r(g)\Big)^{-1}}\, r\big(h \triangleleft b\big)^{-1},\\
&\qquad \big(h \triangleleft br(g)\big)\, g\Big)\\
&= \Big(a\, r(h)\big(h \triangleright b\big) r\big(h \triangleleft b\big)^{-1},\, \underline{\big(h \triangleleft br(g)\big)\, g}\Big)\\
&\overset{(1.63)}{=} \Big(\underline{a\, r(h)\big(h \triangleright b\big) r\big(h \triangleleft b\big)^{-1}},\, \big(h \triangleleft b\big) \bullet g\Big)\\
&\overset{(1.64)}{=} \Big(a\, (h \triangleright^r b),\, (h \triangleleft b) \bullet g\Big) = (a,\, h) \cdot^r (b,\, g)
\end{aligned}$$

where $\cdot^r$ is the multiplication given by (1.18) associated to the new pair of maps $(\triangleright^r, \triangleleft)$. Now we apply Proposition 1.1.8.

(3) First we remark that the isomorphism of groups $\psi : A \bowtie^r H_r \to A \bowtie H$ given by (1.65) stabilizes A. Hence $A \cong \psi(A) = A \times \{1\} \leq A \bowtie H$ and $H_r \cong \psi(\{1\} \times H_r) = \{(r(h), h) \,|\, h \in H\}$ is a subgroup of $A \bowtie H$. Now, $A \bowtie H$ factorizes through A and H_r since in $A \bowtie H$ we have:

$$(a, h) = (ar(h)^{-1}, 1) \cdot (r(h), h)$$

for all $a \in A$ and $h \in H$. Of course, $A \times \{1\}$ and $\{(r(h), h) \,|\, h \in H\} \cong H_r$ have trivial intersection in $A \bowtie H$ as r is a unitary map. The proof is now completely finished. □

Now we prove the converse of Theorem 1.3.6 which gives the description of all A-complements of G in terms of a fixed one H.

Theorem 1.3.7 *Let $A \leq G$ be a subgroup of G and H a given A-complement of G. Then $\mathbb{H}$ is an A-complement of G if and only if there exists an isomorphism of groups $\mathbb{H} \cong H_r$, for some deformation map $r : H \to A$ of the canonical matched pair $(A, H, \triangleright, \triangleleft)$ associated to the factorization $G = AH$.*

Proof: Let $A \bowtie H$ be the bicrossed product of the canonical matched pair $(A, H, \triangleright, \triangleleft)$. Then the multiplication map $m_G : A \bowtie H \to G$ is an isomorphism of groups that stabilizes A. Consider $(A, \mathbb{H}, \triangleright', \triangleleft')$ to be the canonical matched pair associated to the factorization $G = A\mathbb{H}$; hence the multiplication map $m'_G : A \bowtie' \mathbb{H} \to G$ is also an isomorphism of groups that stabilizes A. Then $\psi := m_G^{-1} \circ m'_G : A \bowtie' \mathbb{H} \to A \bowtie H$ is a group isomorphism that stabilizes A as a composition of such morphisms. Now by applying Proposition 1.3.4 it follows that ψ is uniquely determined by a pair of maps $(\overline{r}, \overline{v})$ consisting of a unitary map $\overline{r} : \mathbb{H} \to A$ and a unitary bijective map $\overline{v} : \mathbb{H} \to H$ satisfying the compatibility conditions

$$h' \triangleright' a = \overline{r}(h')\,\big(\overline{v}(h') \triangleright a\big)\,\overline{r}(h' \triangleleft' a)^{-1} \tag{1.66}$$

$$\overline{v}(h' \triangleleft' a) = \overline{v}(h') \triangleleft a \tag{1.67}$$

$$\overline{r}(h'g') = \overline{r}(h')\,\big(\overline{v}(h') \triangleright \overline{r}(g')\big) \tag{1.68}$$

$$\overline{v}(h'g') = \big(\overline{v}(h') \triangleleft \overline{r}(g')\big)\,\overline{v}(g') \tag{1.69}$$

for all h', $g' \in \mathbb{H}$ and $a \in A$. Moreover, $\psi : A \bowtie' \mathbb{H} \to A \bowtie H$ is given by:

$$\psi(a, h') = (a\,\overline{r}(h'), \overline{v}(h'))$$

for all $a \in A$ and $h' \in \mathbb{H}$. We define

$$r : H \to A, \quad r := \overline{r} \circ \overline{v}^{-1}$$

and we will prove that r is a deformation map of the matched pair $(A, H, \triangleright, \triangleleft)$ and $\overline{v} : \mathbb{H} \to H_r$ is an isomorphism of groups. First, notice that r is unitary as $\overline{r}$, $\overline{v}$ are both unitary. We have to show that the compatibility condition (1.62) holds for r. Indeed, from (1.68) and (1.69) we obtain:

$$\overline{r} \circ \overline{v}^{-1}\big[\big(\overline{v}(h') \triangleleft \overline{r}(g')\big)\,\overline{v}(g')\big] = \overline{r}(h')\big(\overline{v}(h') \triangleright \overline{r}(g')\big) \tag{1.70}$$

for all h', $g' \in \mathbb{H}$. Let h, $g \in H$ and write the compatibility condition (1.70) for $h' = \overline{v}^{-1}(h)$ and $g' = \overline{v}^{-1}(g)$. We obtain

$$r\Big(\big(h \triangleleft r(g)\big)\, g\Big) = r(h)\,\big(h \triangleright r(g)\big)$$

that is, (1.62) holds and hence $r : H \to A$ is a deformation map. Finally, $\overline{v} : \mathbb{H} \to H_r$ is a bijective map as $H = H_r$ as sets. Hence, we are left to prove that $\overline{v}$ is also a morphism of groups. Indeed, for any h', $g' \in \mathbb{H}$ we have:

$$\overline{v}(h'g') \overset{(1.69)}{=} \big(\overline{v}(h') \triangleleft \overline{r}(g')\big)\,\overline{v}(g') \overset{(1.63)}{=} \overline{v}(h') \bullet \overline{v}(g')$$

where $\bullet$ is the multiplication on H_r as defined by (1.63). Hence $\overline{v} : \mathbb{H} \to H_r$ is an isomorphism of groups and the proof is finished. □

Remark 1.3.8 *Assume that in Theorem 1.3.6 the deformation map $r : H \to A$ is the trivial one or the right action $\triangleleft$ is the trivial action of A on H. Then $H_r = H$ as groups. In general, the new group H_r may not be isomorphic to*

H as groups. Example 1.4.3 shows how the Klein's group $C_2 \times C_2$ can be constructed as an r-deformation of the cyclic group C_4, for some deformation map $r : C_4 \to S_3$. On the other hand, there are also examples of non-trivial deformation maps, with a non-trivial action $\triangleleft$, such that H_r is a group isomorphic to H. Such an example is provided in Example 1.4.5.

Corollary 1.3.9 *Let A and H be two groups, $A \ltimes H$ an arbitrary semidirect product of A and H. Then the factorization index $[A \ltimes H : A]^f = 1$.*

Proof: Indeed, $H \cong \{1\} \times H$ is an A-complement of the semidirect product $A \ltimes H$. Moreover, the right action $\triangleleft$ of the canonical matched pair $(A, H, \triangleright, \triangleleft)$ constructed in (1.19) for the factorization $A \ltimes H = (A \times \{1\})(\{1\} \times H)$ is the trivial action. Thus, using Remark 1.3.8, any r-deformation of $H \cong \{1\} \times H$ coincides with H. The rest follows from Theorem 1.3.7. □

In order to provide the classification of complements we need one more definition:

Definition 1.3.10 *Let $(A, H, \triangleright, \triangleleft)$ be a matched pair of groups. Two deformation maps r, $R : H \to A$ are called* equivalent *and we denote this by $r \sim R$ if there exists $\sigma : H \to H$ a permutation on the set H such that $\sigma(1_H) = 1_H$ and for all g, $h \in H$ we have:*

$$\sigma\big((h \triangleleft r(g))\, g\big) = \big(\sigma(h) \triangleleft R(\sigma(g))\big)\, \sigma(g) \tag{1.71}$$

As a conclusion of all the above results, the main theorem of this section which gives the classification of all A-complements of a group G now follows.

Theorem 1.3.11 *Let $A \leq G$ be a subgroup of G, H a given A-complement of G and $(A, H, \triangleright, \triangleleft)$ the associated canonical matched pair. Then:*

(1) *$\sim$ is an equivalence relation on $\mathcal{DM}(H, A \,|\, (\triangleright, \triangleleft))$ and the map*

$$\mathcal{D}(H, A \,|\, (\triangleright, \triangleleft)) \to \mathcal{F}(A, G), \quad \overline{r} \mapsto H_r$$

is a bijection between sets, where $\mathcal{D}(H, A \,|\, (\triangleright, \triangleleft)) := \mathcal{DM}(H, A \,|\, (\triangleright, \triangleleft))/\sim$ is the quotient set through the relation $\sim$ and $\overline{r}$ is the equivalence class of r via $\sim$.

(2) *The factorization index $[G : A]^f$ is computed by the formula:*

$$[G : A]^f = |\mathcal{D}(H, A \,|\, (\triangleright, \triangleleft))|$$

Proof: It follows from Theorem 1.3.7 that if $\mathbb{H}$ is an arbitrary A-complement of G, then there exists an isomorphism of groups $\mathbb{H} \cong H_r$, for some deformation map $r : H \to A$ of the matched pair $(A, H, \triangleright, \triangleleft)$. Thus, in order to classify all A-complements on G we can consider only r-deformations of H, for various deformation maps $r : H \to A$. Now let r, $R : H \to A$ be two deformation maps of the matched pair $(A, H, \triangleright, \triangleleft)$. As H_r and H_R coincide as sets, the groups H_r and H_R are isomorphic if and only if there exists $\sigma : H \to H$ a unitary bijective map such that $\sigma : H_r \to H_R$ is a morphism of groups. Taking into account the definition of the multiplication on H_r given by (1.63) it follows that σ is a group morphism if and only if the compatibility condition (1.71) of Definition 1.3.10 holds, i.e., $r \sim R$. Hence, $r \sim R$ if and only if there exists a map σ such that $\sigma : H_r \to H_R$ is an isomorphism of groups. Therefore $\sim$ is an equivalence relation on $\mathcal{DM}(H, A \,|\, (\triangleright, \triangleleft))$ and the map

$$\mathcal{D}(H, A \,|\, (\triangleright, \triangleleft)) \to \mathcal{F}(A, E), \qquad \overline{r} \mapsto H_r$$

is well defined and a bijection between sets, where $\overline{r}$ is the equivalence class of r via the relation $\sim$. (2) follows from (1) and the proof is now finished. □

1.4 Examples: Applications to the structure of finite groups

Let n be a positive integer. In this section we apply the results obtained in Section 1.3 to the factorization $S_n = S_{n-1}C_n$. As a consequence, we derive a combinatorial formula for computing the number of types of groups of order n as well as an explicit description for the multiplication on any group of order n. In what follows, we consider the usual presentation of the symmetric group S_n:

$$S_n = \langle s_1,\, s_2,\, \ldots,\, s_{n-1} \mid s_i^2 = 1,\, s_i\, s_{i+1}\, s_i = s_{i+1}\, s_i\, s_{i+1},\, s_i\, s_j = s_j\, s_i,\, |i-j| > 1 \rangle$$

We shall see the cyclic group C_n as a subgroup of S_n generated by $x := s_1\, s_2\, \ldots\, s_{n-1}$ while S_{n-1} will be seen as the subgroup of S_n generated by $s_1,\, s_2,\, \ldots\, s_{n-2}$. To start with, we describe the canonical matched pair associated to the factorization $S_n = S_{n-1}C_n$. It is enough to define the two actions $\triangleright : C_n \times S_{n-1} \to S_{n-1}$ and $\triangleleft : C_n \times S_{n-1} \to C_n$ on the generators of S_{n-1} and C_n as they can be extended to the entire group by using the compatibilities (1.15) and (1.16).

Proposition 1.4.1 *The canonical matched pair $(S_{n-1}, C_n, \triangleright, \triangleleft)$ associated to the factorization $S_n = S_{n-1}C_n$ is given as follows:*

$$x \triangleright s_i = \begin{cases} s_{i+1}, & \textit{if} \;\; i < n-2 \\ s_{n-2}\, s_{n-3}\, \ldots\, s_1, & \textit{if} \;\; i = n-2 \end{cases}$$

$$x \triangleleft s_i = \begin{cases} x, & \textit{if} \;\; i < n-2 \\ x^2, & \textit{if} \;\; i = n-2 \end{cases}$$

where $x := s_1\, s_2\, \ldots\, s_{n-1}$.

Proof: We compute the canonical matched pair by using the approach highlighted in the proof of Proposition 1.1.9. We start by computing the xs_i's, for al $i \in 1, 2, \dots, n-2$. If $i < n-2$ we have:

$$\begin{aligned} xs_i &= s_1 \dots s_{i-1}\, s_i\, s_{i+1}\, s_{i+2} \dots s_{n-1}\, s_i \\ &= s_1 \dots s_{i-1}\, (s_i\, s_{i+1}\, s_i)\, s_{i+2} \dots s_{n-1} \\ &= s_1 \dots s_{i-1}\, (s_{i+1}\, s_i\, s_{i+1})\, s_{i+2} \dots s_{n-1} \\ &= s_{i+1}\, s_1 \dots\, s_{n-1} = s_{i+1} x \end{aligned}$$

If $i = n-2$ we obtain:

$$\begin{aligned} xs_{n-2} &= s_1 \dots s_{n-3}\, (s_{n-2}\, s_{n-1}\, s_{n-2}) \\ &= s_1 \dots s_{n-3}\, (s_{n-1}\, s_{n-2}\, s_{n-1}) \\ &= s_{n-1}\, s_1 \dots s_{n-1} \\ &= s_{n-2}\, s_{n-3} \dots s_1 (s_1\, s_2 \dots s_{n-1})^2 = x' x^2 \end{aligned}$$

where $x' := s_{n-2}\, s_{n-3}\, \dots\, s_1$ and the conclusion follows easily. □

By applying Theorem 1.3.7 and Theorem 1.3.11 for the factorization $S_n = S_{n-1}C_n$ we obtain the following result concerning the structure and the number of types of groups of finite order.

Corollary 1.4.2 *Let n be a positive integer and $(S_{n-1}, C_n, \triangleright, \triangleleft)$ the canonical matched pair associated to the factorization $S_n = S_{n-1}C_n$. Then:*

(1) *Any group of order n is isomorphic to an r-deformation of the cyclic group C_n, for some deformation map $r : C_n \to S_{n-1}$ of the canonical matched pair $(S_{n-1}, C_n, \triangleright, \triangleleft)$. The multiplication $\bullet$ on $(C_n)_r$ is given by: $x \bullet y = \big(x \triangleleft r(y)\big)\, y$, for all x, $y \in (C_n)_r$, where we denoted by juxtaposition the multiplication in the cyclic group C_n.*

(2) *The number of isomorphism types of all groups of order n is equal to*

$$|\, \mathcal{D}(C_n, S_{n-1} \,|\, (\triangleright, \triangleleft))\, |$$

Proof: It follows from Theorem 1.3.7 and Theorem 1.3.11 taking into account that any group H of order n is an S_{n-1}-complement of S_n according to (3) of Example 1.3.3. □

Now we provide some explicit examples in order to see how Corollary 1.4.2 works.

Example 1.4.3 *Consider the extension $S_3 \leq S_4$ of factorization index 2. Then the canonical matched pair $(S_3, C_4, \triangleright, \triangleleft)$ associated to the factorization $S_4 = S_3C_4$ from Proposition 1.4.1 takes the following form:*

$\triangleright$	1	s_1	$s_1\, s_2$	$s_2\, s_1$	s_2	$s_1\, s_2\, s_1$
1	1	s_1	$s_1\, s_2$	$s_2\, s_1$	s_2	$s_1\, s_2\, s_1$
x	1	s_2	s_1	$s_1\, s_2$	$s_2\, s_1$	$s_1\, s_2\, s_1$
x^2	1	$s_2\, s_1$	s_2	s_1	$s_1\, s_2$	$s_1\, s_2\, s_1$
x^3	1	$s_1\, s_2$	$s_2\, s_1$	s_2	s_1	$s_1\, s_2\, s_1$

$\triangleleft$	1	s_1	$s_1\,s_2$	$s_2\,s_1$	s_2	$s_1\,s_2\,s_1$
1	1	1	1	1	1	1
x	x	x	x^2	x^3	x^2	x^3
x^2	x^2	x^3	x^3	x	x	x^2
x^3	x^3	x^2	x	x^2	x^3	x

By a straightforward computation one can prove that there are two deformation maps for the canonical matched pair $(S_3, C_4, \triangleright, \triangleleft)$*: namely the trivial one* $r' : C_4 \to S_3$, $r'(c) = 1$, *for any* $c \in C_4$ *and the map given by*

$$r : C_4 \to S_3, \qquad r(1) = r(x^2) = 1, \qquad r(x) = r(x^3) = s_1\,s_2\,s_1$$

We consider the following presentation of the Klein's group: $C_2 \times C_2 = \langle a = (12)(34),\, b = (13)(24) \mid a^2 = b^2 = 1,\, ab = ba\rangle$. *Then we can easily prove that the map:*

$$\varphi : C_2 \times C_2 \to (C_4)_r, \quad \varphi(1) = 1, \quad \varphi(a) = x, \quad \varphi(b) = x^2, \quad \varphi(ab) = x^3$$

is an isomorphism of groups, that is $C_2 \times C_2 \cong (C_4)_r$.

Corollary 1.4.2 proves that any finite group of order n is isomorphic to an r-deformation of the cyclic group C_n, for some deformation map $r : C_n \to S_{n-1}$ of the canonical matched pair associated to the factorization $S_n = S_{n-1}C_n$. The next example shows how the symmetric group S_3 appears as an r-deformation of the cyclic group C_6 arising from a given matched pair $(C_3, C_6, \triangleright, \triangleleft)$.

Example 1.4.4 *Let* $C_3 = \langle a \mid a^3 = 1\rangle$ *and* $C_6 = \langle b \mid b^6 = 1\rangle$ *be the cyclic groups of order* 3 *respectively* 6. *As a special case of [14, Proposition 4.2] we have a matched pair of groups* $(C_3, C_6, \triangleright, \triangleleft)$, *where the actions* $(\triangleright, \triangleleft)$ *on generators are defined by:*

$$b \triangleright a := a^2 \qquad b \triangleleft a := b^3$$

By a rather long but straightforward computation it can be seen that the map:

$$r : C_6 \to C_3, \quad r(1) = r(b^3) = 1, \quad r(b) = r(b^4) = a^2, \quad r(b^2) = r(b^5) = a$$

is a deformation map of the matched pair $(C_3, C_6, \triangleright, \triangleleft)$ *and* $\varphi : S_3 \to (C_6)_r$ *given by:*

$$\varphi(1) = 1,\ \varphi(s_1) = b,\ \varphi(s_1\,s_2) = b^2, \varphi(s_2\,s_1) = b^4,\ \varphi(s_2) = b^5,\ \varphi(s_1\,s_2\,s_1) = b^3$$

is an isomorphism of groups. Hence S_3 *is an* r*-deformation of the cyclic group* C_6.

Our last example provides a non-trivial deformation map $r : H \to A$ such that $H_r \cong H$.

Example 1.4.5 *Let* $(C_3, C_6, \triangleright, \triangleleft)$ *be the matched pair of Example 1.4.4. Then the map*

$$R : C_6 \to C_3, \quad R(1) = R(b^2) = R(b^4) = 1, \qquad R(b) = R(b^3) = R(b^5) = a$$

is also a deformation map of $(C_3, C_6, \triangleright, \triangleleft)$. *Then, one can easily check that* $(C_6)_R$ *is a group isomorphic to* C_6.

Bibliographical Notes

The material presented in this chapter is based on the author's papers [14], [11], [12], [18], and respectively, [23].

Chapter 2

Leibniz algebras

Leibniz algebras were introduced by Bloh [51] under the name of D-algebras and rediscovered later on by Loday [166] as non-commutative generalizations of Lie algebras. Several classical theorems known in the context of Lie algebras were extended to Leibniz algebras, there exists a (co)homology theory for them, the classification of certain types of Leibniz algebras of a given (small) dimension was recently performed, their interaction with vertex operator algebras, and the Godbillon-Vey invariants for foliations or differential geometry was highlighted. For more details and motivations we refer to [28], [29], [45], [66], [67], [68], [70], [72], [104], [119], [137], [149], [166], [167], [178], [205], [206] and the references therein. This chapter deals with the extending structure problems for Leibniz algebras which comes down to the following question:

Let $\mathfrak{g}$ be a Leibniz algebra and E a vector space containing $\mathfrak{g}$ as a subspace. Describe and classify the set of all Leibniz algebra structures $[-,-]$ that can be defined on E such that $\mathfrak{g}$ becomes a Leibniz subalgebra of $(E,[-,-])$.

We will provide an answer to the above problem as follows: first we will describe explicitly all Leibniz algebra structures on E which contain $\mathfrak{g}$ as a subalgebra; then we will classify them up to a Leibniz algebra isomorphism $\varphi : E \to E$ that *stabilizes* $\mathfrak{g}$, that is, φ acts as the identity on $\mathfrak{g}$. The extending structures (ES) problem was formulated at the level of groups in [18] and for arbitrary categories in [13] where it was studied for quantum groups. The ES problem is obviously very difficult. For instance, if $\mathfrak{g} = \{0\}$, then the ES problem asks for the classification of all Leibniz algebra structures on an arbitrary vector space E, which is of course a hopeless problem for vector spaces of large dimension: the classification of all 3-dimensional (resp. 4-dimensional) Leibniz algebras was finished only recently in [72] (resp. [68]). For this reason, from now on we will assume that $\mathfrak{g} \neq \{0\}$. Despite the difficulty of the ES problem, we can still provide detailed answers to it in certain special cases which depend both on the choice of the Leibniz algebra $\mathfrak{g}$ and on the codimension of $\mathfrak{g}$ in E. It generalizes and unifies the *extension problem* and the *factorization problem*.

The chapter is organized as follows: in Section 2.1 we introduce the abstract construction of the *unified product* $\mathfrak{g} \ltimes V$ for Leibniz algebras: it is associated to a Leibniz algebra $\mathfrak{g}$, a vector space V, and a system of data $\Omega(\mathfrak{g}, V) = \big(\triangleleft, \triangleright, \leftharpoonup, \rightharpoonup, f, \{-,-\}\big)$ called an extending datum of $\mathfrak{g}$ through V. Theorem 2.1.3 establishes the set of axioms that need to be satisfied by $\Omega(\mathfrak{g}, V)$ such that $\mathfrak{g} \ltimes V$ with a given bracket becomes a Leibniz algebra, i.e., is a unified product. In this case, $\Omega(\mathfrak{g}, V) = \big(\triangleleft, \triangleright, \leftharpoonup, \rightharpoonup, f, \{-,-\}\big)$ will be called a *Leibniz extending structure* of $\mathfrak{g}$ through V. Now let $\mathfrak{g}$ be a Leibniz algebra, E a vector space containing $\mathfrak{g}$ as a subspace and V a given complement of $\mathfrak{g}$ in E. Theorem 2.1.5 provides the answer to the description part of the ES problem: there exists a Leibniz algebra structure $[-,-]$ on E such that $\mathfrak{g}$ is a subalgebra of $(E, [-,-])$ if and only if there exists an isomorphism of Leibniz algebras $(E, [-,-]) \cong \mathfrak{g} \ltimes V$, for some Leibniz extending structure $\Omega(\mathfrak{g}, V) = \big(\triangleleft, \triangleright, \leftharpoonup, \rightharpoonup, f, \{-,-\}\big)$ of $\mathfrak{g}$ through V. The theoretical answer to the classification part of the ES problem is given in Theorem 2.1.9: we will construct explicitly a cohomology 'group', denoted by $\mathcal{HL}^2_{\mathfrak{g}}(V, \mathfrak{g})$, which will be the classifying object of all extending structures of the Leibniz algebra $\mathfrak{g}$ to E—the classification is given up to an isomorphism of Leibniz algebras which stabilizes $\mathfrak{g}$. The construction of a second classifying object, denoted by $\mathcal{HL}^2(V, \mathfrak{g})$ is also given and it parameterizes all extending structures of $\mathfrak{g}$ to E up to an isomorphism which simultaneously stabilizes $\mathfrak{g}$ and co-stabilizes V—i.e., this classification is given from the point of view of the extension problem. In Section 2.2 we give some explicit examples of computing $\mathcal{HL}^2_{\mathfrak{g}}(V, \mathfrak{g})$ and $\mathcal{HL}^2(V, \mathfrak{g})$ in the case of *flag extending structures* as defined in Definition 2.2.1. The main result of this section is Theorem 2.2.9: several special cases of it are discussed and explicit examples are given. Section 2.3 deals with two main special cases of the unified product. First we introduce the *crossed product of two Leibniz algebras* as a special case of the unified product. Corollary 2.3.1 shows that any extension of a given Leibniz algebra $\mathfrak{h}$ by a Leibniz algebra $\mathfrak{g}$ is equivalent to a crossed product extension and the classifying object $\mathbb{HL}^2(\mathfrak{h}, \mathfrak{g})$ of all extensions of $\mathfrak{h}$ by $\mathfrak{g}$ is constructed in Corollary 2.3.2 as a generalization of the second Loday-Pirashvili cohomology group $\mathrm{HL}^2(\mathfrak{h}, \mathfrak{g})$ [167]. The second important special case of the unified product is what we have called the *bicrossed product of Leibniz algebras*. The terminology is obviously inspired by its connections to the factorization problem which is highlighted in Corollary 2.3.6. Section 2.4 deals with the converse of the factorization problem, called the *classifying complements problem*.

In Section 2.5 we discuss Ito's theorem in the context of Leibniz algebras. Itô's theorem provides an important piece of information in group theory and it was the foundation of many structural results for finite groups. The aim of this section is to prove that Itô's theorem remains valid at the level of Leibniz algebras (Corollary 2.5.2): if $\mathfrak{g}$ is a Leibniz algebra such that $\mathfrak{g} = A + B$, for two abelian subalgebras A and B, then $\mathfrak{g}$ is metabelian (2-step solvable); the corresponding theorem for Lie algebras is well known [43, 61, 198]. The converse as well as possible generalizations of Itô's theorem are also considered.

2.1 Unified products for Leibniz algebras

In this section we give the theoretical answer to the ES problem by constructing two cohomological type objects which will parameterize ExtdL $(E,\, \mathfrak{g})$ and ExtdL$'$ $(E,\, \mathfrak{g})$. First we introduce the following:

Definition 2.1.1 *Let $\mathfrak{g}$ be a Leibniz algebra and V a vector space. An extending datum of $\mathfrak{g}$ through V is a system $\Omega(\mathfrak{g}, V) = \big(\triangleleft,\, \triangleright,\, \leftharpoonup,\, \rightharpoonup,\, f,\, \{-,\, -\}\big)$ consisting of six bilinear maps*

$$\triangleleft : V \times \mathfrak{g} \to V, \qquad \triangleright : V \times \mathfrak{g} \to \mathfrak{g}, \quad \leftharpoonup : \mathfrak{g} \times V \to \mathfrak{g}, \quad \rightharpoonup : \mathfrak{g} \times V \to V$$
$$f : V \times V \to \mathfrak{g}, \quad \{-,\, -\} : V \times V \to V$$

Let $\Omega(\mathfrak{g}, V) = \big(\triangleleft,\, \triangleright,\, \leftharpoonup,\, \rightharpoonup,\, f,\, \{-,\, -\}\big)$ be an extending datum. We denote by $\mathfrak{g} \ltimes_{\Omega(\mathfrak{g},V)} V = \mathfrak{g} \ltimes V$ the vector space $\mathfrak{g} \times V$ together with the bilinear map $[-,\, -] : (\mathfrak{g} \times V) \times (\mathfrak{g} \times V) \to \mathfrak{g} \times V$ defined by:

$$[(g, x),\, (h, y)] := \big([g,\, h] + x \triangleright h + g \leftharpoonup y + f(x, y),\; \{x,\, y\} + x \triangleleft h + g \rightharpoonup y\big) \quad (2.1)$$

for all $g,\, h \in \mathfrak{g}$ and $x,\, y \in V$. The object $\mathfrak{g} \ltimes V$ is called the unified product of $\mathfrak{g}$ and $\Omega(\mathfrak{g}, V)$ if it is a Leibniz algebra with the bracket given by (2.1). In this case the extending datum $\Omega(\mathfrak{g}, V) = \big(\triangleleft,\, \triangleright,\, \leftharpoonup,\, \rightharpoonup,\, f,\, \{-,\, -\}\big)$ is called a Leibniz extending structure of $\mathfrak{g}$ through V. The maps $\triangleleft$, $\triangleright$, $\rightharpoonup$ and $\leftharpoonup$ are called the actions of $\Omega(\mathfrak{g}, V)$ and f is called the cocycle of $\Omega(\mathfrak{g}, V)$.

Example 2.1.2 *The unified product is a very general construction: in particular, the* hemisemidirect product *introduced in differential geometry [149, Example 2.2] is a special case of it. Let $\mathfrak{g}$ be a Lie algebra and $(V, \triangleleft)$ be a right $\mathfrak{g}$-module. Then $\mathfrak{g} \times V$ is a Leibniz algebra with the bracket $[(g, x),\, (h, y)] := \big([g, h],\, x \triangleleft h\big)$, for all $g,\, h \in \mathfrak{g}$, $x,\, y \in V$ called the hemisemidirect product of $\mathfrak{g}$ and V. This Leibniz algebra is not a Lie algebra if $\mathfrak{g}$ acts nontrivially on V. The hemisemidirect product is a special case of the unified product if we let $\triangleright$, $\leftharpoonup$, $\rightharpoonup$, f and $\{-, -\}$ be the trivial maps and $\mathfrak{g}$ be a Lie algebra.*

Let $\Omega(\mathfrak{g}, V)$ be an extending datum of $\mathfrak{g}$ through V. The bracket defined by (2.1) has a rather complicated formula; however, for some specific elements we obtain easier forms which will be very useful for future computations. More precisely, the following relations hold in $\mathfrak{g} \ltimes V$ for any $g,\, h \in \mathfrak{g}$, $x,\, y \in V$:

$$[(g, 0),\, (h, 0)] = \big([g,\, h]\,,\, 0\big), \qquad [(g, 0),\, (0, y)] = \big(g \leftharpoonup y,\, g \rightharpoonup y\big) \quad (2.2)$$
$$[(0, x),\, (h, 0)] = \big(x \triangleright h,\, x \triangleleft h\big), \quad [(0, x),\, (0, y)] = \big(f(x, y),\, \{x, y\}\big) \quad (2.3)$$

The next theorem provides the set of axioms that need to be fulfilled by an extending datum $\Omega(\mathfrak{g}, V)$ such that $\mathfrak{g} \ltimes V$ is a unified product.

Theorem 2.1.3 *Let $\mathfrak{g}$ be a Leibniz algebra, V a vector space and $\Omega(\mathfrak{g}, V)$ an extending datum of $\mathfrak{g}$ by V. Then $\mathfrak{g} \ltimes V$ is a unified product if and only if the following compatibility conditions hold for any $g,\ h \in \mathfrak{g}$, $x,\ y,\ z \in V$:*

(L1) $(V, \triangleleft)$ *is a right* $\mathfrak{g}$*-module, i.e.,* $x \triangleleft [g,\ h] = (x \triangleleft g) \triangleleft h - (x \triangleleft h) \triangleleft g$

(L2) $x \triangleright [g,\ h] = [x \triangleright g,\ h] - [x \triangleright h,\ g] + (x \triangleleft g) \triangleright h - (x \triangleleft h) \triangleright g$

(L3) $[g,\ h] \rightharpoonup x = g \rightharpoonup (h \rightharpoonup x) + (g \rightharpoonup x) \triangleleft h$

(L4) $[g,\ h] \leftharpoonup x = [g,\ h \leftharpoonup x] + [g \leftharpoonup x,\ h] + g \leftharpoonup (h \rightharpoonup x) + (g \rightharpoonup x) \triangleright h$

(L5) $x \triangleright f(y,\ z) = f(x,\ y) \leftharpoonup z - f(x,\ z) \leftharpoonup y + f(\{x,\ y\},\ z) - f(\{x,\ z\},\ y) - f(x,\ \{y,\ z\})$

(L6) $x \triangleleft f(y,\ z) = f(x,\ y) \rightharpoonup z - f(x,\ z) \rightharpoonup y + \{\{x,\ y\},\ z\} - \{\{x,\ z\},\ y\} - \{x,\ \{y,\ z\}\}$

(L7) $\{x,\ y\} \triangleright g = x \triangleright (y \triangleright g) + (x \triangleright g) \leftharpoonup y + f(x,\ y \triangleleft g) + f(x \triangleleft g,\ y) - [f(x,\ y),\ g]$

(L8) $\{x,\ y\} \triangleleft g = x \triangleleft (y \triangleright g) + (x \triangleright g) \rightharpoonup y + \{x,\ y \triangleleft g\} + \{x \triangleleft g,\ y\}$

(L9) $g \rightharpoonup \{x,\ y\} = (g \leftharpoonup x) \rightharpoonup y - (g \leftharpoonup y) \rightharpoonup x + \{g \rightharpoonup x,\ y\} - \{g \rightharpoonup y,\ x\}$

(L10) $g \leftharpoonup \{x,\ y\} = (g \leftharpoonup x) \leftharpoonup y - (g \leftharpoonup y) \leftharpoonup x + f(g \rightharpoonup x,\ y) - f(g \rightharpoonup y,\ x) - [g,\ f(x, y)]$

(L11) $[g,\ h \leftharpoonup x] + [g,\ x \triangleright h] + g \leftharpoonup (h \rightharpoonup x) + g \leftharpoonup (x \triangleleft h) = 0$

(L12) $x \triangleright (y \triangleright g) + x \triangleright (g \leftharpoonup y) + f(x,\ y \triangleleft g) + f(x,\ g \rightharpoonup y) = 0$

(L13) $x \triangleleft (y \triangleright g) + x \triangleleft (g \leftharpoonup y) + \{x,\ y \triangleleft g\} + \{x,\ g \rightharpoonup y\} = 0$

(L14) $g \rightharpoonup (h \rightharpoonup x) + g \rightharpoonup (x \triangleleft h) = 0$

Proof: The proof relies on a detailed analysis of the Leibniz identity for the bracket given by (2.1). As the computations are rather long but straightforward we will only indicate the essential steps of the proof, the details being left to the reader. To start with, we note that $\mathfrak{g} \ltimes V$ is a Leibniz algebra if and only if Leibniz's identity holds, i.e.,:

$$\big[(g, x),\ [(h, y),\ (l, z)]\big] = \big[\,[(g, x),\ (h, y)],\ (l, z)\big] - \big[\,[(g, x),\ (l, z)],\ (h, y)\big] \quad (2.4)$$

for all $g,\ h,\ l \in \mathfrak{g}$ and $x,\ y,\ z \in V$. Since in $\mathfrak{g} \ltimes V$ we have $(g, x) = (g, 0) + (0, x)$ it follows that (2.4) holds if and only if it holds for all generators of $\mathfrak{g} \ltimes V$, i.e., the set $\{(g,\ 0) \mid g \in \mathfrak{g}\} \cup \{(0,\ x) \mid x \in V\}$. Hence we are left to deal with eight cases which are necessary and sufficient for testing the compatibility condition (2.4). First, we should notice that (2.4) holds for the triple $(g, 0)$, $(h, 0)$, $(l, 0)$, since in $\mathfrak{g} \ltimes V$ we have that $[(g, 0),\ (h, 0)] = ([g, h], 0)$. Now, taking into account (2.2), we obtain that (2.4) holds for $(g,0)$, $(h,0)$, $(0,x)$ if and only if

$$\big([g,\, h] \leftharpoonup x - [g \leftharpoonup x,\, h] - (g \rightharpoonup x) \rhd h,\; [g,\, h] \rightharpoonup x - (g \rightharpoonup x) \lhd h\big) = \big([g,\, h \leftharpoonup x] + g \leftharpoonup (h \rightharpoonup x),\; g \rightharpoonup (h \rightharpoonup x)\big)$$

i.e., if and only if ($L3$) and ($L4$) hold. A similar computation proves that the compatibility condition (2.4) is fulfilled for the triple $(g,0)$, $(0,x)$, $(h,0)$ if and only if:

$$\begin{aligned} [g,\, h] \leftharpoonup x &= [g \leftharpoonup x,\, h] - [g,\, x \rhd h] + (g \rightharpoonup x) \rhd h - g \leftharpoonup (x \lhd h) \\ [g,\, h] \rightharpoonup x &= (g \rightharpoonup x) \lhd h - g \rightharpoonup (x \lhd h) \end{aligned}$$

Taking into account that we are looking for a minimal and independent set of axioms we obtain, assuming that ($L3$) and ($L4$) hold, that (2.4) is fulfilled for the triple $(g,0)$, $(0,x)$, $(h,0)$ if and only if ($L11$) and ($L14$) hold. Next, it is straightforward to see that (2.4) holds for $(g,0)$, $(0,x)$, $(0,y)$ if and only if ($L9$) and ($L10$) hold. In a similar manner, one can show that (2.4) holds for $(0,x)$, $(0,y)$, $(0,z)$ if and only if axiom ($L5$) and ($L6$) are fulfilled. We are left with three more cases to study. First, observe that (2.4) holds for $(0,x)$, $(g,0)$, $(h,0)$ if and only if

$$\big(x \rhd [g,\, h],\; x \lhd [g,\, h]\big) = \big([x \rhd g,\, h] - [x \rhd h,\, g] + (x \lhd g) \rhd h - (x \lhd h) \rhd g,\; (x \lhd g) \lhd h - (x \lhd h) \lhd g\big)$$

which is equivalent to the fact that axioms ($L1$) and ($L2$) hold. Analogously, we can show that (2.4) holds for $(0,x)$, $(0,y)$, $(g,0)$ if and only if ($L7$) and ($L8$) hold. Finally, it is straightforward to see that (2.4) holds for $(0,x)$, $(g,0)$, $(0,y)$ if and only if the following two compatibilities are fulfilled:

$$\begin{aligned} \{x,\, y\} \rhd g &= (x \rhd g) \leftharpoonup y + f(x \lhd g,\, y) - [f(x,\, y),\, g] - x \rhd (g \leftharpoonup y) \\ &\quad - f(x,\, g \rightharpoonup y) \\ \{x,\, y\} \lhd g &= \{x \lhd g,\, y\} + (x \rhd g) \rightharpoonup y - x \lhd (g \leftharpoonup y) - \{x,\, g \rightharpoonup y\} \end{aligned}$$

Since we are looking for a minimal set of axioms we obtain, assuming that ($L7$) and ($L8$) hold, that (2.4) is fulfilled for the triple $(0,x)$, $(g,0)$, $(0,y)$ if and only if ($L12$) and ($L13$) hold. The proof is now finished. □

From now on, a Leibniz extending structure of $\mathfrak{g}$ through V will be viewed as a system $\Omega(\mathfrak{g}, V) = \big(\lhd,\, \rhd,\, \leftharpoonup,\, \rightharpoonup,\, f,\, \{-,\, -\}\big)$ satisfying the compatibility conditions ($L1$)–($L14$). We denote by $\mathcal{LE}(\mathfrak{g}, V)$ the set of all Leibniz extending structures of $\mathfrak{g}$ through V.

Example 2.1.4 *We provide the first example of a Leibniz extending structure and the corresponding unified product. More examples will be given in Section 2.2 and Section 2.3.*

Let $\Omega(\mathfrak{g}, V) = \big(\triangleleft, \triangleright, \leftharpoonup, \rightharpoonup, f, \{-, -\}\big)$ *be an extending datum of a Leibniz algebra* $\mathfrak{g}$ *through a vector space* V *such that its actions are all the trivial maps, i.e.,* $x \triangleleft g = x \triangleright g = g \rightharpoonup x = g \leftharpoonup x = 0$, *for all* $x \in V$ *and* $g \in \mathfrak{g}$. *Then,* $\Omega(\mathfrak{g}, V) = \big(f, \{-, -\}\big)$ *is a Leibniz extending structure of* $\mathfrak{g}$ *through* V *if and only if* $(V, \{-, -\})$ *is a Leibniz algebra and* $f : V \times V \to \mathfrak{g}$ *is an abelian 2-cocycle, that is*

$$[g,\, f(x,y)] = [f(x,y),\, g] = 0, \quad f(x,\, \{y,\, z\}) - f(\{x,\, y\},\, z) + f(\{x,\, z\},\, y) = 0 \tag{2.5}$$

for all $g \in \mathfrak{g}$, x, y *and* $z \in V$. *The first part of (2.5) shows that the image of* f *is contained in the center of* $\mathfrak{g}$, *while the second part is a 2-cocycle condition (see [167, Section 1.7]) which follows from the axiom (L5). In this case, the associated unified product* $\mathfrak{g} \ltimes_{\Omega(\mathfrak{g},V)} V$ *will be called the* twisted product *of the Leibniz algebras* $\mathfrak{g}$ *and* V.

Let $\Omega(\mathfrak{g}, V) = \big(\triangleleft, \triangleright, \leftharpoonup, \rightharpoonup, f, \cdot\big) \in \mathcal{LZ}(\mathfrak{g}, V)$ be a Leibniz extending structure and $\mathfrak{g} \ltimes V$ the associated unified product. Then the canonical inclusion

$$i_{\mathfrak{g}} : \mathfrak{g} \to \mathfrak{g} \ltimes V, \qquad i_{\mathfrak{g}}(g) = (g,\, 0)$$

is an injective Leibniz algebra map. Therefore, we can see $\mathfrak{g}$ as a subalgebra of $\mathfrak{g} \ltimes V$ through the identification $\mathfrak{g} \cong i_{\mathfrak{g}}(\mathfrak{g}) \cong \mathfrak{g} \times \{0\}$.

In order to prove the converse of the above result we need to introduce a few concepts. Let $\mathfrak{g}$ be a Leibniz algebra, E a vector space such that $\mathfrak{g}$ is a subspace of E and V a complement of $\mathfrak{g}$ in E, i.e., V is a subspace of E such that $E = \mathfrak{g} + V$ and $V \cap \mathfrak{g} = 0$. For a linear map $\varphi : E \to E$ we consider the diagram:

$$\begin{array}{ccccc} \mathfrak{g} & \xrightarrow{\ i\ } & E & \xrightarrow{\ \pi\ } & V \\ \Big\downarrow{\scriptstyle Id} & & \Big\downarrow{\scriptstyle \varphi} & & \Big\downarrow{\scriptstyle Id} \\ \mathfrak{g} & \xrightarrow{\ i\ } & E & \xrightarrow{\ \pi\ } & V \end{array} \tag{2.6}$$

where $\pi : E \to V$ is the canonical projection of $E = \mathfrak{g} + V$ on V and $i : \mathfrak{g} \to E$ is the inclusion map. We say that $\varphi : E \to E$ *stabilizes* $\mathfrak{g}$ (resp. *co-stabilizes* V) if the left square (resp. the right square) of the diagram (2.6) is commutative.

Two Leibniz algebra structures $\{-, -\}$ and $\{-, -\}'$ on E containing $\mathfrak{g}$ as a Leibniz subalgebra are called *equivalent* and we denote this by $(E, \{-, -\}) \equiv (E, \{-, -\}')$, if there exists a Leibniz algebra isomorphism $\varphi : (E, \{-, -\}) \to (E, \{-, -\}')$ which stabilizes $\mathfrak{g}$. $\{-, -\}$ and $\{-, -\}'$ are called *cohomologous*, and we denote this by $(E, \{-, -\}) \approx (E, \{-, -\}')$, if there exists a Leibniz algebra isomorphism $\varphi : (E, \{-, -\}) \to (E, \{-, -\}')$ which stabilizes $\mathfrak{g}$ and co-stabilizes V, i.e., the diagram (2.6) is commutative.

$\equiv$ and $\approx$ are both equivalence relations on the set of all Leibniz algebra structures on E containing $\mathfrak{g}$ as subalgebra and we denote by $\mathrm{ExtdL}\,(E, \mathfrak{g})$ (resp. $\mathrm{ExtdL}'\,(E, \mathfrak{g})$) the set of all equivalence classes via $\equiv$ (resp. $\approx$).

$\mathrm{ExtdL}\,(E,\mathfrak{g})$ is the classifying object of the ES problem: by explicitly computing $\mathrm{ExtdL}\,(E,\mathfrak{g})$ we obtain a parameterization of the set of all isomorphism classes of Leibniz algebra structures on E that stabilize $\mathfrak{g}$. $\mathrm{ExtdL}'\,(E,\mathfrak{g})$ gives a classification of the ES problem from the point of view of the extension problem. Any two cohomologous brackets on E are of course equivalent, hence there exists a canonical projection:

$$\mathrm{ExtdL}'\,(E,\mathfrak{g}) \twoheadrightarrow \mathrm{ExtdL}\,(E,\mathfrak{g})$$

We are now in a position to provide an answer to the description part of the ES problem:

Theorem 2.1.5 *Let $\mathfrak{g}$ be a Leibniz algebra, E a vector space containing $\mathfrak{g}$ as a subspace and $[-,\,-]$ a Leibniz algebra structure on E such that $\mathfrak{g}$ is a Leibniz subalgebra in $(E,[-,\,-])$. Then there exists a Leibniz extending structure $\Omega(\mathfrak{g},V) = \big(\triangleleft,\,\triangleright,\,\leftharpoonup,\,\rightharpoonup,\,f,\,[-,-]\big)$ of $\mathfrak{g}$ through a subspace V of E and an isomorphism of Leibniz algebras $(E,[-,\,-]) \cong \mathfrak{g} \ltimes V$ that stabilizes $\mathfrak{g}$ and co-stabilizes V.*

Proof: Since k is a field, there exists a linear map $p: E \to \mathfrak{g}$ such that $p(g) = g$, for all $g \in \mathfrak{g}$. Then $V := \mathrm{Ker}(p)$ is a complement of $\mathfrak{g}$ in E. We define the extending datum $\Omega(\mathfrak{g},V) = \big(\triangleleft = \triangleleft_p,\,\triangleright = \triangleright_p,\,\leftharpoonup = \leftharpoonup_p,\,\rightharpoonup = \rightharpoonup_p,\,f = f_p,\,[-,-] = [-,-]_p\big)$ of $\mathfrak{g}$ through V by the following formulas:

$$\begin{aligned}
\triangleright : V\times\mathfrak{g}\to\mathfrak{g}, &\quad x\triangleright g = p\big([x,\,g]\big),\\
\triangleleft : V\times\mathfrak{g}\to V, &\quad x\triangleleft g = [x,\,g] - p\big([x,\,g]\big)\\
\leftharpoonup : \mathfrak{g}\times V\to\mathfrak{g}, &\quad g\leftharpoonup x = p\big([g,\,x]\big),\\
\rightharpoonup : \mathfrak{g}\times V\to V, &\quad g\rightharpoonup x = [g,\,x] - p\big([g,\,x]\big)\\
f : V\times V\to\mathfrak{g}, &\quad f(x,y) = p\big([x,\,y]\big),\\
\{\,,\,\} : V\times V\to V, &\quad \{x,y\} = [x,\,y] - p\big([x,\,y]\big)
\end{aligned}$$

for all $g \in \mathfrak{g}$, x, $y \in V$. Now, the map $\varphi : \mathfrak{g}\times V \to E$, $\varphi(g,x) := g+x$, is a linear isomorphism between the direct product of vector spaces $\mathfrak{g}\times V$ and the Leibniz algebra $(E,[-,-])$ with the inverse given by $\varphi^{-1}(y) := \big(p(y),\,y - p(y)\big)$, for all $y \in E$. Hence, there exists a unique Leibniz algebra structure on $\mathfrak{g}\times V$ such that φ is an isomorphism of Leibniz algebras and this unique bracket is given by:

$$[(g,x),\,(h,y)] := \varphi^{-1}\big([\varphi(g,x),\,\varphi(h,y)]\big)$$

for all g, $h \in \mathfrak{g}$ and x, $y \in V$. Using Theorem 2.1.3, the proof will be finished if we prove that this bracket is the one defined by (2.1) associated to the system $\big(\triangleleft_p,\,\triangleright_p,\,\leftharpoonup_p,\,\rightharpoonup_p,\,f_p,\{-,\,-\}_p\big)$ constructed above. Indeed, for any g, $h \in \mathfrak{g}$, x,

$y \in V$ we have:

$$\begin{aligned}
[(g,x),\,(h,y)] &= \varphi^{-1}\big([\varphi(g,x),\,\varphi(h,y)]\big) = \varphi^{-1}\big([g+x,\,h+y]\big)\\
&= \varphi^{-1}\big([g,h]+[g,y]+[x,h]+[x,y]\big)\\
&= \Big([g,h]+p([g,y])+p([x,h])+p([x,y]),\\
&\quad [g,y]+[x,h]+[x,y]-p([g,y])-p([x,h])-p([x,y])\Big)\\
&= \big([g,\,h]+x\triangleright h+g\leftharpoonup y+f(x,y),\ \{x,\,y\}+x\triangleleft h+g\rightharpoonup y\big)
\end{aligned}$$

as needed. Thus, $\varphi : \mathfrak{g} \ltimes V \to E$ is an isomorphism of Leibniz algebras and the following diagram is commutative

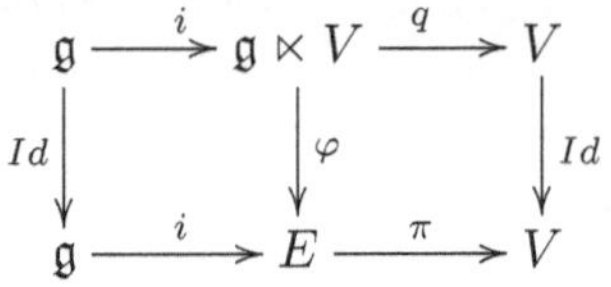

where $\pi : E \to V$ is the projection of $E = A + V$ on the vector space V and $q : A \ltimes V \to V$, $q(g,x) := x$ is the canonical projection. □

Based on Theorem 2.1.5, the classification of all Leibniz algebra structures on E that contain $\mathfrak{g}$ as a Leibniz subalgebra reduces to the classification of all unified products $\mathfrak{g} \ltimes V$, associated to all Leibniz extending structures $\Omega(\mathfrak{g}, V) = \big(\triangleleft, \triangleright, \leftharpoonup, \rightharpoonup, f, \{-,\,-\}\big)$, for a given complement V of $\mathfrak{g}$ in E. In order to construct the cohomological objects $\mathcal{HL}^2_{\mathfrak{g}}\,(V,\,\mathfrak{g})$ and $\mathcal{HL}^2\,(V,\,\mathfrak{g})$ which will parameterize the classifying sets $\mathrm{ExtdL}\,(E, \mathfrak{g})$ and respectively $\mathrm{ExtdL}'\,(E, \mathfrak{g})$, we need the following technical result:

Lemma 2.1.6 *Let $\Omega(\mathfrak{g}, V) = \big(\triangleleft, \triangleright, \leftharpoonup, \rightharpoonup, f, \{-,\,-\}\big)$ and $\Omega'(\mathfrak{g}, V) = \big(\triangleleft', \triangleright', \leftharpoonup', \rightharpoonup', f', \{-,\,-\}'\big)$ be two Leibniz extending structures of $\mathfrak{g}$ through V and $\mathfrak{g} \ltimes V$, $\mathfrak{g} \ltimes' V$ the associated unified products. Then there exists a bijection between the set of all morphisms of Leibniz algebras $\psi : \mathfrak{g} \ltimes V \to \mathfrak{g} \ltimes' V$ which stabilizes $\mathfrak{g}$ and the set of pairs (r, v), where $r : V \to \mathfrak{g}$, $v : V \to V$ are two linear maps satisfying the following compatibility conditions for any $g \in \mathfrak{g}$, $x,\, y \in V$:*

(ML1) $v(g \rightharpoonup x) = g \rightharpoonup' v(x)$;

(ML2) $v(x \triangleleft g) = v(x) \triangleleft' g$;

(ML3) $x \triangleright g + r(x \triangleleft g) = [r(x),\, g] + v(x) \triangleright' g$;

(ML4) $g \leftharpoonup x + r(g \rightharpoonup x) = [g,\, r(x)] + g \leftharpoonup' v(x)$;

(ML5) $v(\{x,\, y\}) = r(x) \rightharpoonup' v(y) + v(x) \triangleleft' r(y) + \{v(x),\, v(y)\}'$;

(ML6) $f(x, y) + r(\{x, y\}) = [r(x), r(y)] + r(x) \leftharpoonup' v(y) + v(x) \triangleright' r(y) + f'(v(x), v(y))$

Under the above bijection, the morphism of Leibniz algebras $\psi = \psi_{(r,v)} : \mathfrak{g} \ltimes V \to \mathfrak{g} \ltimes' V$ *corresponding to* (r, v) *is given for any* $g \in \mathfrak{g}$ *and* $x \in V$ *by:*

$$\psi(g, x) = (g + r(x), v(x))$$

Moreover, $\psi = \psi_{(r,v)}$ *is an isomorphism if and only if* $v : V \to V$ *is an isomorphism and* $\psi = \psi_{(r,v)}$ *co-stabilizes* V *if and only if* $v = \mathrm{Id}_V$.

Proof: A linear map $\psi : \mathfrak{g} \ltimes V \to \mathfrak{g} \ltimes' V$ which stabilizes $\mathfrak{g}$ is uniquely determined by two linear maps $r : V \to \mathfrak{g}$, $v : V \to V$ such that $\psi(g, x) = (g + r(x), v(x))$, for all $g \in \mathfrak{g}$, and $x \in V$. Indeed, by denoting $\psi(0, x) = (r(x), v(x)) \in \mathfrak{g} \times V$ for all $x \in V$, we obtain:

$$\psi(g, x) \quad = \quad \psi\big((g, 0) + \psi(0, x)\big) = \psi(g, 0) + \psi(0, x) = \big(g + r(x), v(x)\big)$$

Let $\psi = \psi_{(r,v)}$ be such a linear map, i.e., $\psi(g, x) = (g + r(x), v(x))$, for some linear maps $r : V \to \mathfrak{g}$, $v : V \to V$. We will prove that ψ is a morphism of Leibniz algebras if and only if the compatibility conditions $(ML1)$–$(ML6)$ hold. It is enough to prove that the compatibility

$$\psi\big([(g, x), (h, y)]\big) = [\psi(g, x), \psi(h, y)] \tag{2.7}$$

holds for all generators of $\mathfrak{g} \ltimes V$. First of all, it is easy to see that (2.7) holds for the pair $(g, 0)$, $(h, 0)$, for all g, $h \in \mathfrak{g}$. Now we prove that (2.7) holds for the pair $(g, 0)$, $(0, x)$ if and only if $(ML1)$ and $(ML4)$ hold. Indeed, $\psi\big([(g, 0), (0, x)]\big) = [\psi(g, 0), \psi(0, x)]$ is equivalent to $\psi(g \leftharpoonup x, g \rightharpoonup x) = [(g, 0), (r(x), v(x))]$ and hence to $(g \leftharpoonup x + r(g \rightharpoonup x), v(g \rightharpoonup x)) = ([g, r(x)] + g \leftharpoonup' v(x), g \rightharpoonup' v(x))$, i.e., due to the fact that $(ML1)$ and $(ML4)$ hold.

Next we prove that (2.7) holds for the pair $(0, x)$, $(g, 0)$ if and only if $(ML2)$ and $(ML3)$ hold. Indeed, $\psi\big([(0, x), (g, 0)]\big) = [\psi(0, x), \psi(g, 0)]$ is equivalent to $\psi(x \triangleright g, x \triangleleft g) = [(r(x), v(x)), (g, 0)]$ and therefore to $(x \triangleright g + r(x \triangleleft g), v(x \triangleleft g)) = ([r(x), g] + v(x) \triangleright' g, v(x) \triangleleft' g)$, i.e., due to the fact that $(ML2)$ and $(ML3)$ hold.

To this end, we prove that (2.7) holds for the pair $(0, x)$, $(0, y)$ if and only if $(ML5)$ and $(ML6)$ hold. Indeed, $\psi\big([(0, x), (0, y)]\big) = [\psi(0, x), \psi(0, y)]$ is equivalent to $\psi(f(x, y), \{x, y\}) = [(r(x), v(x)), (r(y), v(y))]$; thus it is equivalent to: $\big(f(x, y) + r(\{x, y\}), v(\{x, y\})\big) = \big([r(x), r(y)] + r(x) \leftharpoonup' v(y) + v(x) \triangleright' r(y) + f'(v(x), v(y)), r(x) \rightharpoonup' v(y) + v(x) \triangleleft' r(y) + \{v(x), v(y)\}'\big)$, i.e., due to the fact that $(ML5)$ and $(ML6)$ hold.

Assume now that $v : V \to V$ is bijective. Then $\psi_{(r,v)}$ is an isomorphism of Leibniz algebras with the inverse given by $\psi^{-1}_{(r,v)}(h, y) = (h - r(v^{-1}(y)), v^{-1}(y))$, for all $h \in \mathfrak{g}$ and $y \in V$. Conversely, assume that $\psi_{(r,v)}$ is bijective. It follows easily that v is surjective. Thus, we are left to prove that v is injective. Indeed, let $x \in V$ such that $v(x) = 0$. We have

$\psi_{(r,v)}(0,0) = (0,0) = (0, v(x)) = \psi_{(r,v)}(-r(x), x)$, and hence we obtain $x = 0$, i.e., v is a bijection. The last assertion is trivial and the proof is now finished. □

Definition 2.1.7 *Two Leibniz extending structures* $\Omega(\mathfrak{g}, V) = (\triangleleft, \triangleright, \leftharpoonup, \rightharpoonup, f, \{-, -\})$ *and* $\Omega'(\mathfrak{g}, V) = (\triangleleft', \triangleright', \leftharpoonup', \rightharpoonup', f', \{-, -\}')$ *are called* equivalent *and we denote this by* $\Omega(\mathfrak{g}, V) \equiv \Omega'(\mathfrak{g}, V)$, *if there exists a pair* (r, v) *of linear maps, where* $r : V \to \mathfrak{g}$ *and* $v \in \mathrm{Aut}_k(V)$ *such that* $(\triangleleft, \triangleright, \leftharpoonup, \rightharpoonup, f, \{-, -\})$ *is implemented from* $(\triangleleft', \triangleright', \leftharpoonup', \rightharpoonup', f', \{-, -\}')$ *using* (r, v) *via:*

$$\begin{aligned}
x \triangleleft g &= v^{-1}\big(v(x) \triangleleft' g\big)\\
g \rightharpoonup x &= v^{-1}\big(g \rightharpoonup' v(x)\big)\\
x \triangleright g &= [r(x),\, g] + v(x) \triangleright' g - r \circ v^{-1}\big(v(x) \triangleleft' g\big)\\
g \leftharpoonup x &= [g,\, r(x)] + g \leftharpoonup' v(x) - r \circ v^{-1}\big(g \rightharpoonup' v(x)\big)\\
\{x,\, y\} &= v^{-1}\big(r(x) \rightharpoonup' v(y) + v(x) \triangleleft' r(y) + \{v(x),\, v(y)\}'\big)\\
f(x,\, y) &= [r(x),\, r(y)] + r(x) \leftharpoonup' v(y) + v(x) \triangleright' r(y) + f'\big(v(x),\, v(y)\big) -\\
&\quad r \circ v^{-1}\big(r(x) \rightharpoonup' v(y) + v(x) \triangleleft' r(y) + \{v(x),\, v(y)\}'\big)
\end{aligned}$$

for all $g \in \mathfrak{g}$, x, $y \in V$.

Using Lemma 2.1.6, we obtain that $\Omega(\mathfrak{g}, V) \equiv \Omega'(\mathfrak{g}, V)$ if and only if there exists $\psi : \mathfrak{g} \ltimes V \to \mathfrak{g} \ltimes' V$ an isomorphism of Leibniz algebras that stabilizes $\mathfrak{g}$, where $\mathfrak{g} \ltimes V$ and $\mathfrak{g} \ltimes' V$ are the corresponding unified products. On the other hand, the isomorphisms between two unified products that stabilize $\mathfrak{g}$ and co-stabilize V are decoded by the following:

Definition 2.1.8 *Two Leibniz extending structures* $\Omega(\mathfrak{g}, V) = (\triangleleft, \triangleright, \leftharpoonup, \rightharpoonup, f, \{-, -\})$ *and* $\Omega'(\mathfrak{g}, V) = (\triangleleft', \triangleright', \leftharpoonup', \rightharpoonup', f', \{-, -\}')$ *are called* cohomologous *and we denote this by* $\Omega(\mathfrak{g}, V) \approx \Omega'(\mathfrak{g}, V)$ *if and only if* $\triangleleft = \triangleleft'$, $\rightharpoonup = \rightharpoonup'$ *and there exists a linear map* $r : V \to \mathfrak{g}$ *such that for any* $g \in \mathfrak{g}$, x, $y \in V$*:*

$$\begin{aligned}
x \triangleright g &= [r(x),\, g] + x \triangleright' g - r(x \triangleleft' g)\\
g \leftharpoonup x &= [g,\, r(x)] + g \leftharpoonup' x - r(g \rightharpoonup' x)\\
\{x,\, y\} &= r(x) \rightharpoonup' y + x \triangleleft' r(y) + \{x,\, y\}'\\
f(x,\, y) &= [r(x),\, r(y)] + r(x) \leftharpoonup' y + x \triangleright' r(y) + f'(x,\, y) -\\
&\quad r\big(r(x) \rightharpoonup' y + x \triangleleft' r(y) + \{x,\, y\}'\big)
\end{aligned}$$

As a conclusion of this section, the theoretical answer to the ES problem follows:

Theorem 2.1.9 *Let* $\mathfrak{g}$ *be a Leibniz algebra,* E *a vector space that contains* $\mathfrak{g}$ *as a subspace and* V *a complement of* $\mathfrak{g}$ *in* E*. Then:*

(1) $\equiv$ *is an equivalence relation on the set* $\mathcal{LZ}(\mathfrak{g}, V)$ *of all Leibniz extending structures of* $\mathfrak{g}$ *through* V. *If we denote* $\mathcal{HL}^2_{\mathfrak{g}}(V, \mathfrak{g}) := \mathcal{LZ}(\mathfrak{g}, V)/\equiv$, *then the map*

$$\mathcal{HL}^2_{\mathfrak{g}}(V, \mathfrak{g}) \to \mathrm{ExtdL}(E, \mathfrak{g}), \qquad \overline{(\triangleleft, \triangleright, \leftharpoonup, \rightharpoonup, f, \{-,-\})} \mapsto (\mathfrak{g} \ltimes V, [-,-])$$

is bijective, where $\overline{(\triangleleft, \triangleright, \leftharpoonup, \rightharpoonup, f, \{-,-\})}$ *is the equivalence class of* $(\triangleleft, \triangleright, \leftharpoonup, \rightharpoonup, f, \{-,-\})$ *via* $\equiv$.

(2) $\approx$ *is an equivalence relation on the set* $\mathcal{LZ}(\mathfrak{g}, V)$ *of all Leibniz extending structures of* $\mathfrak{g}$ *through* V. *If we denote* $\mathcal{HL}^2(V, \mathfrak{g}) := \mathcal{LZ}(\mathfrak{g}, V)/\approx$, *then the map*

$$\mathcal{HL}^2(V, \mathfrak{g}) \to \mathrm{ExtdL}'(E, \mathfrak{g}), \qquad \overline{\overline{(\triangleleft, \triangleright, \leftharpoonup, \rightharpoonup, f, \{-,-\})}} \mapsto (\mathfrak{g} \ltimes V, [-,-])$$

is bijective, where $\overline{\overline{(\triangleleft, \triangleright, \leftharpoonup, \rightharpoonup, f, \{-,-\})}}$ *is the equivalence class of* $(\triangleleft, \triangleright, \leftharpoonup, \rightharpoonup, f, \{-,-\})$ *via* $\approx$.

2.2 Flag extending structures of Leibniz algebras: Examples

After we have provided a theoretical answer to the ES problem in Theorem 2.1.9, we are left to compute the classifying object $\mathcal{HL}^2_{\mathfrak{g}}(V, \mathfrak{g})$ for a given Leibniz algebra $\mathfrak{g}$ that is a subspace in a vector space E with a complement V and then to describe all Leibniz algebra structures on E which extend the one of $\mathfrak{g}$. In this section we propose an algorithm to tackle the problem for a large class of such structures called *flag extending structures.*

Definition 2.2.1 *Let* $\mathfrak{g}$ *be a Leibniz algebra and* E *a vector space containing* $\mathfrak{g}$ *as a subspace. A Leibniz algebra structure on* E *is called a* flag extending structure *of* $\mathfrak{g}$ *if there exists a finite chain of Leibniz subalgebras of* E

$$\mathfrak{g} = E_0 \subset E_1 \subset \cdots \subset E_m = E \tag{2.8}$$

such that E_i *has codimension* 1 *in* E_{i+1}, *for all* $i = 0, \cdots, m-1$.

As an easy consequence of Definition 2.2.1 we have that $\dim_k(V) = m$, where V is the complement of $\mathfrak{g}$ in E. In what follows we will provide a way of describing all flag extending structures of $\mathfrak{g}$ to E in a recursive manner which relies on the first step, namely $m = 1$. Therefore, we start by describing and classifying all unified products $\mathfrak{g} \ltimes V_1$, for a 1-dimensional vector space V_1. This procedure can be iterated by replacing the initial Leibniz algebra $\mathfrak{g}$ with a unified product $\mathfrak{g} \ltimes V_1$ obtained in the previous step. After m steps we arrive at the description of all flag extending structures of $\mathfrak{g}$ to E. We start by introducing the following concepts which play a key role in the construction of the different non-abelian cohomological objects arising from the classification part of the ES problem.

Definition 2.2.2 *An* anti-derivation *of a Leibniz algebra* $\mathfrak{g}$ *is a linear map* $D : \mathfrak{g} \to \mathfrak{g}$ *such that*

$$D([g, h]) = [D(g), h] - [D(h), g] \tag{2.9}$$

for all $g, h \in \mathfrak{g}$. *We denote by* $\mathrm{ADer}(\mathfrak{g})$ *the space of all anti-derivations of* $\mathfrak{g}$.

Example 2.2.3 *For a Lie algebra* $\mathfrak{g}$ *we have that* $\mathrm{ADer}(\mathfrak{g}) = \mathrm{Der}(\mathfrak{g})$ *but in the case of Leibniz algebras the two spaces are, in general, not equal. The next example illustrates this. Let* $\mathfrak{g}$ *be the 3-dimensional Leibniz algebra with the basis* $\{e_1, e_2, e_3\}$ *and the bracket defined by:* $[e_1, e_3] = e_2$, $[e_3, e_3] = e_1$. *A straightforward computation shows that the set* $\mathrm{Der}(g)$ *(resp.* $\mathrm{ADer}(g)$*) coincides with the set of all arrays* Δ *(resp.* D*) of the form:*

$$\Delta = \begin{pmatrix} 2b_1 & 0 & b_2 \\ b_2 & 3b_1 & b_3 \\ 0 & 0 & b_1 \end{pmatrix} \qquad (\text{resp.} \quad D = \begin{pmatrix} 0 & 0 & d_1 \\ 0 & 0 & d_2 \\ 0 & 0 & d_3 \end{pmatrix} \quad) \tag{2.10}$$

for all $b_1, b_2, b_3, d_1, d_2, d_3 \in k$.

The following new concept will play an important role in the construction of the two non-abelian cohomological objects introduced in Section 2.2 and Section 2.3.

Definition 2.2.4 *Let* $\mathfrak{g}$ *be a Leibniz algebra. A* pointed double derivation *of* $\mathfrak{g}$ *is a triple* (g_0, D, Δ), *where* $g_0 \in \mathfrak{g}$ *and* $D, \Delta : \mathfrak{g} \to \mathfrak{g}$ *are linear maps satisfying the following compatibilities for any* $g, h \in \mathfrak{g}$:

$$D(g_0) = [g, g_0] = [g, D(h) + \Delta(h)] = D^2(g) + D\big(\Delta(g)\big) = 0 \tag{2.11}$$

$$D^2(g) + \Delta\big(D(g)\big) = [g_0, g] \tag{2.12}$$

$$\Delta\big([g, h]\big) = [\Delta(g), h] + [g, \Delta(h)] \tag{2.13}$$

$$D\big([g, h]\big) = [D(g), h] - [D(h), g] \tag{2.14}$$

We denote by $\mathcal{D}(\mathfrak{g})$ *the space of all pointed double derivations.*

The compatibility conditions (2.13)-(2.14) show that Δ (resp. D) is a derivation (resp. an antiderivation) of $\mathfrak{g}$, hence $\mathcal{D}(\mathfrak{g}) \subseteq \mathfrak{g} \times \mathrm{ADer}(\mathfrak{g}) \times \mathrm{Der}(\mathfrak{g})$. If $\mathfrak{g}$ is a Lie algebra then we can easily see that the space $\mathcal{D}(\mathfrak{g})$ coincides with the set of all pairs $(g_0, D) \in Z(\mathfrak{g}) \times \mathrm{Der}(\mathfrak{g})$ such that $D(g_0) = 0$. An example of computing the space $\mathcal{D}(\mathfrak{g})$ is given in Example 2.3.4.

Definition 2.2.5 *Let* $\mathfrak{g}$ *be a Leibniz algebra. A* flag datum of the first kind *of* $\mathfrak{g}$ *is a 5-tuple* $(g_0, \alpha, \lambda, D, \Delta)$, *where* $g_0 \in \mathfrak{g}$, $\alpha \in k$, $\lambda : \mathfrak{g} \to k$, D *and* $\Delta : \mathfrak{g} \to \mathfrak{g}$ *are linear maps satisfying the following compatibilities for any* g, $h \in \mathfrak{g}$:

(F1) $\lambda([g, h]) = 0, \quad \lambda(D(g)) + \alpha\,\lambda(g) = 0, \quad \lambda(\Delta(g)) = 0;$

(F2) $D(g_0) = -\,\alpha\, g_0, \quad \lambda(g_0) = -\,\alpha^2, \quad \alpha\,\Delta(g) = -[g, g_0];$

(F3) $[g, \Delta(h)] + [g, D(h)] = -\lambda(h)\,\Delta(g);$

(F4) $D^2(g) + D(\Delta(g)) = -\lambda(g)\, g_0;$

(F5) $D^2(g) + \Delta(D(g)) = \alpha\, D(g) + [g_0, g] - 2\,\lambda(g)\, g_0;$

(F6) $\Delta([g, h]) = [\Delta(g), h] + [g, \Delta(h)];$

(F7) $D([g, h]) = [D(g), h] - [D(h), g] + \lambda(g)D(h) - \lambda(h)D(g)$

We denote by $\mathcal{F}_1(\mathfrak{g})$ the set of all flag datums of the first kind of $\mathfrak{g}$.

Definition 2.2.6 *Let $\mathfrak{g}$ be a Leibniz algebra. A* flag datum of the second kind *of $\mathfrak{g}$ is a quadruple (g_0, ν, D, Δ), where $g_0 \in \mathfrak{g}$, $\nu : \mathfrak{g} \to k$, $\nu \neq 0$ is a non-trivial map, D and $\Delta : \mathfrak{g} \to \mathfrak{g}$ are linear maps satisfying the following compatibilities for any $g, h \in \mathfrak{g}$:*

(G1) $\nu([g, h]) = 0, \quad [g, g_0] = 0, \quad D(g_0) = 0, \quad \nu(g_0) = 0;$

(G2) $[g, \Delta(h)] + [g, D(h)] = 0, \quad \nu(D(g)) + \nu(\Delta(g)) = 0;$

(G3) $D^2(g) + D(\Delta(g)) = 0, \quad D^2(g) + \Delta(D(g)) = [g_0, g] + 2\,\nu(g)\, g_0;$

(G4) $\Delta([g, h]) = [\Delta(g), h] + [g, \Delta(h)] + \nu(h)\Delta(g) + \nu(g)D(h);$

(G5) $D([g, h]) = [D(g), h] - [D(h), g] - \nu(g)D(h) + \nu(h)D(g).$

We denote by $\mathcal{F}_2(\mathfrak{g})$ the set of all flag datums of the second kind of $\mathfrak{g}$ and by $\mathcal{F}(\mathfrak{g}) := \mathcal{F}_1(\mathfrak{g}) \sqcup \mathcal{F}_2(\mathfrak{g})$, the disjoint union of the two sets. The elements of $\mathcal{F}(\mathfrak{g})$ will be called *flag datums* of $\mathfrak{g}$. $\mathcal{F}(\mathfrak{g})$ contains the space of pointed double derivations $\mathcal{D}(\mathfrak{g})$ of $\mathfrak{g}$ via the canonical embedding: $\mathcal{D}(\mathfrak{g}) \hookrightarrow \mathcal{F}_1(\mathfrak{g})$, $(g_0, D, \Delta) \mapsto (g_0, 0, 0, D, \Delta)$. The next proposition shows that the space $\mathcal{F}(\mathfrak{g})$ is the counterpart for Leibniz algebras of the space of twisted derivations of a Lie algebra:

Proposition 2.2.7 *Let $\mathfrak{g}$ be a Leibniz algebra and V a vector space of dimension 1 with a basis $\{x\}$. Then there exists a bijection between the set $\mathcal{LZ}(\mathfrak{g}, V)$ of all Leibniz extending structures of $\mathfrak{g}$ through V and $\mathcal{F}(\mathfrak{g}) = \mathcal{F}_1(\mathfrak{g}) \sqcup \mathcal{F}_2(\mathfrak{g})$.*

Under the above bijective correspondence, the Leibniz extending structure $\Omega(\mathfrak{g}, V) = (\triangleleft, \triangleright, \leftharpoonup, \rightharpoonup, f, \{-,-\})$ corresponding to $(g_0, \alpha, \lambda, D, \Delta) \in \mathcal{F}_1(\mathfrak{g})$ is given by:

$$x \triangleleft g = \lambda(g)x, \qquad x \triangleright g = D(g), \qquad f(x, x) = g_0 \tag{2.15}$$

$$g \leftharpoonup x = \Delta(g), \qquad g \rightharpoonup x = 0, \qquad \{x, x\} = \alpha\, x \tag{2.16}$$

while the Leibniz extending structure $\Omega(\mathfrak{g}, V) = \big(\triangleleft, \triangleright, \leftharpoonup, \rightharpoonup, f, \{-,-\}\big)$ *corresponding to* $(g_0, \nu, D, \Delta) \in \mathcal{F}_2(\mathfrak{g})$ *is given by:*

$$x \triangleleft g = -\nu(g)x, \quad x \triangleright g = D(g), \quad f(x,x) = g_0 \tag{2.17}$$

$$g \leftharpoonup x = \Delta(g), \quad g \rightharpoonup x = \nu(g)x, \quad \{x, x\} = 0 \tag{2.18}$$

for all $g \in \mathfrak{g}$.

Proof: We have to compute the set of all bilinear maps $\triangleleft : V \times \mathfrak{g} \to V$, $\triangleright : V \times \mathfrak{g} \to \mathfrak{g}$, $\leftharpoonup: \mathfrak{g} \times V \to \mathfrak{g}$, $\rightharpoonup: \mathfrak{g} \times V \to V$, $f : V \times V \to \mathfrak{g}$ and $\{-, -\} : V \times V \to V$ satisfying the compatibility conditions $(L1)$–$(L14)$ of Theorem 2.1.3. Since V has dimension 1 there exists a bijection between the set of all bilinear maps $\triangleleft : V \times \mathfrak{g} \to V$ and the set of all linear maps $\lambda : \mathfrak{g} \to k$ and the bijection is given such that the action $\triangleleft : V \times \mathfrak{g} \to V$ associated to λ is given by the formula: $x \triangleleft g := \lambda(g)x$, for all $g \in \mathfrak{g}$. In the same manner, the action $\rightharpoonup: \mathfrak{g} \times V \to V$ is uniquely determined by a linear map $\nu : \mathfrak{g} \to k$ such that $g \rightharpoonup x = \nu(g)x$, for all $g \in \mathfrak{g}$. Similarly, the bilinear maps $\triangleright : V \times \mathfrak{g} \to \mathfrak{g}$, $\leftharpoonup: \mathfrak{g} \times V \to \mathfrak{g}$ are uniquely implemented by linear maps $D = D_{\triangleright} : \mathfrak{g} \to \mathfrak{g}$ respectively $\Delta = \Delta_{\leftharpoonup} : \mathfrak{g} \to \mathfrak{g}$ via the formulas: $x \triangleright g := D(g)$ and $g \leftharpoonup x := \Delta(g)$, for all $g \in \mathfrak{g}$. Finally, any bilinear map $f : V \times V \to \mathfrak{g}$ is uniquely implemented by an element $g_0 \in \mathfrak{g}$ such that $f(x, x) = g_0$ and any bracket $\{-, -\} : V \times V \to V$ is uniquely determined by a scalar $\alpha \in k$ such that $\{x, x\} = \alpha x$.

Now, the compatibility condition $(L9)$ is equivalent to $\alpha\,\nu(g) = 0$, for all $g \in \mathfrak{g}$, while $(L14)$ gives $\nu(g)\big(\nu(h) + \lambda(h)\big) = 0$, for all $g, h \in \mathfrak{g}$. If $\nu = 0$, the trivial map on $\mathfrak{g}$, then $(L9)$ and $(L14)$ are trivially fulfilled and the rest of the compatibility conditions of Theorem 2.1.3 came down to $(F1)$-$(F7)$ from the definition of $\mathcal{F}_1(\mathfrak{g})$. This can be proved by a routine computation: for instance, axiom $(L1)$ is equivalent to $\lambda\big([g, h]\big) = 0$ while axiom $(L2)$ is equivalent to $(F7)$. Otherwise, if $\nu \neq 0$ implies that $\alpha = 0$ and $\lambda = -\nu$. Based on this, it is straightforward to see that the compatibility conditions $(L1)$–$(L14)$ of Theorem 2.1.3 are equivalent to $(G1)$–$(G5)$ from the definition of $\mathcal{F}_2(\mathfrak{g})$. □

Let $(g_0, \alpha, \lambda, D, \Delta) \in \mathcal{F}_1(\mathfrak{g})$. The unified product $\mathfrak{g} \ltimes_{(g_0,\, \alpha,\, \lambda,\, D,\, \Delta)} V$ associated to the Leibniz extending structure given by (2.15)–(2.16) will be denoted by $\mathfrak{g}_1\left(x \,|\, (g_0, \alpha, \lambda, D, \Delta)\right)$ and has the bracket defined by:

$$[(g,0), (h,0)] = ([g,h], 0), \quad [(g,0), (0,x)] = (\Delta(g), 0) \tag{2.19}$$

$$[(0,x), (0,x)] = (g_0, \alpha x), \quad [(0,x), (g,0)] = (D(g), \lambda(g)x) \tag{2.20}$$

for all $g, h \in \mathfrak{g}$. On the other hand, for $(g_0, \nu, D, \Delta) \in \mathcal{F}_2(\mathfrak{g})$, the unified product $\mathfrak{g} \ltimes_{(g_0,\, \nu,\, D,\, \Delta)} V$ associated to the Leibniz extending structure given by (2.17)-(2.18) will be denoted by $\mathfrak{g}_2\left(x \,|\, (g_0, \nu, D, \Delta)\right)$ and has the bracket defined by:

$$[(g,0), (h,0)] = ([g,h], 0), \quad [(g,0), (0,x)] = (\Delta(g), \nu(g)x) \tag{2.21}$$

$$[(0,x), (0,x)] = (g_0, 0), \quad [(0,x), (g,0)] = (D(g), -\nu(g)x) \tag{2.22}$$

for all g, $h \in \mathfrak{g}$. Thus, we have obtained the following:

Corollary 2.2.8 *Any Leibniz algebra that contains a given Leibniz algebra $\mathfrak{g}$ as a subalgebra of codimension 1 is isomorphic to a Leibniz algebra of type $\mathfrak{g}_1\,(x \,|\, (g_0,\, \alpha,\, \lambda,\, D,\, \Delta))$, for some $(g_0,\, \alpha,\, \lambda,\, D,\, \Delta) \in \mathcal{F}_1\,(\mathfrak{g})$ or to a Leibniz algebra of type $\mathfrak{g}_2\,(x \,|\, (g_0,\, \nu,\, D,\, \Delta))$, for some $(g_0,\, \nu,\, D,\, \Delta) \in \mathcal{F}_2\,(\mathfrak{g})$.*

Now, we shall classify these Leibniz algebras up to an isomorphism that stabilizes $\mathfrak{g}$, i.e., we give the first explicit classification result for the ES problem. This is the key step in the classification of all flag extending structures of $\mathfrak{g}$.

Theorem 2.2.9 *Let $\mathfrak{g}$ be a Leibniz algebra of codimension 1 in the vector space E. Then:*

$$\mathrm{ExtdL}\,(E, \mathfrak{g}) \cong \mathcal{HL}^2_{\mathfrak{g}}(k, \mathfrak{g}) \cong (\mathcal{F}_1\,(\mathfrak{g})/\equiv_1) \sqcup (\mathcal{F}_2\,(\mathfrak{g})/\equiv_2), \qquad \text{where :}$$

$\equiv_1$ is the equivalence relation on the set $\mathcal{F}_1\,(\mathfrak{g})$ defined as follows: $(g_0,\, \alpha,\, \lambda,\, D,\, \Delta) \equiv_1 (g_0',\, \alpha',\, \lambda',\, D',\, \Delta')$ if and only if $\lambda = \lambda'$ and there exists a pair $(q, G) \in k^* \times \mathfrak{g}$ such that for any $g \in \mathfrak{g}$:

$$g_0 = q^2\, g_0' + [G, G] + q\, D'(G) + q\, E'(G) - q\, \alpha' G - \lambda'(G) G \tag{2.23}$$
$$\alpha = q\, \alpha' + \lambda'(G) \tag{2.24}$$
$$D(g) = q\, D'(g) + [G, g] - \lambda'(g) G \tag{2.25}$$
$$\Delta(g) = q\, \Delta'(g) + [g, G] \tag{2.26}$$

$\equiv_2$ is the equivalence relation on $\mathcal{F}_2\,(\mathfrak{g})$ given by: $(g_0,\, \nu,\, D,\, \Delta) \equiv_2 (g_0',\, \nu',\, D',\, \Delta')$ if and only if $\nu = \nu'$ and there exists a pair $(q, G) \in k^* \times \mathfrak{g}$ such that for any $g \in \mathfrak{g}$:

$$g_0 = q^2\, g_0' + [G, G] + q\, D'(G) + q\, E'(G) \tag{2.27}$$
$$D(g) = q\, D'(g) + [G, g] + \nu'(g)\, G \tag{2.28}$$
$$\Delta(g) = q\, \Delta'(g) + [g, G] - \nu'(g)\, G \tag{2.29}$$

The bijection between $(\mathcal{F}_1\,(\mathfrak{g})/\equiv_1) \sqcup (\mathcal{F}_2\,(\mathfrak{g})/\equiv_2)$ and $\mathrm{ExtdL}\,(E, \mathfrak{g})$ is given by: [1]

$$\overline{(g_0,\, \alpha,\, \lambda,\, D,\, \Delta)}^{1} \mapsto \mathfrak{g}_1\,(x \,|\, (g_0,\, \alpha,\, \lambda,\, D,\, \Delta)) \text{ and}$$
$$\overline{(g_0,\, \nu,\, D,\, \Delta)}^{2} \mapsto \mathfrak{g}_2\,(x \,|\, (g_0,\, \nu,\, D,\, \Delta))$$

Proof: The proof relies on Proposition 2.2.7 and Theorem 2.1.9. Let V be a complement of $\mathfrak{g}$ in E having $\{x\}$ as a basis. Since $\dim_k(V) = 1$, any linear map $r : V \to \mathfrak{g}$ is uniquely determined by an element $G \in \mathfrak{g}$ such that $r(x) = G$, where $\{x\}$ is a basis in V. On the other hand, any automorphism v of V is

[1] As usual we denote by $\overline{y}^i$ the equivalence class of y via the relation $\equiv_i$, $i = 1, 2$.

uniquely determined by a non-zero scalar $q \in k^*$ such that $v(x) = qx$. Based on these facts, a little computation shows that the compatibility conditions from Definition 2.1.7, imposed for the Leibniz extending structures (2.15)–(2.16) and respectively (2.17)–(2.18), take precisely the form given in the statement of the theorem. We should mention here that a Leibniz extending structure given by (2.15)–(2.16) is never equivalent in the sense of Definition 2.1.7 to a Leibniz extending structure given by (2.17)-(2.18), thanks to the compatibility condition (ML1) of Lemma 2.1.6. Therefore, we obtain the disjoint union from the statement and the proof is finished. □

Remark 2.2.10 *In the context of Theorem 2.2.9 we also have that*

$$\mathrm{ExtdL}'(E,\mathfrak{g}) \cong \mathcal{HL}^2(k,\mathfrak{g}) \cong (\mathcal{F}_1(\mathfrak{g})/\approx_1) \sqcup (\mathcal{F}_2(\mathfrak{g})/\approx_2), \qquad \text{where :}$$

$\approx_i$ *is the following relation on* $\mathcal{F}_i(\mathfrak{g})$*:* $(g_0, \alpha, \lambda, D, \Delta) \approx_1 (g_0', \alpha', \lambda', D', \Delta')$ *if and only if* $\lambda = \lambda'$ *and there exists* $G \in \mathfrak{g}$ *such that relations (2.23)-(2.26) hold for* $q = 1$ *and respectively* $(g_0, \nu, D, \Delta) \approx_2 (g_0', \nu', D', \Delta')$ *if and only if* $\nu = \nu'$ *and there exists* $G \in \mathfrak{g}$ *such that (2.27)-(2.29) hold for* $q = 1$.

Theorem 2.2.9 takes a simplified form for perfect Leibniz algebras. Indeed, let $\mathfrak{g}$ be a perfect Leibniz algebra, i.e., $\mathfrak{g}$ is generated as a vector space by all brackets $[x,y]$. Then (G1) shows that $\mathcal{F}_2(\mathfrak{g})$ is the empty set since, by definition, an element ν of a quadruple $(g_0, \nu, D, \Delta) \in \mathcal{F}_2(\mathfrak{g})$ is a non-trivial map. Thus, we have that $\mathcal{F}(\mathfrak{g}) = \mathcal{F}_1(\mathfrak{g})$. Let now $(g_0, \alpha, \lambda, D, \Delta) \in \mathcal{F}_1(\mathfrak{g})$; it follows from (F1) and (F2) that $\lambda = 0$, the trivial map, and $\alpha = 0$. Furthermore, we can easily see that for a perfect Leibniz algebra $\mathfrak{g}$, $\mathcal{F}(\mathfrak{g})$ identifies with the set of all triples (g_0, D, Δ), where $g_0 \in \mathfrak{g}$, $D, \Delta : \mathfrak{g} \to \mathfrak{g}$ are linear maps satisfying the compatibilities (2.11)-(2.14), that is $\mathcal{F}(\mathfrak{g}) \cong \mathcal{D}(\mathfrak{g})$, where $\mathcal{D}(\mathfrak{g})$ is the space of all pointed double derivations of $\mathfrak{g}$ as defined in Definition 2.2.4.

Two pointed double derivations (g_0, D, Δ) and (g_0', D', Δ') are equivalent and we write
$(g_0, D, \Delta) \equiv (g_0', D', \Delta')$ if and only if there exists a pair $(q, G) \in k^* \times \mathfrak{g}$ such that:

$$g_0 = q^2 g_0' + [G,G] + q\,D'(G) + q\,\Delta'(G) \tag{2.30}$$
$$D - q\,D' = [G,-], \qquad \Delta - q\,\Delta' = [-,G] \tag{2.31}$$

On the other hand, two pointed double derivations (g_0, D, Δ) and (g_0', D', Δ') are cohomologous and we write $(g_0, D, \Delta) \approx (g_0', D', \Delta')$ if and only if there exists $G \in \mathfrak{g}$ such that:

$$g_0 = g_0' + [G,G] + D'(G) + \Delta'(G) \tag{2.32}$$
$$D - D' = [G,-], \qquad \Delta - \Delta' = [-,G] \tag{2.33}$$

Taking into account the unified product defined by (2.19)-(2.20) we obtain:

Corollary 2.2.11 *Let $\mathfrak{g}$ be a perfect Leibniz algebra having $\{e_i \,|\, i \in I\}$ as a basis. Then any Leibniz algebra E containing $\mathfrak{g}$ as a subalgebra of codimension 1 has the bracket $[-,-]_E$ defined on the basis $\{x,\, e_i \,|\, i \in I\}$ by:*

$$[e_i,\, e_j]_E := [e_i,\, e_j], \quad [x,\, x]_E := g_0, \quad [e_i,\, x]_E := \Delta(e_i), \quad [x,\, e_i]_E := D(e_i)$$

for all $(g_0,\, D,\, \Delta) \in \mathcal{D}(\mathfrak{g})$. Furthermore, $\mathcal{HL}^2_{\mathfrak{g}}(k,\mathfrak{g}) \cong \mathcal{D}(\mathfrak{g})/\equiv$ and $\mathcal{HL}^2(k,\mathfrak{g}) \cong \mathcal{D}(\mathfrak{g})/\approx$, where $\equiv$ (resp. $\approx$) is the relation defined by (2.30)-(2.31) (resp. (2.32)-(2.33)).

On the other hand, we have the following result for abelian Leibniz algebras:

Example 2.2.12 *Let $\mathfrak{g}$ be a vector space with $\{e_i \,|\, i \in I\}$ as a basis viewed as an abelian Leibniz algebra. Then, there exist three families of Leibniz algebras that contain $\mathfrak{g}$ as a subalgebra of codimension 1. They have $\{x,\, e_i \,|\, i \in I\}$ as a basis and the bracket given for any $i \in I$ as follows:*

$$\mathfrak{g}_{11}^{(g_0,\, D,\, \Delta)}: \qquad [e_i,\, e_j] = 0, \quad [e_i,\, x] = \Delta(e_i), \quad [x,\, x] = g_0, \quad [x,\, e_i] = D(e_i)$$

for all triples $(g_0, D, \Delta) \in \mathfrak{g} \times \mathrm{Hom}_k(\mathfrak{g},\mathfrak{g})^2$ such that $g_0 \in \mathrm{Ker}(D)$ and $D \circ \Delta = \Delta \circ D = -D^2$. The Leibniz algebra $\mathfrak{g}_{11}^{(g_0,\, D,\, \Delta)}$ is the unified product associated to the flag datum of the first kind $(g_0,\, \alpha,\, \lambda,\, D,\, \Delta)$ for which $\alpha := 0$ and $\lambda := 0$.

The second family of Leibniz algebras has the bracket given as follows:

$$\mathfrak{g}_{12}^{(u,\, h_0,\, \lambda)}: \qquad [e_i,\, e_j] = 0, \quad [e_i,\, x] = 0, \quad [x,\, e_i] = u\,\lambda(e_i)\, h_0 + \lambda(e_i)\, x$$
$$[x,\, x] = -u^2\, \lambda(h_0)\, h_0 - u\, \lambda(h_0)\, x$$

for all triples $(u,\, h_0,\, \lambda) \in k^* \times \mathfrak{g} \times \mathrm{Hom}_k(\mathfrak{g}, k)$ such that $\lambda \neq 0$. The Leibniz algebra $\mathfrak{g}_{12}^{(u,\, h_0,\, \lambda)}$ is the unified product associated to the flag datum of the first kind $(g_0,\, \alpha,\, \lambda,\, D,\, \Delta)$ for which $g_0 := u^2\lambda(h_0)\, h_0$, $\alpha := -u\lambda(h_0)$, $\Delta := 0$ and $D(g) := u\lambda(g)\, h_0$, for all $g \in \mathfrak{g}$.

Finally, for the last family of Leibniz algebras the bracket is given as follows:

$$\mathfrak{g}_{2}^{(u,\, g_0,\, h_0,\, \nu)}: \qquad [e_i,\, e_j] = 0, \quad [e_i,\, x] = -\,[x,\, e_i] = -u\,\nu(e_i)\, h_0 + \nu(e_i)\, x,$$
$$[x,\, x] = g_0$$

for all $(u,\, g_0,\, h_0,\, \nu) \in k^* \times \mathfrak{g}^2 \times \mathrm{Hom}_k(\mathfrak{g}, k)$ such that: $2g_0 = 0$, $\nu(g_0) = 0$ and $\nu \neq 0$. The Leibniz algebra $\mathfrak{g}_{2}^{(u,\, g_0,\, h_0,\, \nu)}$ is the unified product associated to the flag datum of the second kind $(g_0,\, \nu,\, D,\, \Delta)$ for which $D(g) := u\,\nu(g)h_0$ and $\Delta(g) := -u\,\nu(g)h_0$, for all $g \in \mathfrak{g}$.

The above results are obtained by a straightforward computation which relies on the explicit description of the set $\mathcal{F}(\mathfrak{g})$ for an abelian Leibniz algebra. For instance, axiom (F3) from the flag datum of first kind, for an abelian Leibniz algebra, takes the form $\lambda(h)\Delta(g) = 0$, for all $g,\, h \in \mathfrak{g}$. Therefore, we *have to consider two cases in order to describe the set $\mathcal{F}_1(\mathfrak{g})$: the first one corresponds to $\lambda = 0$, while for the second one we have $\lambda \neq 0$. The corresponding unified products are the ones given by the first two families of Leibniz algebras. In order to describe the set $\mathcal{F}_2(\mathfrak{g})$ we mention that the condition $2g_0 = 0$ is derived from axiom (G2). In this case the corresponding unified product is the one given by the last family of Leibniz algebras.*

Next we provide some explicit examples. First, we prove that any Leibniz algebra which contains a semisimple Lie algebra $\mathfrak{g}$ as a subalgebra of codimension 1 is in fact a Lie algebra and the classifying objects $\mathcal{H}^2_{\mathfrak{g}}(k, \mathfrak{g})$ and $\mathcal{H}^2(k, \mathfrak{g})$ are both singletons.

Example 2.2.13 *Let $\mathfrak{g}$ be a semisimple Lie algebra of codimension 1 in the vector space E and $\{e_i \,|\, i = 1, \cdots, n\}$ a basis of $\mathfrak{g}$. Then any Leibniz algebra structure on E that contains $\mathfrak{g}$ as a subalgebra is isomorphic to the Lie algebra having $\{x, e_i \,|\, i = 1, \cdots, n\}$ as a basis and the bracket $[-,-]_E$ defined by for any $i = 1, \cdots, n$ by:*

$$[e_i, e_j]_E := [e_i, e_j], \quad [x, e_i]_E = -[e_i, x]_E := [h_0, e_i], \quad [x, x] = 0$$

for some $h_0 \in \mathfrak{g}$. Furthermore, $\mathcal{HL}^2_{\mathfrak{g}}(k, \mathfrak{g}) = \mathcal{HL}^2(k, \mathfrak{g}) = 0$.

Indeed, we apply Corollary 2.2.11 taking into account that any semisimple Lie algebra $\mathfrak{g}$ is perfect, $\mathrm{Inn}(\mathfrak{g}) = \mathrm{Der}(\mathfrak{g})$ *and $Z(\mathfrak{g}) = 0$. Let $(g_0, D, \Delta) \in \mathcal{F}(\mathfrak{g})$ be a flag datum of $\mathfrak{g}$; then, since $\mathfrak{g}$ has a trivial center we obtain from (2.11) that $g_0 = 0$. Moreover, as $\mathfrak{g}$ is perfect it follows again from (2.11) that $\Delta = -D$. Thus, $\mathcal{F}(\mathfrak{g}) = \mathrm{Der}(\mathfrak{g})$. Since $\mathfrak{g}$ is semisimple, any derivation $D \in \mathrm{Der}(\mathfrak{g})$ is inner, i.e., there exists $h_0 \in \mathfrak{g}$ such that $D = [h_0, -]$. Thus, the Leibniz algebra $\mathfrak{g}_1(x \,|\, (g_0, \alpha, \lambda, D, \Delta)) = \mathfrak{g}_1(x \,|\, D)$ defined by (2.19) and (2.20) takes the form given in the statement. Moreover, two derivations $D = [h_0, -]$ and $D' = [h'_0, -]$ are equivalent in the sense of (2.33) if and only if there exists $G \in \mathfrak{g}$ such that $h_0 = h'_0 + G$, i.e., any two derivations are cohomologous. This shows that $\mathcal{HL}^2(k, \mathfrak{g})$ is a singleton having only 0 as an element and so is $\mathcal{HL}^2_{\mathfrak{g}}(k, \mathfrak{g})$ being a quotient of it.*

Now we will provide an explicit example which highlights the efficiency of Theorem 2.2.9. More precisely, we will describe all 4-dimensional Leibniz algebras that contain a given non-perfect 3-dimensional Leibniz algebra $\mathfrak{g}$ as a subalgebra. Then we will be able to compute the classifying object $\mathrm{ExtdL}(k^4, \mathfrak{g}) \cong \mathcal{HL}^2_{\mathfrak{g}}(k, \mathfrak{g})$. The detailed computations are rather long but straightforward.

Example 2.2.14 *Let $\mathfrak{g}$ be the 3-dimensional Leibniz algebra with the basis $\{e_1, e_2, e_3\}$ and the bracket defined by: $[e_1, e_3] = e_2$, $[e_3, e_3] = e_1$.*

Then, there exist four families of 4-dimensional Leibniz algebras which contain $\mathfrak{g}$ as a subalgebra: they have $\{e_1, e_2, e_3, x\}$ as a basis and the bracket is given as follows (the first three families of Leibniz algebras can be defined *over any field k while in case of the fourth family we need to distinguish between fields of characteristic 2 and those of characteristic different than 2):*

(1) *If* $\mathrm{char}(k) \neq 2$ *then the four families of Leibniz algebras that contain* $\mathfrak{g}$ *as a subalgebra are the following:*

$$\mathfrak{g}_{11}^{(b_1,\, b_2,\, c,\, d_1,\, d_2)}: \quad \begin{aligned} &[e_1,\, e_3] = e_2, \quad [e_3,\, e_3] = e_1, \\ &[e_1,\, x] = b_1\, e_2, \quad [e_3,\, x] = b_1\, e_1 + b_2\, e_2, \\ &[x,\, x] = b_1 d_1\, e_1 + c\, e_2, \quad [x,\, e_3] = d_1\, e_1 + d_2\, e_2 \end{aligned}$$

for all $b_1,\, b_2,\, c,\, d_1,\, d_2 \in k$. *The Leibniz algebra* $\mathfrak{g}_{11}^{(b_1,\, b_2,\, c,\, d_1,\, d_2)}$ *is the unified product associated to the flag datum of the first kind* $(g_0,\, \alpha,\, \lambda,\, D,\, \Delta)$ *defined as follows:* $\alpha := 0$, $\lambda := 0$, $g_0 := b_1 d_1\, e_1 + c\, e_2$ *and* D, Δ *are given by*

$$D := \begin{pmatrix} 0 & 0 & d_1 \\ 0 & 0 & d_2 \\ 0 & 0 & 0 \end{pmatrix} \qquad \Delta := \begin{pmatrix} 0 & 0 & b_1 \\ b_1 & 0 & b_2 \\ 0 & 0 & 0 \end{pmatrix}$$

The second family of Leibniz algebras has the bracket given by:

$$\mathfrak{g}_{12}^{(b_1,\, b_2,\, b_3,\, c,\, d)}: \quad \begin{aligned} &[e_1,\, e_3] = e_2, \quad [e_3,\, e_3] = e_1, \quad [e_1,\, x] = 2b_1\, e_1 + b_2\, e_2, \\ &[e_2,\, x] = 3b_1\, e_2, \quad [e_3,\, x] = b_2\, e_1 + b_3\, e_2 + b_1\, e_3, \\ &[x,\, x] = (2b_1 d + b_2^2 - b_1 b_3)\, e_1 + c\, e_2, \\ &[x,\, e_3] = b_2\, e_1 + d\, e_2 - b_1\, e_3 \end{aligned}$$

for all $b_1 \in k^*$ *and* $b_2,\, b_3,\, c,\, d \in k$. *The Leibniz algebra* $\mathfrak{g}_{12}^{(b_1,\, b_2,\, b_3,\, c,\, d)}$ *is the unified product associated to the flag datum of the first kind* $(g_0,\, \alpha,\, \lambda,\, D,\, \Delta)$ *defined as follows:* $\alpha := 0$, $\lambda := 0$, $g_0 := (2b_1 d + b_2^2 - b_1 b_3)\, e_1 + c\, e_2$ *and* D, Δ *are given by:*

$$D := \begin{pmatrix} 0 & 0 & b_2 \\ 0 & 0 & d \\ 0 & 0 & -b_1 \end{pmatrix} \qquad \Delta := \begin{pmatrix} 2b_1 & 0 & b_2 \\ b_2 & 3b_1 & b_3 \\ 0 & 0 & b_1 \end{pmatrix}$$

The third family of Leibniz algebras has the bracket given by:

$$\mathfrak{g}_{13}^{(\alpha,\, \lambda_0,\, d_1,\, d_2)}: \quad \begin{aligned} &[e_1,\, e_3] = e_2, \quad [e_3,\, e_3] = e_1, \quad [e_1,\, x] = \alpha\, \lambda_0^{-1}\, e_2, \\ &[e_3,\, x] = \alpha\, \lambda_0^{-1}\, e_1, \\ &[x,\, x] = \alpha\, \lambda_0^{-1}\, (d_1\, e_1 + d_2\, e_2 - \alpha\, e_3) + \alpha\, x, \\ &[x,\, e_3] = d_1\, e_1 + d_2\, e_2 - \alpha\, e_3 + \lambda_0\, x \end{aligned}$$

for all $\lambda_0 \in k^*$ *and* $\alpha,\, d_1,\, d_2 \in k$. *The Leibniz algebra* $\mathfrak{g}_{13}^{(\alpha,\, \lambda_0,\, d_1,\, d_2)}$ *is the unified product associated to the flag datum of the first kind* $(g_0,\, \alpha,\, \lambda,\, D,\, \Delta)$ *defined as follows:* $\lambda(e_1) = \lambda(e_2) := 0$, $\lambda(e_3) := \lambda_0 \neq 0$, $g_0 := \alpha\, \lambda_0^{-1} d_1\, e_1 + \alpha\, \lambda_0^{-1} d_2\, e_2 - \alpha^2\, \lambda_0^{-1} e_3$ *and* D, Δ *are given by*

$$D := \begin{pmatrix} 0 & 0 & d_1 \\ 0 & 0 & d_2 \\ 0 & 0 & -\alpha \end{pmatrix} \qquad \Delta := \begin{pmatrix} 0 & 0 & \alpha\, \lambda_0^{-1} \\ \alpha\, \lambda_0^{-1} & 0 & 0 \\ 0 & 0 & 0 \end{pmatrix}$$

Finally, the last family of Leibniz algebras has the bracket defined as follows:

$$\mathfrak{g}_{21}^{(\nu_0,\, d_1,\, d_2,,d_3)}: \quad \begin{aligned} &[e_1,\, e_3] = e_2, \quad [e_3,\, e_3] = e_1, \quad [e_1,\, x] = \nu_0^{-1}\, d_3\, e_2,\\ &[e_3,\, x] = (-d_1 + 2\nu_0^{-1} d_3) e_1 - (d_2 - \nu_0^{-1} d_1 + \nu_0^{-2} d_3) e_2\\ &\qquad -d_3 e_3 + \nu_0 x,\\ &[x,\, x] = \nu_0^{-2}\, d_3^2\, e_1 + (\nu_0^{-2}\, d_1 d_3 - \nu_0^{-3}\, d_3^2)\, e_2,\\ &[x,\, e_3] = d_1\, e_1 + d_2\, e_2 + d_3\, e_3 - \nu_0\, x \end{aligned}$$

for all $\nu_0 \in k^*$ and d_1, d_2, $d_3 \in k$. The Leibniz algebra $\mathfrak{g}_{21}^{(\nu_0,\, d_1,\, d_2,\, d_3)}$ is the unified product associated to the flag datum of the second kind $(g_0,\, \nu,\, D,\, \Delta)$ defined as follows: $\nu(e_1) = \nu(e_2) := 0$, $\nu(e_3) := \nu_0 \neq 0$, $g_0 := \nu_0^{-2}\, d_3^2\, e_1 + (\nu_0^{-2}\, d_1 d_3 - \nu_0^{-3}\, d_3^2)\, e_2$ and D, Δ are given by

$$D := \begin{pmatrix} 0 & 0 & d_1 \\ 0 & 0 & d_2 \\ 0 & 0 & d_3 \end{pmatrix} \qquad \Delta := \begin{pmatrix} 0 & 0 & -d_1 + 2\nu_0^{-1} d_3 \\ \nu_0^{-1} d_3 & 0 & -d_2 + \nu_0^{-1} d_1 - \nu_0^{-2} d_3 \\ 0 & 0 & -d_3 \end{pmatrix} \tag{2.34}$$

(2) *If* $\mathrm{char}(k) = 2$, *then the four families of Leibniz algebras that contain $\mathfrak{g}$ as a subalgebra are the following: $\mathfrak{g}_{11}^{(b_1,\, b_2,\, c,\, d_1,\, d_2)}$, $\mathfrak{g}_{12}^{(b_1,\, b_2,\, b_3,\, c,\, d)}$, $\mathfrak{g}_{13}^{(\alpha,\, \lambda_0,\, d_1,\, d_2)}$ defined above together with the family of Leibniz algebras defined as follows:*

$$\mathfrak{g}_{22}^{(c,\, \nu_0,\, d_1,\, d_2,,d_3)}: \quad \begin{aligned} &[e_1,\, e_3] = e_2, \quad [e_3,\, e_3] = e_1, \quad [e_1,\, x] = \nu_0^{-1}\, d_3\, e_2,\\ &[e_3,\, x] = -d_1\, e_1 - (d_2 - \nu_0^{-1} d_1 + \nu_0^{-2} d_3) e_2 - d_3 e_3 + \nu_0 x,\\ &[x,\, x] = \nu_0^{-2}\, d_3^2\, e_1 + c\, e_2,\\ &[x,\, e_3] = d_1\, e_1 + d_2\, e_2 + d_3\, e_3 - \nu_0\, x \end{aligned}$$

for all $\nu_0 \in k^*$ and c, d_1, d_2, $d_3 \in k$. The Leibniz algebra $\mathfrak{g}_{22}^{(c,\, \nu_0,\, d_1,\, d_2,\, d_3)}$ is the unified product associated to the flag datum of the second kind $(g_0,\, \nu,\, D,\, \Delta)$ defined as follows: $\nu(e_1) = \nu(e_2) := 0$, $\nu(e_3) := \nu_0 \neq 0$, $g_0 := \nu_0^{-2}\, d_3^2\, e_1 + c\, e_2$ and D, Δ are given by (2.34).

The proof is a purely computational one and we will only indicate the main steps. We start by computing $\mathcal{F}_1\,(\mathfrak{g})$. First, notice that a linear map $\lambda : \mathfrak{g} \to k$ satisfies the first compatibility of (F1), i.e., $\lambda([g, h]) = 0$ if and only if λ is given by $\lambda(e_1) = \lambda(e_2) = 0$ and $\lambda(e_3) = \lambda_0$, for some $\lambda_0 \in k$. For such a λ we can easily show that a pair (D, Δ) satisfies the compatibilities (F6) and (F7) if and only if we have:

$$D = \begin{pmatrix} 0 & 0 & d_1 \\ 0 & 0 & d_2 \\ 0 & 0 & d_3 \end{pmatrix} \qquad \Delta = \begin{pmatrix} 2b_1 & 0 & b_2 \\ b_2 & 3b_1 & b_3 \\ 0 & 0 & b_1 \end{pmatrix}$$

for some d_1, d_2, d_3, b_1, b_2, $b_3 \in k$. Let now $\alpha \in k$ and consider $g_0 = c_1 e_1 + c_2 e_2 + c_3 e_3$, for some c_1, c_2, $c_3 \in k$. We can easily see that the 5-tuple $(g_0,\, \alpha,\, \lambda,\, D,\, \Delta)$ satisfies the compatibilities (F1)–(F7), i.e., it is a flag *datum of the first kind if and only if it coincides with one of the three flag datums described in* (1). *For instance, the compatibility (F1) is fulfilled if and only if* $\lambda_0(\alpha + d_3) = \lambda_0 b_1 = 0$. *This last equality leads us to consider two cases, namely* $\lambda_0 = 0$ *or* $\lambda_0 \neq 0$. *It is now straightforward to describe* $\mathcal{F}_1(\mathfrak{g})$ *(without depending on the characteristic of* k*).*

Analogously, we can describe $\mathcal{F}_2(\mathfrak{g})$. *A non-trivial map* $\nu : \mathfrak{g} \to k$ *satisfies the first compatibility of (G1) if and only if* ν *is given by* $\nu(e_1) = \nu(e_2) = 0$ *and* $\nu(e_3) = \nu_0$, *for some* $\nu_0 \in k^*$. *For such a map* ν, *we can easily show that* (D, Δ) *satisfies the compatibilities (G4) and (G5) if and only if* D *and* Δ *are given by (2.34). By considering again* $g_0 = c_1e_1 + c_2e_2 + c_3e_3$, *for some* c_1, c_2, $c_3 \in k$ *we see that the compatibility (G1) is fulfilled if and only if* $c_3 = 0$. *The last compatibility of (G3) is equivalent to:*

$$2\,c_1 = 2\,\nu_0^{-2}\,d_3^2, \qquad c_1 + 2\,\nu_0 c_2 = 2\,\nu_0^{-1} d_1 d_3 - \nu_0^{-1} d_3^2$$

The above two compatibilities are the ones that lead us to the description of $\mathcal{F}_2(\mathfrak{g})$ *depending on the characteristic of* k. *Moreover, these computations provide also the description of the classifying object* $\mathcal{HL}^2_{\mathfrak{g}}(k, \mathfrak{g})$. *If* $\mathrm{char}(k) \neq 2$ *then*

$$\mathcal{HL}^2_{\mathfrak{g}}(k, \mathfrak{g}) \cong (k^5/\equiv_{11}) \sqcup ((k^* \times k^4)/\equiv_{12}) \sqcup ((k^* \times k^3)/\equiv_{13}) \sqcup ((k^* \times k^3)/\equiv_2) \tag{2.35}$$

where $\equiv_{1i}$ *are the equivalence relations (2.23)-(2.26) while* $\equiv_2$ *is the equivalence relation (2.27)-(2.29). In the case when* $\mathrm{char}(k) = 2$, *then the last term of (2.35) is replaced by* $(k^* \times k^4)/\equiv_2$.

2.3 Special cases of unified products for Leibniz algebras

In this section we deal with two special cases of the unified product, namely the crossed product and the bicrossed product of two Leibniz algebras. Along the way we emphasize the problem to which each of these products is connected. We use the following convention: if one of the maps $\triangleleft$, $\triangleright$, $\leftharpoonup$, $\rightharpoonup$, f or $\{-,-\}$ of an extending datum $\Omega(\mathfrak{g}, V) = \big(\triangleleft, \triangleright, \leftharpoonup, \rightharpoonup, f, \{-,-\}\big)$ is trivial, we will omit it from the 6-tuple $\big(\triangleleft, \triangleright, \leftharpoonup, \rightharpoonup, f, \{-,-\}\big)$.

Crossed products and the extension problem for Leibniz algebras

We shall highlight a first special case of the unified product, namely the crossed product of Leibniz algebras, which plays the key role in the study of the extension problem in its full generality. The extension problem asks for the classification of all extensions of $\mathfrak{h}$ by $\mathfrak{g}$ and it was first studied in [167] for $\mathfrak{g}$

abelian; in this case all such extensions are classified by the second cohomology group $\mathrm{HL}^2(\mathfrak{h}, \mathfrak{g})$ [167, Proposition 1.9]. The fact that $\mathfrak{g}$ is abelian is essential in proving the aforementioned classification result. However, a classification result can still be proved in the non-abelian case and the classification object, denoted by $\mathbb{HL}^2(\mathfrak{h}, \mathfrak{g})$, arises as a special case of the ES problem and will generalize the second cohomology group $\mathrm{HL}^2(\mathfrak{h}, \mathfrak{g})$. The main drawback of this construction is the fact that $\mathbb{HL}^2(\mathfrak{h}, \mathfrak{g})$ does not arise as a cohomology group of a certain complex, it will be constructed using the theory of crossed products for Leibniz algebras. To conclude, the extension problem appears as a special case of the ES problem as follows: if in the ES problem we replace the condition "$\mathfrak{g}$ is a Leibniz subalgebra of $(E, [-,-])$" by a more restrictive one, namely "$\mathfrak{g}$ is a two-sided ideal of E and the quotient $E/\mathfrak{g}$ is isomorphic to a given Leibniz algebra $\mathfrak{h}$", then what we obtain is in fact the extension problem.

Let $\Omega(\mathfrak{g}, V) = \big(\triangleleft, \triangleright, \leftharpoonup, \rightharpoonup, f, \{-, -\}\big)$ be an extending datum of a Leibniz algebra $\mathfrak{g}$ through V such that $\triangleleft$ and $\rightharpoonup$ are both trivial, i.e., $x \triangleleft g = g \rightharpoonup x = 0$, for all $x \in V$ and $g \in \mathfrak{g}$. Then, it follows from Theorem 2.1.3 that $\Omega(\mathfrak{g}, V) = \big(\triangleright, \leftharpoonup, f, \{-, -\}\big)$ is a Leibniz extending structure of $\mathfrak{g}$ through V if and only if $(V, \{-,-\})$ is a Leibniz algebra and $(\mathfrak{g}, V, \triangleright, \leftharpoonup, f)$ is a *crossed system* of Leibniz algebras, i.e., the following compatibilities hold for any g, $h \in \mathfrak{g}$ and x, y, $z \in V$:

(CS1) $[g, h] \leftharpoonup x = [g, h \leftharpoonup x] + [g \leftharpoonup x, h]$;

(CS2) $g \leftharpoonup \{x, y\} = (g \leftharpoonup x) \leftharpoonup y - (g \leftharpoonup y) \leftharpoonup x - [g, f(x,y)]$;

(CS3) $x \triangleright f(y, z) = f(x, y) \leftharpoonup z - f(x, z) \leftharpoonup y + f(\{x, y\}, z) - f(\{x, z\}, y) - f(x, \{y, z\})$;

(CS4) $x \triangleright [g, h] = [x \triangleright g, h] - [x \triangleright h, g]$;

(CS5) $\{x, y\} \triangleright g = x \triangleright (y \triangleright g) + (x \triangleright g) \leftharpoonup y - [f(x, y), g]$;

(CS6) $[g, h \leftharpoonup x] + [g, x \triangleright h] = 0$;

(CS7) $x \triangleright (y \triangleright g) + x \triangleright (g \leftharpoonup y) = 0$.

In this case, the associated unified product $\mathfrak{g} \ltimes_{\Omega(\mathfrak{g}, V)} V$ will be denoted by $\mathfrak{g} \#_{\triangleright, \leftharpoonup}^{f} V$ and we shall call it the *crossed product* of the Leibniz algebras $\mathfrak{g}$ and V. Hence, the crossed product associated to the crossed system $(\mathfrak{g}, V, \triangleright, \leftharpoonup, f)$ is the Leibniz algebra defined as follows: $\mathfrak{g} \#_{\triangleright, \leftharpoonup}^{f} V = \mathfrak{g} \times V$ with the bracket given for any g, $h \in \mathfrak{g}$ and x, $y \in V$ by:

$$[(g, x), (h, y)] := \big([g, h] + x \triangleright h + g \leftharpoonup y + f(x, y), \{x, y\}\big) \tag{2.36}$$

The crossed product of Leibniz algebras is the object responsible for answering the following special case of the ES problem, which is a generalization of the extension problem: *Let $\mathfrak{g}$ be a Leibniz algebra, and E a vector space* *containing $\mathfrak{g}$ as a subspace. Describe and classify all Leibniz algebra structures on E such that $\mathfrak{g}$ is a two-sided ideal of E.* The classical extension problem first appeared in [167] and is a special case of this question if we require the additional assumption on the quotient $E/\mathfrak{g}$ to be isomorphic to a given Leibniz algebra $\mathfrak{h}$.

Indeed, let $(\mathfrak{g}, V, \triangleright, \leftharpoonup, f)$ be a crossed system of two Leibniz algebras. Then, $\mathfrak{g} \cong \mathfrak{g} \times \{0\}$ is a two-sided ideal in the crossed product $\mathfrak{g}\#_{\triangleright,\leftharpoonup}^{f} V$ since $[(g,0),\, (h,y)] := \big([g,\, h] + g \leftharpoonup y,\, 0\big)$ and $[(g,x),\, (h,0)] := \big([g,\, h] + x \triangleright h,\, 0\big)$. Conversely, we have:

Corollary 2.3.1 *Let $\mathfrak{g}$ be a Leibniz algebra, E a vector space containing $\mathfrak{g}$ as a subspace. Then any Leibniz algebra structure on E that contains $\mathfrak{g}$ as a two-sided ideal is isomorphic to a crossed product of Leibniz algebras $\mathfrak{g}\#_{\triangleright,\leftharpoonup}^{f} V$ and the isomorphism can be chosen to stabilize $\mathfrak{g}$ and co-stabilize V.*

Proof: Let $[-,\,-]$ be a Leibniz algebra structure on E such that $\mathfrak{g}$ is a two-sided ideal in E. In particular, $\mathfrak{g}$ is a subalgebra of E and hence we can apply Theorem 2.1.5. In this case the actions $\triangleleft = \triangleleft_p$ and $\rightharpoonup = \rightharpoonup_p$ of the Leibniz extending structure $\Omega(\mathfrak{g}, V) = \big(\triangleleft_p, \triangleright_p, f_p, \{-,\,-\}_p\big)$ constructed in the proof of Theorem 2.1.5 are both trivial since for any $x \in V$ and $g \in \mathfrak{g}$ we have that $[x,g]$, $[g,x] \in \mathfrak{g}$ and hence $p([x,g]) = [x,g]$ and $p([g,x]) = [g,x]$. Thus, $x \triangleleft_p g = g \rightharpoonup_p x = 0$ and hence the Leibniz extending structure $\Omega(\mathfrak{g}, V) = \big(\triangleleft, \triangleright, \leftharpoonup, \rightharpoonup, f, \{-,\,-\}\big)$ constructed in the proof of Theorem 2.1.5 is precisely a crossed system of Leibniz algebras and the unified product $\mathfrak{g} \ltimes_{\Omega(\mathfrak{g},V)} V = \mathfrak{g}\#_{\triangleright,\leftharpoonup}^{f} V$ is the crossed product of $\mathfrak{g}$ and $V = \mathrm{Ker}(p)$. □

Let $\mathfrak{g}$ and $\mathfrak{h}$ be two given Leibniz algebras. The extension problem asks for the classification of all extensions of $\mathfrak{h}$ by $\mathfrak{g}$, i.e., of all Leibniz algebras $\mathfrak{E}$ that fit into an exact sequence

$$0 \longrightarrow \mathfrak{g} \xrightarrow{\ i\ } \mathfrak{E} \xrightarrow{\ \pi\ } \mathfrak{h} \longrightarrow 0 \tag{2.37}$$

The classification is up to an isomorphism of Leibniz algebras that stabilizes $\mathfrak{g}$ and co-stabilizes $\mathfrak{h}$ and we denote by $\mathcal{EP}(\mathfrak{h},\, \mathfrak{g})$ the isomorphism classes of all extensions of $\mathfrak{h}$ by $\mathfrak{g}$ up to this equivalence relation. If $\mathfrak{g}$ is abelian, then $\mathcal{EP}(\mathfrak{h},\, \mathfrak{g}) \cong \mathrm{HL}^2(\mathfrak{h},\, \mathfrak{g})$, where $\mathrm{HL}^2(\mathfrak{h},\, \mathfrak{g})$ is the second cohomology group [167, Proposition 1.9]. The crossed product is the tool to approach the extension problem in its full generality, leaving aside the abelian case. Let us explain this briefly. Consider $\mathfrak{g}$ and $\mathfrak{h}$ to be two Leibniz algebras and we denote by $\mathcal{CS}\,(\mathfrak{h},\, \mathfrak{g})$ the set of all triples $(\triangleright,\, \leftharpoonup,\, f)$ such that $(\mathfrak{g},\, \mathfrak{h},\, \triangleright,\, \leftharpoonup,\, f)$ is a crossed system of Leibniz algebras. First we remark that, if $(\mathfrak{g},\, \mathfrak{h},\, \triangleright,\, \leftharpoonup,\, f)$ is a crossed system, then the crossed product $\mathfrak{g}\#_{\triangleright,\leftharpoonup}^{f} \mathfrak{h}$ is an extension of $\mathfrak{h}$ by $\mathfrak{g}$ via

$$0 \longrightarrow \mathfrak{g} \xrightarrow{\ i_{\mathfrak{g}}\ } \mathfrak{g}\#_{\triangleright,\leftharpoonup}^{f} \mathfrak{h} \xrightarrow{\ \pi_{\mathfrak{h}}\ } \mathfrak{h} \longrightarrow 0 \tag{2.38}$$

where $i_{\mathfrak{g}}(g) = (g, 0)$ and $\pi_{\mathfrak{h}}(g, h) = h$ are the canonical maps. Conversely,

Corollary 2.3.1 shows that any extension $\mathfrak{E}$ of $\mathfrak{h}$ by $\mathfrak{g}$ is equivalent to a crossed product extension of the form (2.38). Thus, the classification of all extensions of $\mathfrak{h}$ by $\mathfrak{g}$ reduces to the classification of all crossed products $\mathfrak{g}\#_{\triangleright,\leftharpoonup}^{f}\mathfrak{h}$ associated to all crossed systems of Leibniz algebras $(\mathfrak{g}, \mathfrak{h}, \triangleright, \leftharpoonup, f)$. Definition 2.1.8, in the special case of crossed systems, takes the following simplified form: two triples $(\triangleright, \leftharpoonup, f)$ and $(\triangleright', \leftharpoonup', f')$ of $\mathcal{CS}(\mathfrak{h}, \mathfrak{g})$ are *cohomologous* and we denote this by $(\triangleright, \leftharpoonup, f) \approx (\triangleright', \leftharpoonup', f')$ if there exists a linear map $r : \mathfrak{h} \to \mathfrak{g}$ such that:

$$\begin{aligned} x \triangleright g &= x \triangleright' g + [r(x), g] \\ g \leftharpoonup x &= g \leftharpoonup' x + [g, r(x)] \\ f(x,y) &= f'(x,y) + [r(x), r(y)] - r(\{x, y\}) + r(x) \leftharpoonup' y + x \triangleright' r(y) \end{aligned}$$

for all $g \in \mathfrak{g}$, $x, y \in \mathfrak{h}$. Then, as we mentioned before Definition 2.1.8, $(\triangleright, \leftharpoonup, f) \approx (\triangleright', \leftharpoonup', f')$ if and only if there exists $\psi : \mathfrak{g}\#_{\triangleright,\leftharpoonup}^{f}\mathfrak{h} \to \mathfrak{g}\#_{\triangleright',\leftharpoonup'}^{f'}\mathfrak{h}$ an isomorphism of Leibniz algebras that stabilizes $\mathfrak{g}$ and co-stabilizes $\mathfrak{h}$. As a special case of Theorem 2.1.9, we obtain the theoretical answer to the extension problem in the general (non-abelian) case:

Corollary 2.3.2 *Let $\mathfrak{g}$ and $\mathfrak{h}$ be two arbitrary Leibniz algebras. Then $\approx$ is an equivalence relation on the set $\mathcal{CS}(\mathfrak{h}, \mathfrak{g})$ of all crossed systems and the map*

$$\mathbb{HL}^2(\mathfrak{h}, \mathfrak{g}) := \mathcal{CS}(\mathfrak{h}, \mathfrak{g})/\approx \longrightarrow \mathcal{EP}(\mathfrak{h}, \mathfrak{g}), \qquad \overline{(\triangleright, \leftharpoonup, f)} \mapsto \mathfrak{g}\#_{\triangleright,\leftharpoonup}^{f}\mathfrak{h}$$

is a bijection between sets, where $\overline{(\triangleright, \leftharpoonup, f)}$ is the equivalence class of $(\triangleright, \leftharpoonup, f)$ via $\approx$.

If $\mathfrak{g}$ is an abelian Leibniz algebra, then $\mathbb{HL}^2(\mathfrak{h}, \mathfrak{g})$ coincides with the second cohomology group $\mathrm{HL}^2(\mathfrak{h}, \mathfrak{g})$ constructed in [167]. The explicit answer to the extension problem for two given Leibniz algebras $\mathfrak{g}$ and $\mathfrak{h}$ will be given once we compute the non-abelian cohomological object $\mathbb{HL}^2(\mathfrak{h}, \mathfrak{g})$ which in general is a highly non-trivial problem. A detailed study of this object for various Leibniz algebras will be given elsewhere; here we give only one example that corresponds to the case when $\mathfrak{h} := k$, the abelian Leibniz algebra of dimension 1, as this is a special case of Theorem 2.2.9 and Remark 2.2.10.

Corollary 2.3.3 *Let $\mathfrak{g}$ be a Leibniz algebra with $\{e_i \,|\, i \in I\}$ as a basis. Then*

$$\mathbb{HL}^2(k, \mathfrak{g}) \cong \mathcal{D}(g)/\approx$$

where $\mathcal{D}(g)$ is the space of all pointed double derivations of $\mathfrak{g}$ and $\approx$ is the equivalence relation defined by (2.32)-(2.33). In particular, any extension of k by $\mathfrak{g}$ is isomorphic to the Leibniz algebra having $\{x, e_i \,|\, i \in I\}$ as a basis and the bracket $[-, -]_{(g_0, D, \Delta)}$ defined for any $i \in I$ by:

$$\begin{aligned} [e_i, e_j]_{(g_0, D, \Delta)} &:= [e_i, e_j], \quad [x, x]_{(g_0, D, \Delta)} := g_0 \\ [e_i, x]_{(g_0, D, \Delta)} &:= \Delta(e_i), \quad [x, e_i]_{(g_0, D, \Delta)} := D(e_i) \end{aligned}$$

for some $(g_0, D, \Delta) \in \mathcal{D}(g)$.

Proof: Follows from Theorem 2.2.9 since the set of crossed systems $\mathcal{CS}(k,\, \mathfrak{g})$ is precisely the set $\Omega(\mathfrak{g}, k) = \big(\triangleleft, \triangleright, \leftharpoonup, \rightharpoonup, f, \{-,\, -\}\big)$ of all Leibniz extending structures of $\mathfrak{g}$ through k having the actions $\triangleleft$ and $\rightharpoonup$ both trivial. Moreover, any extension E of k by $\mathfrak{g}$ is a Leibniz algebra containing $\mathfrak{g}$ as a subalgebra of codimension 1. In this context, the compatibility conditions (F1)-(F7) that define a flag datum of the first kind collapses to (2.11)-(2.14). The fact that $\triangleleft$ is the trivial action implies that $\lambda = 0$. The Leibniz algebra from the statement is the unified (crossed) product defined by (2.19)-(2.20). □

In the next example we compute explicitly the object $\mathbb{HL}^2(k,\, \mathfrak{g})$ for a certain Leibniz algebra $\mathfrak{g}$.

Example 2.3.4 *Let $\mathfrak{g}$ be the 3-dimensional Leibniz algebra with the basis $\{e_1, e_2, e_3\}$ and the bracket defined by: $[e_1,\, e_3] = e_2$, $[e_3,\, e_3] = e_1$. A little computation, similar to one performed in Example 2.2.14, shows that the set $\mathcal{D}(g)$ identifies with the set of all 6-tuples $(c,\, b_1,\, b_2,\, b_3,\, d_1,\, d_2)$ $\in k^6$ which satisfy:*

$$b_1\,(d_1 - b_2) = 0$$

The bijection is defined such that $(g_0,\, D,\, \Delta) \in \mathcal{D}(g)$ corresponding to $(c,\, b_1,\, b_2,\, b_3,\, d_1,\, d_2)$ is given by

$$g_0 := (2\,b_1 d_2 + b_2 d_1 - b_1 b_3)\, e_1 + c\, e_2,$$

$$D := \begin{pmatrix} 0 & 0 & d_1 \\ 0 & 0 & d_2 \\ 0 & 0 & -b_1 \end{pmatrix} \qquad \Delta := \begin{pmatrix} 2b_1 & 0 & b_2 \\ b_2 & 3b_1 & b_3 \\ 0 & 0 & b_1 \end{pmatrix}$$

The compatibility condition $b_1\,(d_1 - b_2) = 0$ imposes a discussion on whether $b_1 = 0$ or $b_1 \neq 0$. This leads to the description of $\mathbb{L}^2(k,\, \mathfrak{g})$ as the following coproduct of sets:

$$\mathbb{HL}^2(k,\, \mathfrak{g}) \cong (k^5/\approx_1) \sqcup (k^* \times k^4/\approx_2), \qquad \text{where :}$$

$\approx_1$ is the following relation on k^5: $(c,\, b_2,\, b_3,\, d_1,\, d_2) \approx_1 (c',\, b_2',\, b_3',\, d_1',\, d_2')$ if and only if there exist $u,\, v \in k$ such that

$$c = c' + uv, \quad b_2 = b_2' + v, \quad b_3 = b_3' + v, \quad d_1 = d_1' + v, \quad d_2 = d_2' + u$$

and $\approx_2$ is the relation on $k^* \times k^4$ defined by: $(b_1,\, c,\, b_3,\, d_1,\, d_2) \approx_2 (b_1',\, c',\, b_3',\, d_1',\, d_2')$ if and only if $b_1 = b_1'$, $b_3 = b_3'$, $d_2 = d_2'$ and there exist $v,\, w \in k$ such that

$$c = c' + v d_1' + 3 w b_1' + (d_1 - d_1')\,(v + d_2' + b_3')$$

Another important special case of the unified product is the bicrossed product which provides the answer to the factorization problem. We present it in the sequel in more detail.

Bicrossed products and the factorization problem for Leibniz algebras

This section is devoted to introducing the concept of a matched pair of Leibniz algebras. These are obtained from extending datums $\Omega(\mathfrak{g}, V) = (\triangleleft, \triangleright, \leftharpoonup, \rightharpoonup, f, \{-, -\})$ for which f is the trivial map.

Definition 2.3.5 *A matched pair of Leibniz algebras is a system* $(\mathfrak{g}, \mathfrak{h}, \triangleleft, \triangleright, \leftharpoonup, \rightharpoonup)$ *consisting of two Leibniz algebras* $(\mathfrak{g}, [-,-])$, $(\mathfrak{h}, \{-,-\})$ *and four bilinear maps* $\triangleleft : \mathfrak{h} \times \mathfrak{g} \to \mathfrak{h}$, $\triangleright : \mathfrak{h} \times \mathfrak{g} \to \mathfrak{g}$, $\leftharpoonup : \mathfrak{g} \times \mathfrak{h} \to \mathfrak{g}$, $\rightharpoonup : \mathfrak{g} \times \mathfrak{h} \to \mathfrak{h}$ *satisfying the following compatibilities for any* $g, h \in \mathfrak{g}$, $x, y \in \mathfrak{h}$:

(MP1) $(\mathfrak{h}, \triangleleft)$ *is a right* $\mathfrak{g}$*-module, i.e.,* $x \triangleleft [g, h] = (x \triangleleft g) \triangleleft h - (x \triangleleft h) \triangleleft g$;

(MP2) $(\mathfrak{g}, \leftharpoonup)$ *is a right* $\mathfrak{h}$*-module, i.e.,* $g \leftharpoonup \{x, y\} = (g \leftharpoonup x) \leftharpoonup y - (g \leftharpoonup y) \leftharpoonup x$;

(MP3) $x \triangleright [g, h] = [x \triangleright g, h] - [x \triangleright h, g] + (x \triangleleft g) \triangleright h - (x \triangleleft h) \triangleright g$;

(MP4) $\{x, y\} \triangleleft g = x \triangleleft (y \triangleright g) + (x \triangleright g) \rightharpoonup y + \{x, y \triangleleft g\} + \{x \triangleleft g, y\}$;

(MP5) $\{x, y\} \triangleright g = x \triangleright (y \triangleright g) + (x \triangleright g) \leftharpoonup y$;

(MP6) $[g, h] \leftharpoonup x = [g, h \leftharpoonup x] + [g \leftharpoonup x, h] + g \leftharpoonup (h \rightharpoonup x) + (g \rightharpoonup x) \triangleright h$;

(MP7) $[g, h] \rightharpoonup x = g \rightharpoonup (h \rightharpoonup x) + (g \rightharpoonup x) \triangleleft h$;

(MP8) $g \rightharpoonup \{x, y\} = (g \leftharpoonup x) \rightharpoonup y - (g \leftharpoonup y) \rightharpoonup x + \{g \rightharpoonup x, y\} - \{g \rightharpoonup y, x\}$;

(MP9) $[g, h \leftharpoonup x] + [g, x \triangleright h] + g \leftharpoonup (h \rightharpoonup x) + g \leftharpoonup (x \triangleleft h) = 0$;

(MP10) $x \triangleright (y \triangleright g) + x \triangleright (g \leftharpoonup y) = 0$;

(MP11) $x \triangleleft (y \triangleright g) + x \triangleleft (g \leftharpoonup y) + \{x, y \triangleleft g\} + \{x, g \rightharpoonup y\} = 0$;

(MP12) $g \rightharpoonup (h \rightharpoonup x) + g \rightharpoonup (x \triangleleft h) = 0$.

Let $(\mathfrak{g}, \mathfrak{h}, \triangleleft, \triangleright, \leftharpoonup, \rightharpoonup)$ be a matched pair of Leibniz algebras. Then $\mathfrak{g} \bowtie \mathfrak{h} := \mathfrak{g} \times \mathfrak{h}$, as a vector space, with the bracket defined for any $g, h \in \mathfrak{g}$ and $x, y \in \mathfrak{h}$ by

$$[(g, x), (h, y)] := \big([g, h] + x \triangleright h + g \leftharpoonup y, \ \{x, y\} + x \triangleleft h + g \rightharpoonup y\big) \tag{2.39}$$

is a Leibniz algebra called the *bicrossed product* associated to the matched pair of Leibniz algebras $(\mathfrak{g}, \mathfrak{h}, \triangleleft, \triangleright, \leftharpoonup, \rightharpoonup)$.

This fact can be proved directly, but it can also be derived as a special case of Theorem 2.1.3. Indeed, let $\mathfrak{g}$ be a Leibniz algebra and $\Omega(\mathfrak{g}, V) = (\triangleleft, \triangleright, \leftharpoonup, \rightharpoonup, f, \{-, -\})$ an extending datum of $\mathfrak{g}$ through V such that f is the trivial map, i.e., $f(x, y) = 0$, for all $x, y \in V$. Then, we can easily see that

$\Omega(\mathfrak{g}, V) = \big(\triangleleft, \triangleright, \leftharpoonup, \rightharpoonup, \{-,-\}\big)$ is a Leibniz extending structure of $\mathfrak{g}$ through V if and only if $(V, \{-,-\})$ is a Leibniz algebra and $(\mathfrak{g}, V, \triangleleft, \triangleright, \leftharpoonup, \rightharpoonup)$ is a matched pair of Leibniz algebras in the sense of Definition 2.3.5. In this case, the associated unified product $\mathfrak{g} \ltimes_{\Omega(\mathfrak{g},V)} V = \mathfrak{g} \bowtie V$ is the bicrossed product of the matched pair $(\mathfrak{g}, V, \triangleleft, \triangleright, \leftharpoonup, \rightharpoonup)$ as defined by (2.39).

The bicrossed product of two Leibniz algebras is the construction responsible for the *factorization problem*, which is a special case of the ES problem and can be stated as follows: *Let* $\mathfrak{g}$ *and* $\mathfrak{h}$ *be two given Leibniz algebras. Describe and classify all Leibniz algebras* Ξ *that factorize through* $\mathfrak{g}$ *and* $\mathfrak{h}$*, i.e.,* Ξ *contains* $\mathfrak{g}$ *and* $\mathfrak{h}$ *as Leibniz subalgebras such that* $\Xi = \mathfrak{g} + \mathfrak{h}$ *and* $\mathfrak{g} \cap \mathfrak{h} = \{0\}$. Indeed, using Theorem 2.1.5 we can prove the following:

Corollary 2.3.6 *A Leibniz algebra* Ξ *factorizes through* $\mathfrak{g}$ *and* $\mathfrak{h}$ *if and only if there exists a matched pair of Leibniz algebras* $(\mathfrak{g}, \mathfrak{h}, \triangleleft, \triangleright, \leftharpoonup, \rightharpoonup)$ *such that* $\Xi \cong \mathfrak{g} \bowtie \mathfrak{h}$.

Proof: To start with, notice that any bicrossed product $\mathfrak{g} \bowtie \mathfrak{h}$ factorizes through $\mathfrak{g} \cong \mathfrak{g} \times \{0\}$ and $\mathfrak{h} \cong \{0\} \times \mathfrak{h}$. Conversely, assume that Ξ factorizes through $\mathfrak{g}$ and $\mathfrak{h}$. Let $p : \Xi \to \mathfrak{g}$ be the k-linear projection of Ξ on $\mathfrak{g}$, i.e., $p(g + x) := g$, for all $g \in \mathfrak{g}$ and $x \in \mathfrak{h}$. Now, we apply Theorem 2.1.5 for $V := \mathrm{Ker}(p) = \mathfrak{h}$. Since V is a Leibniz subalgebra of $E := \Xi$, the map $f = f_p$ constructed in the proof of Theorem 2.1.5 is the trivial map as $[x, y] \in V = \mathrm{Ker}(p)$. Thus, the Leibniz extending structure $\Omega(\mathfrak{g}, V) = \big(\triangleleft, \triangleright, \leftharpoonup, \rightharpoonup, f, \{-,-\}\big)$ constructed in the proof of Theorem 2.1.5 is precisely a matched pair of Leibniz algebras and the unified product $\mathfrak{g} \ltimes_{\Omega(\mathfrak{g},V)} V = \mathfrak{g} \bowtie V$ is the bicrossed product of the matched pair $(\mathfrak{g}, V, \triangleleft, \triangleright, \leftharpoonup, \rightharpoonup)$. Explicitly, the matched pair $(\mathfrak{g}, \mathfrak{h}, \triangleleft = \triangleleft_p, \triangleright = \triangleright_p, \leftharpoonup = \leftharpoonup_p, \rightharpoonup = \rightharpoonup_p)$ is given by:

$$x \triangleright g := p\big([x,\, g]\big), \qquad x \triangleleft g := [x,\, g] - p\big([x,\, g]\big) \tag{2.40}$$

$$g \leftharpoonup x := p\big([g,\, x]\big), \qquad g \rightharpoonup x := [g,\, x] - p\big([g,\, x]\big) \tag{2.41}$$

for all $x \in \mathfrak{h}$ and $g \in \mathfrak{g}$. □

From now on, the matched pair constructed in (2.40) and (2.41) will be called the *canonical matched pair* associated to the factorization $\Xi = \mathfrak{g} + \mathfrak{h}$ of Ξ through $\mathfrak{g}$ and $\mathfrak{h}$. Based on Corollary 2.3.6 the factorization problem can be restated in a computational manner as follows: Let $\mathfrak{g}$ and $\mathfrak{h}$ be two given Leibniz algebras. Describe the set of all matched pairs $(\mathfrak{g}, \mathfrak{h}, \triangleleft, \triangleright, \leftharpoonup, \rightharpoonup)$ and classify up to an isomorphism all bicrossed products $\mathfrak{g} \bowtie \mathfrak{h}$.

Example 2.3.7 *Let* $\mathfrak{g}$ *be the* 3*-dimensional Leibniz algebra considered in Example 2.2.14 and* k *be the* 1*-dimensional (abelian) Leibniz algebra. Then all bicrossed products* $\mathfrak{g} \bowtie k$ *can be explicitly described as a special case of Example 2.2.14. To this end we need to consider* $g_0 := 0$ *and* $\alpha := 0$ *in the unified products associated to all flag datums of the first kind provided in Example 2.2.14 and* $g_0 := 0$ *in the unified products associated to all flag datums* *of the second kind. For instance, by taking $d_3 = 0$ in the Leibniz algebra $\mathfrak{g}_{21}^{(\nu_0, d_1, d_2, d_3)}$ of Example 2.2.14 we obtain the bicrossed product $\mathfrak{g} \bowtie k$ which is a 4-dimensional Leibniz algebra with the basis $\{e_1, e_2, e_3, x\}$ and the bracket given by:*

$$\begin{aligned} [e_1, e_3] &= e_2, \quad [e_3, e_3] = e_1, \quad [e_3, x] = -d_1 e_1 - (d_2 - \nu_0^{-1} d_1) e_2 + \nu_0 x, \\ [x, e_3] &= d_1 e_1 + d_2 e_2 - \nu_0 x \end{aligned}$$

for all $\nu_0 \in k^*$ and d_1, $d_2 \in k$. This Leibniz algebra is the bicrossed product associated to the following matched pair $(\mathfrak{g}, k, \triangleleft, \triangleright, \leftharpoonup, \rightharpoonup)$:

$$\begin{aligned} x \triangleleft e_3 &:= -\nu_0 x, \quad x \triangleright e_3 := d_1 e_1 + d_2 e_2, \\ e_3 \rightharpoonup x &:= \nu_0 x, \quad e_3 \leftharpoonup x := -d_1 e_1 + (-d_2 + \nu_0^{-1} d_1) e_2 \end{aligned}$$

where the undefined actions are zero and x is a basis of k. Another example of a matched pair of Leibniz algebras and the corresponding bicrossed product will be given in Example 2.4.6.

2.4 Classifying complements for extensions of Leibniz algebras

This section is devoted to the classifying complements (CC) problem whose statement is given below. We first introduce the notion of a *complement* of a Leibniz subalgebra $\mathfrak{g}$ of Ξ: a Leibniz subalgebra $\mathfrak{h}$ of Ξ is called a *complement* of $\mathfrak{g}$ in Ξ (or a $\mathfrak{g}$-*complement* of Ξ) if $\Xi = \mathfrak{g} + \mathfrak{h}$ and $\mathfrak{g} \cap \mathfrak{h} = \{0\}$.

Classifying complements problem. *Let $\mathfrak{g} \subseteq \mathfrak{E}$ be an extension of Leibniz algebras. If a complement of $\mathfrak{g}$ in $\mathfrak{E}$ exists, describe explicitly and classify all complements of $\mathfrak{g}$ in $\mathfrak{E}$, i.e., Leibniz subalgebras $\mathfrak{h}$ of $\mathfrak{E}$ such that $\mathfrak{E} = \mathfrak{g} + \mathfrak{h}$ and $\mathfrak{g} \cap \mathfrak{h} = \{0\}$.*

Let $\mathfrak{g} \subseteq \Xi$ be a Leibniz subalgebra of Ξ. If $\mathfrak{h}$ is a complement of $\mathfrak{g}$ in Ξ, Corollary 2.3.6 shows that $\Xi \cong \mathfrak{g} \bowtie \mathfrak{h}$, where $\mathfrak{g} \bowtie \mathfrak{h}$ is the bicrossed product associated to the canonical matched pair of the factorization $\Xi = \mathfrak{g} + \mathfrak{h}$ as constructed in (2.40) and (2.41).

We denote by $\mathcal{F}(\mathfrak{g}, \Xi)$ the (possibly empty) isomorphism classes of all $\mathfrak{g}$-complements of Ξ. The *factorization index* of $\mathfrak{g}$ in Ξ is defined by $[\Xi : \mathfrak{g}]^f := |\,\mathcal{F}(\mathfrak{g}, \Xi)\,|$ as a numerical measure of the (CC) problem.

Definition 2.4.1 *Let $(\mathfrak{g}, \mathfrak{h}, \triangleright, \triangleleft, \leftharpoonup, \rightharpoonup)$ be a matched pair of Leibniz algebras. A linear map $r : \mathfrak{h} \to \mathfrak{g}$ is called a* deformation map *of the matched pair $(\mathfrak{g}, \mathfrak{h}, \triangleright, \triangleleft, \leftharpoonup, \rightharpoonup)$ if the following compatibility holds for any x, $y \in \mathfrak{h}$:*

$$r\big([x, y]\big) - \big[r(x), r(y)\big] = x \triangleright r(y) + r(x) \leftharpoonup y - r\big(x \triangleleft r(y) + r(x) \rightharpoonup y\big) \quad (2.42)$$

We denote by $\mathcal{DM}\,(\mathfrak{h}, \mathfrak{g}\,|\,(\triangleright, \triangleleft, \leftharpoonup, \rightharpoonup))$ the set of all deformation maps of the matched pair $(\mathfrak{g}, \mathfrak{h}, \triangleright, \triangleleft, \leftharpoonup, \rightharpoonup)$. The trivial map $r(x) = 0$, for all $x \in \mathfrak{h}$, is of course a deformation map. The right-hand side of (2.42) measures how far $r : \mathfrak{h} \to \mathfrak{g}$ is from being a Leibniz algebra map. Using this concept which will play a key role in solving the (CC) problem, we introduce the following deformation of a Leibniz algebra:

Theorem 2.4.2 *Let $\mathfrak{g}$ be a Leibniz subalgebra of Ξ, $\mathfrak{h}$ a given $\mathfrak{g}$-complement of Ξ and $r : \mathfrak{h} \to \mathfrak{g}$ a deformation map of the associated canonical matched pair $(\mathfrak{g}, \mathfrak{h}, \triangleright, \triangleleft, \leftharpoonup, \rightharpoonup)$.*

(1) *Let $f_r : \mathfrak{h} \to \Xi = \mathfrak{g} \bowtie \mathfrak{h}$ be the k-linear map defined for any $x \in \mathfrak{h}$ by:*

$$f_r(x) = (r(x),\, x)$$

Then $\widetilde{\mathfrak{h}} := \mathrm{Im}(f_r)$ is a $\mathfrak{g}$-complement of Ξ.

(2) *$\mathfrak{h}_r := \mathfrak{h}$, as a vector space, with the new bracket defined for any $x,\, y \in \mathfrak{h}$ by:*

$$[x,\, y]_r := [x,\, y] + x \triangleleft r(y) + r(x) \rightharpoonup y \qquad (2.43)$$

is a Leibniz algebra called the r-deformation of $\mathfrak{h}$. Furthermore, $\mathfrak{h}_r \cong \widetilde{\mathfrak{h}}$, as Leibniz algebras.

Proof: (1) To start with, we will prove that $\widetilde{\mathfrak{h}} = \{(r(x),\, x) \mid x \in \mathfrak{h}\}$ is a Leibniz subalgebra of $\mathfrak{g} \bowtie \mathfrak{h} = \Xi$. Indeed, for all $x,\, y \in \mathfrak{h}$ we have:

$$\begin{aligned}
\big[(r(x), x), (r(y), y)\big] &\overset{(2.39)}{=} \Big(\underline{[r(x), r(y)] + x \triangleright r(y) + r(x) \leftharpoonup y,} \\
&\qquad [x, y] + x \triangleleft r(y) + r(x) \rightharpoonup y\Big) \\
&\overset{(2.42)}{=} \Big(r([x,\, y] + x \triangleleft r(y) + r(x) \rightharpoonup y), \\
&\qquad [x,\, y] + x \triangleleft r(y) + r(x) \rightharpoonup y\Big)
\end{aligned}$$

i.e., $\big[(r(x),\, x),\, (r(y),\, y)\big] \in \widetilde{\mathfrak{h}}$. Moreover, it is straightforward to see that $\mathfrak{g} \cap \widetilde{\mathfrak{h}} = \{0\}$ and $(g,\, x) = (g - r(x),\, 0) + (r(x),\, x) \in \mathfrak{g} + \widetilde{\mathfrak{h}}$ for all $g \in \mathfrak{g}$, $x \in \mathfrak{h}$. Here, we view $\mathfrak{g} \cong \mathfrak{g} \times \{0\}$ as a subalgebra of $\mathfrak{g} \bowtie \mathfrak{h}$. Therefore, $\widetilde{\mathfrak{h}}$ is a $\mathfrak{g}$-complement of $\Xi = \mathfrak{g} \bowtie \mathfrak{h}$.

(2) We denote by $\widetilde{f_r} : \mathfrak{h} \to \widetilde{\mathfrak{h}}$ the linear isomorphism induced by f_r. We will prove that $\widetilde{f_r}$ is also a Leibniz algebra map if we consider on $\mathfrak{h}$ the bracket

given by (2.43). Indeed, for any $x,\ y \in \mathfrak{h}$ we have:

$$\begin{aligned}
\widetilde{f}_r([x,\ y]_r) &\stackrel{(2.43)}{=} \widetilde{f}_r\big([x,\ y] + x \triangleleft r(y) + r(x) \rightharpoonup y\big) \\
&= \Big(\underline{r\big([x,\ y] + x \triangleleft r(y) + r(x) \rightharpoonup y\big)}, \\
&\qquad [x,\ y] + x \triangleleft r(y) + r(x) \rightharpoonup y\Big) \\
&\stackrel{(2.42)}{=} \Big([r(x),\ r(y)] + x \triangleright r(y) + r((x) \leftharpoondown y, \\
&\qquad [x,\ y] + x \triangleleft r(y) + r(x) \rightharpoonup y\Big) \\
&\stackrel{(2.39)}{=} [(r(x),\ x),\ (r(y),\ y)] = [\widetilde{f}_r(x),\ \widetilde{f}_r(y)]
\end{aligned}$$

Therefore, $\mathfrak{h}_r$ is a Leibniz algebra and the proof is now finished. □

The following is the converse of Theorem 2.4.2: it proves that all $\mathfrak{g}$-complements of Ξ are r-deformations of a given complement.

Theorem 2.4.3 *Let $\mathfrak{g}$ be a Leibniz subalgebra of Ξ, and $\mathfrak{h}$ a given $\mathfrak{g}$-complement of Ξ with the associated canonical matched pair of Leibniz algebras $(\mathfrak{g}, \mathfrak{h}, \triangleright, \triangleleft, \leftharpoondown, \rightharpoonup)$. Then $\overline{\mathfrak{h}}$ is a $\mathfrak{g}$-complement of Ξ if and only if there exists an isomorphism of Leibniz algebras $\overline{\mathfrak{h}} \cong \mathfrak{h}_r$, for some deformation map $r : \mathfrak{h} \to \mathfrak{g}$ of the matched pair $(\mathfrak{g}, \mathfrak{h}, \triangleright, \triangleleft, \leftharpoondown, \rightharpoonup)$.*

Proof: Let $\overline{\mathfrak{h}}$ be an arbitrary $\mathfrak{g}$-complement of Ξ. Since $\Xi = \mathfrak{g} \oplus \mathfrak{h} = \mathfrak{g} \oplus \overline{\mathfrak{h}}$ we can find four k-linear maps:

$$u : \mathfrak{h} \to \mathfrak{g}, \quad v : \mathfrak{h} \to \overline{\mathfrak{h}}, \quad t : \overline{\mathfrak{h}} \to \mathfrak{g}, \quad w : \overline{\mathfrak{h}} \to \mathfrak{h}$$

such that for all $x \in \mathfrak{h}$ and $y \in \overline{\mathfrak{h}}$ we have:

$$x = u(x) \oplus v(x), \qquad y = t(y) \oplus w(y) \tag{2.44}$$

By an easy computation it follows that $v : \mathfrak{h} \to \overline{\mathfrak{h}}$ is a linear isomorphism of vector spaces. We denote by $\tilde{v} : \mathfrak{h} \to \mathfrak{g} \bowtie \mathfrak{h}$ the composition:

$$\tilde{v} : \mathfrak{h} \xrightarrow{v} \overline{\mathfrak{h}} \stackrel{i}{\hookrightarrow} \Xi = \mathfrak{g} \bowtie \mathfrak{h}$$

Therefore, we have $\tilde{v}(x) \stackrel{(2.44)}{=} \big(-u(x),\ x\big)$, for all $x \in \mathfrak{h}$. Then we shall prove that $r := -u$ is a deformation map and $\overline{\mathfrak{h}} \cong \mathfrak{h}_r$. Indeed, $\overline{\mathfrak{h}} = \mathrm{Im}(v) = \mathrm{Im}(\tilde{v})$ is a Leibniz subalgebra of $\Xi = \mathfrak{g} \bowtie \mathfrak{h}$ and we have:

$$\begin{aligned}
\big[(r(x),\ x),\ (r(y),\ y)\big] &\stackrel{(2.39)}{=} \big([r(x),\ r(y)] + x \triangleright r(y) + r(x) \leftharpoondown y,\ [x,\ y] + \\
&\qquad x \triangleleft r(y) + r(x) \rightharpoonup y\big) = \big(r(z),\ z\big)
\end{aligned}$$

for some $z \in \mathfrak{h}$. Thus, we obtain:

$$r(z) = [r(x),\ r(y)] + x \triangleright r(y) + r(x) \leftharpoondown y, \quad z = [x,\ y] + x \triangleleft r(y) + r(x) \rightharpoonup y \tag{2.45}$$

By applying r to the second part of (2.45) it follows that r is a deformation map of the matched pair $(\mathfrak{g}, \mathfrak{h}, \triangleright, \triangleleft, \leftharpoonup, \rightharpoonup)$. Furthermore, (2.45) and (2.43) show that $v : \mathfrak{h}_r \to \overline{\mathfrak{h}}$ is also a Leibniz algebra map which finishes the proof. □

In order to provide the classification of all complements we introduce the following:

Definition 2.4.4 *Let $(\mathfrak{g}, \mathfrak{h}, \triangleright, \triangleleft, \leftharpoonup, \rightharpoonup)$ be a matched pair of Leibniz algebras. Two deformation maps r, $R : \mathfrak{h} \to \mathfrak{g}$ are called* equivalent *and we denote this by $r \sim R$ if there exists $\sigma : \mathfrak{h} \to \mathfrak{h}$ a k-linear automorphism of $\mathfrak{h}$ such that for any x, $y \in \mathfrak{h}$:*

$$\sigma([x, y]) - [\sigma(x), \sigma(y)] = \sigma(x) \triangleleft R(\sigma(y)) + R(\sigma(x)) \rightharpoonup \sigma(y) - \sigma(x \triangleleft r(y)) \\ - \sigma(r(x) \rightharpoonup y)$$

To conclude this section, the following result provides the answer to the (CC) problem for Leibniz algebras:

Theorem 2.4.5 *Let $\mathfrak{g}$ be a Leibniz subalgebra of Ξ, $\mathfrak{h}$ a $\mathfrak{g}$-complement of Ξ and $(\mathfrak{g}, \mathfrak{h}, \triangleright, \triangleleft, \leftharpoonup, \rightharpoonup)$ the associated canonical matched pair. Then $\sim$ is an equivalence relation on the set $\mathcal{DM}(\mathfrak{h}, \mathfrak{g} \,|\, (\triangleright, \triangleleft, \leftharpoonup, \rightharpoonup))$ and the map*

$$\mathcal{HA}^2(\mathfrak{h}, \mathfrak{g} \,|\, (\triangleright, \triangleleft, \leftharpoonup, \rightharpoonup)) := \mathcal{DM}(\mathfrak{h}, \mathfrak{g} \,|\, (\triangleright, \triangleleft, \leftharpoonup, \rightharpoonup))/\sim \longrightarrow \mathcal{F}(\mathfrak{g}, \Xi), \quad \overline{r} \mapsto \mathfrak{h}_r$$

is a bijection between $\mathcal{HA}^2(\mathfrak{h}, \mathfrak{g} \,|\, (\triangleright, \triangleleft, \leftharpoonup, \rightharpoonup))$ and the isomorphism classes of all $\mathfrak{g}$-complements of Ξ. In particular, the factorization index of $\mathfrak{g}$ in Ξ is computed by the formula:

$$[\Xi : \mathfrak{g}]^f = |\mathcal{HA}^2(\mathfrak{h}, \mathfrak{g} \,|\, (\triangleright, \triangleleft, \leftharpoonup, \rightharpoonup))|$$

Proof: Follows from Theorem 2.4.3 taking into account the fact that two deformation maps r and R are equivalent in the sense of Definition 2.4.4 if and only if the corresponding Leibniz algebras $\mathfrak{h}_r$ and $\mathfrak{h}_R$ are isomorphic. □

Example 2.4.6 *Let $\mathfrak{h}$ be the abelian Lie algebra of dimension 2 with basis $\{f_1, f_2\}$ and $\mathfrak{g}$ the Lie algebra with basis $\{e_1, e_2\}$ and the bracket: $[e_2, e_1] = -[e_1, e_2] = e_2$. Then there exists a matched pair of Leibniz algebras $(\mathfrak{g}, \mathfrak{h}, \triangleleft, \triangleright, \leftharpoonup, \rightharpoonup)$, where the non-zero values of the actions are given as follows:*

$$f_1 \triangleleft e_1 := f_1, \quad f_2 \triangleleft e_1 := f_2, \quad f_1 \triangleright e_1 := e_2, \quad e_1 \leftharpoonup f_1 := -e_2, \\ e_1 \rightharpoonup f_1 := -f_1$$

The bicrossed product $\Xi = \mathfrak{g} \bowtie \mathfrak{h}$ associated to this matched pair is the following 4-dimensional Leibniz algebra having $\{e_1, e_2, f_1, f_2\}$ as a basis and the bracket given by:

$$[e_2, e_1] = -[e_1, e_2] = e_2, \ [f_1, e_1] = f_1 + e_2, \ [e_1, f_1] = -f_1 - e_2, \ [f_2, e_1] = f_2$$

Furthermore, the deformation maps associated with the above matched pair of Leibniz algebras are given as follows:

$$\begin{aligned} \overline{r}_{(\gamma,\,\delta)} &: \mathfrak{h} \to \mathfrak{g}, \quad \overline{r}(f_1) = \gamma e_2, \quad \overline{r}(f_2) = \delta e_2 \\ r_{(\alpha,\,\beta)} &: \mathfrak{h} \to \mathfrak{g}, \quad r(f_1) = \alpha e_2 + \beta e_1, \quad r(f_2) = 0 \end{aligned}$$

for some scalars α, β, γ, $\delta \in k$. One can easily see that $\mathfrak{h}_{\overline{r}_{(\gamma,\,\delta)}}$ coincides with the Lie algebra $\mathfrak{h}$ for all γ, $\delta \in k$ while $\mathfrak{h}_{r_{(\alpha,\,\beta)}}$ has the bracket given by:

$$[f_1,\, f_2]_{r_{(\alpha,\,\beta)}} = [f_1,\, f_1]_{r_{(\alpha,\,\beta)}} = [f_2,\, f_2]_{r_{(\alpha,\,\beta)}} = 0, \quad [f_2,\, f_1]_{r_{(\alpha,\,\beta)}} = \beta f_2$$

Therefore, if $\beta = 0$ then $\mathfrak{h}_{r_{(\alpha,\,\beta)}}$ again coincides with $\mathfrak{h}$. If $\beta \neq 0$, then for any $\alpha \in k$ and $\beta \in k^*$, the Leibniz algebra $\mathfrak{h}_{r_{(\alpha,\,\beta)}}$ is isomorphic to the Leibniz algebra $\mathfrak{k}$ with basis $\{F_1,\, F_2\}$ and the bracket given by: $[F_1,\, F_1] = F_2$, $[F_2,\, F_1] = F_2$. The isomorphism $\psi : \mathfrak{k} \to \mathfrak{h}_{r_{(\alpha,\,\beta)}}$ is given by: $\psi(F_1) := f_1 + f_2$, $\psi(F_2) := \beta f_2$. Since obviously $\mathfrak{k}$ is not isomorphic to the abelian Lie algebra $\mathfrak{h}$ we obtain that the extension $\mathfrak{g} \subseteq \Xi$ has factorization index $[\Xi : \mathfrak{g}]^f = 2$.

2.5 Itô's theorem for Leibniz algebras

The aim of this section is to prove that Itô's theorem remains valid at the level of Leibniz algebras (Corollary 2.5.2): if $\mathfrak{g}$ is a Leibniz algebra such that $\mathfrak{g} = A + B$, for two abelian subalgebras A and B, then $\mathfrak{g}$ is metabelian (2-step solvable). The converse as well as possible generalizations of Itô's theorem are also investigated.

We start by pointing out that for any Leibniz algebra $\mathfrak{g}$ there exists a partial skew-symmetry on the derived algebra. More precisely, the following formula which will be used in the proof of Theorem 2.5.1, holds for any x, y, z, $t \in \mathfrak{g}$:

$$\left[\, [x,\, y]\,,\, [z,\, t]\,\right] = -\left[\, [x,\, y]\,,\, [t,\, z]\,\right]. \tag{2.46}$$

Indeed, the equality (2.46) follows by applying two times the Leibniz law as follows:

$$\left[\, [x,\, y]\,,\, [z,\, t]\,\right] \stackrel{(0.7)}{=} \left[\,\left[\, [x,\, y]\,,\, z\right],\, t\right] - \left[\,\left[\, [x,\, y]\,,\, t\right],\, z\right] \stackrel{(0.7)}{=} -\left[\, [x,\, y]\,,\, [t,\, z]\,\right].$$

In the proof of Theorem 2.5.1 we use intensively the Leibniz law as given in (0.7) as well as the following equivalent form of it:

$$\left[[x,\, z]\,,\, y\right] = \left[[x,\, y]\,,\, z\right] - \left[x,\, [y,\, z]\right]. \tag{2.47}$$

The next result holds for Leibniz algebras defined over any commutative ring k.

Theorem 2.5.1 *Let $\mathfrak{g}$ be a Leibniz algebra and A, B two abelian subalgebras of $\mathfrak{g}$. Then any two-sided ideal $\mathfrak{h}$ of $\mathfrak{g}$ contained in $A+B$ is a metabelian Leibniz subalgebra of $\mathfrak{g}$.*

Proof: We have to prove that $[[x_1, x_2], [x_3, x_4]] = 0$, for all x_1, x_2, x_3, $x_4 \in \mathfrak{h}$. Since $\mathfrak{h} \subseteq A+B$ we can find $a_i \in A$, $b_i \in B$ such that $x_i = a_i + b_i$, for all $i = 1, \cdots, 4$. Using that A and B are both abelian subalgebras of $\mathfrak{g}$ we obtain:

$$\begin{aligned}
&[[x_1, x_2], [x_3, x_4]] = [[a_1+b_1, a_2+b_2], [a_3+b_3, a_4+b_4]] \\
&= [[a_1, b_2] + [b_1, a_2], [a_3, b_4] + [b_3, a_4]] \\
&= [[a_1, b_2], [a_3, b_4]] + [[a_1, b_2], [b_3, a_4]] + [[b_1, a_2], [a_3, b_4]] \\
&\quad + [[b_1, a_2], [b_3, a_4]].
\end{aligned}$$

The proof will be finished once we show that any member of the last sum is equal to zero. Indeed, using that A and B are abelian subalgebras and $\mathfrak{h}$ is a two-sided ideal of $\mathfrak{g}$ we obtain: $[a_3, b_2] = [a_3, a_2+b_2] = [a_3, x_2] \in h$ and similarly $[a_1, b_4] = [x_1, b_4]$, $[b_1, a_3] = [x_1, a_3]$ and $[b_4, a_2] = [b_4, x_2]$ are all elements of $\mathfrak{h}$. Thus, we can find some elements $a_j \in A$, $b_j \in B$, $j = 5, \cdots, 8$ such that:

$$\begin{aligned}
&[a_3, b_2] = a_5 + b_5, \quad [a_1, b_4] = a_6 + b_6, \quad [b_1, a_3] = a_7 + b_7, \\
&[b_4, a_2] = a_8 + b_8.
\end{aligned} \tag{2.48}$$

Then, using the fact that A and B are both abelian, we obtain:

$$\begin{aligned}
[[a_1, b_2], [a_3, b_4]] &\overset{(0.7)}{=} \Big[\underline{[[a_1, b_2], a_3]}, b_4\Big] - \Big[\underline{[[a_1, b_2], b_4]}, a_3\Big] \\
&\overset{(2.47)}{=} [[[a_1, a_3], b_2] - [a_1, [a_3, b_2]], b_4] \\
&\quad - [[[a_1, b_4], b_2] - [a_1, [b_4, b_2]], a_3] \\
&= -\Big[\Big[a_1, \underline{[a_3, b_2]}\Big], b_4\Big] - \Big[\Big[\underline{[a_1, b_4]}, b_2\Big], a_3\Big] \\
&\overset{(2.48)}{=} -[[a_1, a_5+b_5], b_4] - [[a_6+b_6, b_2], a_3] \\
&= -\underline{[[a_1, b_5], b_4]} - \underline{[[a_6, b_2], a_3]} \\
&\overset{(2.47)}{=} [a_1, [b_4, b_5]] - [[a_1, b_4], b_5] + [a_6, [a_3, b_2]] \\
&\quad - [[a_6, a_3], b_2] \\
&= -\Big[\underline{[a_1, b_4]}, b_5\Big] + \Big[a_6, \underline{[a_3, b_2]}\Big] \\
&\overset{(2.48)}{=} -[a_6+b_6, b_5] + [a_6, a_5+b_5] \\
&= -[a_6, b_5] + [a_6, b_5] = 0.
\end{aligned}$$

Thus, $[[a_1, b_2], [a_3, b_4]] = 0$. A similar computation shows that $[[a_1, b_2], [a_4, b_3]] = 0$. This implies, using (2.46), that $[[a_1, b_2], [b_3, a_4]] =$

$-[[a_1, b_2], [a_4, b_3]] = 0$, that is the second term of the above sum is also equal to zero. Furthermore, we have:

$$
\begin{aligned}
[[b_1, a_2], [a_3, b_4]] &\overset{(0.7)}{=} \Big[\underline{[[b_1, a_2], a_3]}, b_4\Big] - \Big[\underline{[[b_1, a_2], b_4]}, a_3\Big] \\
&\overset{(2.47)}{=} [[[b_1, a_3], a_2] - [b_1, [a_3, a_2]], b_4] \\
&\quad - [[[b_1, b_4], a_2] - [b_1, [b_4, a_2]], a_3] \\
&= \Big[\Big[\underline{[b_1, a_3]}, a_2\Big], b_4\Big] + \Big[\Big[b_1, \underline{[b_4, a_2]}\Big], a_3\Big] \\
&\overset{(2.48)}{=} [[a_7 + b_7, a_2], b_4] + [[b_1, a_8 + b_8], a_3] \\
&= \underline{[[b_7, a_2], b_4]} + \underline{[[b_1, a_8], a_3]} \\
&\overset{(2.47)}{=} [[b_7, b_4], a_2] - [b_7, [b_4, a_2]] + [[b_1, a_3], a_8] \\
&\quad - [b_1, [a_3, a_8]] \\
&= -\Big[b_7, \underline{[b_4, a_2]}\Big] + \Big[\underline{[b_1, a_3]}, a_8\Big] \\
&\overset{(2.48)}{=} -[b_7, a_8 + b_8] + [a_7 + b_7, a_8] \\
&= -[b_7, a_8] + [b_7, a_8] = 0
\end{aligned}
$$

i.e., $[[b_1, a_2], [a_3, b_4]] = 0$, that is the third term of the above sum is also equal to zero. A similar computation will prove that $[[b_1, a_2], [a_4, b_3]] = 0$. Finally, applying once again (2.46), we obtain that $[[b_1, a_2], [b_3, a_4]] = -[[b_1, a_2], [a_4, b_3]] = 0$ and the proof is now finished. □

As a special case of Theorem 2.5.1 we obtain:

Corollary 2.5.2 ***(Itô's theorem for Leibniz algebras)*** *Let $\mathfrak{g}$ be a Leibniz algebra over a commutative ring k such that $\mathfrak{g} = A + B$, for two abelian subalgebras A and B of $\mathfrak{g}$. Then $\mathfrak{g}$ is metabelian.*

Example 2.5.3 *Itô's theorem (Corollary 2.5.2) cannot be generalized in the following direction: if $\mathfrak{g}$ is a Lie algebra such that $\mathfrak{g} = A + B$, where A is a metabelian subalgebra and B an abelian subalgebra, then $\mathfrak{g}$ is not necessarily a metabelian Lie algebra. Indeed, let $\mathfrak{g}$ be the 4-dimensional Lie algebra (denoted by L_5 in [3]) having the basis $\{e_1, \cdots, e_4\}$ and the bracket*

$$[e_1, e_2] = e_2, \quad [e_1, e_3] = e_3, \quad [e_1, e_4] = 2\, e_4, \quad [e_2, e_3] = e_4.$$

Then, $[[e_1, e_2], [e_1, e_3]] = e_4$, i.e., $\mathfrak{g}$ is not metabelian. On the other hand, we have the decomposition $\mathfrak{g} = A + B$, where A is the metabelian subalgebra of $\mathfrak{g}$ having $\{e_1, e_2\}$ as a basis and B is the abelian subalgebra of $\mathfrak{g}$ with the basis $\{e_3, e_4\}$.

We end the section by looking at that the converse of Corollary 2.5.2. More precisely, we prove by counterexamples that the converse of Itô's theorem fails to be true for both Lie algebras and Leibniz algebras.

Example 2.5.4 *Let $\mathfrak{g}$ be the 3-dimensional metabelian Leibniz algebra over the field $\mathbb{R}$ having the basis $\{e_1, e_2, e_3\}$ and the bracket:*

$$[e_2, e_2] = e_1, \qquad [e_3, e_3] = e_1.$$

We will prove that $\mathfrak{g}$ cannot be written as a sum of two abelian subalgebras. Indeed, suppose that A is an abelian subalgebra of dimension 1 with basis $x = ae_1 + be_2 + ce_3$, $a, b, c \in k$. As A is abelian we have $[x, x] = 0$ and this implies $b^2 + c^2 = 0$. Therefore, $b = c = 0$ and thus A is generated by e_1. Assume now that B is an abelian subalgebra of dimension 2 with basis $\{x = a_1e_1 + a_2e_2 + a_3e_3, y = b_1e_1 + b_2e_2 + b_3e_3\}$. Again by the fact that B is abelian we obtain $a_2 = a_3 = b_2 = b_3 = 0$. Hence $\mathfrak{g}$ has no abelian subalgebras of dimension 2 and the conclusion follows.

Over the complex field $\mathbb{C}$, the metabelian Lie algebra of smallest dimension which cannot be written as a sum of two proper abelian subalgebras has dimension 5. Indeed, if we consider the classification of 3- and 4-dimensional complex Lie algebras (see for instance [3, Table 1 and 2]) one can easily see that any metabelian Lie algebra in those lists can be written as a direct sum of two proper abelian subalgebras.

Example 2.5.5 *Let $\mathfrak{g}$ be the 5-dimensional metabelian Lie algebra having the basis $\{e_1, e_2, e_3, e_4, e_5\}$ and the bracket given by:*

$$[e_1, e_2] = e_3, \quad [e_1, e_3] = e_4, \quad [e_2, e_3] = e_5. \tag{2.49}$$

We will prove that $\mathfrak{g}$ cannot be written as a sum of two abelian subalgebras based on the following remark. Let $a = \sum_{i=1}^{5} a_i\, e_i$ and $b = \sum_{i=1}^{5} b_i\, e_i$, be two non-zero elements of an abelian Lie subalgebra of $\mathfrak{g}$. Since $[a, b] = 0$, it follows from (2.49), that

$$a_1\, b_2 = a_2\, b_1, \quad a_1\, b_3 = a_3\, b_1, \quad a_2\, b_3 = a_3\, b_2, \tag{2.50}$$

that is, the rank of the following matrix

$$\begin{pmatrix} a_1 & a_2 & a_3 \\ b_1 & b_2 & b_3 \end{pmatrix}$$

is at most 1. Based on this observation we will prove that $\dim_{\mathrm{k}}(A + B) \leq 4$, for any A, B abelian subalgebras of $\mathfrak{g}$. Assume that $\mathfrak{g}$ is the sum of two abelian subalgebras, say, $\mathfrak{g} = A + B$. As $\mathfrak{g}$ is not abelian we have $\dim_{\mathrm{k}}(A) \geq 1$ and $\dim_{\mathrm{k}}(B) \geq 1$. Let $l = \dim_{\mathrm{k}}(A)$ and $t = \dim_{\mathrm{k}}(B)$ and consider $\mathcal{B}_A = \{X_1, ..., X_l\}$, respectively $\mathcal{B}_B = \{Y_1, ..., Y_t\}$, k-bases for A and respectively B. As $\dim_{\mathrm{k}}(A + B) = 5$, we can find 5 linearly independent vectors among the generators $\{X_1, ..., X_l, Y_1, ..., Y_t\}$ which form a basis of $A + B$. Again by the fact that $\mathfrak{g}$ is not abelian, this basis of $A + B$ will contain either three vectors from $\mathcal{B}_A$ and two vectors from $\mathcal{B}_B$ or four vectors from $\mathcal{B}_A$ and one *vector from* $\mathcal{B}_B$. *Therefore, we are left to prove that* $\mathfrak{g}$ *cannot be written as the sum of two abelian subalgebras of dimensions* 3 *and* 2*, respectively* 4 *and* 1. *Suppose first that* $\mathfrak{g} = A + B$*, where* A *and* B *are abelian subalgebras such that* $\dim_k(A) = 3$ *and* $\dim_k(B) = 2$. *Consider* $\mathcal{B}_A = \{X_1 = \sum_{i=1}^5 \alpha_i\, e_i,\, X_2 = \sum_{i=1}^5 \beta_i\, e_i,\, X_3 = \sum_{i=1}^5 \gamma_i\, e_i\}$ *and* $\mathcal{B}_B = \{Y_1 = \sum_{i=1}^5 \delta_i\, e_i,\, Y_2 = \sum_{i=1}^5 \mu_i\, e_i\}$ k*-bases of* A*, respectively* B*, and let* V *be the matrix of coefficients*

$$V = \begin{pmatrix} \alpha_1 & \alpha_2 & \alpha_3 & \alpha_4 & \alpha_5 \\ \beta_1 & \beta_2 & \beta_3 & \beta_4 & \beta_5 \\ \gamma_1 & \gamma_2 & \gamma_3 & \gamma_4 & \gamma_5 \\ \delta_1 & \delta_2 & \delta_3 & \delta_4 & \delta_5 \\ \mu_1 & \mu_2 & \mu_3 & \mu_4 & \mu_5 \end{pmatrix}. \qquad (2.51)$$

Let $W := V_{1,2,3}^{1,2,3}$ *(resp.* $T := V_{4,5}^{1,2,3}$*) be the matrix formed with the first three columns and the first three rows of* V *(resp. the first three columns and the last two rows of* V*). By applying (2.50) to the vectors in* $\mathcal{B}_A$*, respectively* $\mathcal{B}_B$*, it follows that* $\operatorname{rank}(W) = 1$ *and* $\operatorname{rank}(T) \leq 1$. *We obtain that* $\operatorname{rank}(V) \leq 4$*, since* $\det(V) = 0$. *The last assertion follows for instance by first expanding the determinant of* V *along the last column and then by expanding again the resulting determinants along the last column. Thus* $\mathfrak{g}$ *cannot be written as a sum of abelian subalgebras of dimensions* 3 *and* 2. *Finally, the second case will be settled in the negative by proving that* $\mathfrak{g}$ *does not have abelian subalgebras of dimension* 4. *Suppose* A *is such a subalgebra with basis* $\mathcal{B}_A = \{X_1 = \sum_{i=1}^5 \alpha_i\, e_i,\, X_2 = \sum_{i=1}^5 \beta_i\, e_i,\, X_3 = \sum_{i=1}^5 \gamma_i\, e_i,\, X_4 = \sum_{i=1}^5 \delta_i\, e_i\}$ *and denote by* V' *the matrix consisting of the first four rows of* V. *We will reach a contradiction by proving that* $\operatorname{rank}(V') < 4$. *By applying (2.50) to the vectors in* $\mathcal{B}_A$ *we obtain that* $\operatorname{rank}(V^{1,2,3}) = 1$*, where* $V^{1,2,3}$ *is the matrix consisting of the first three columns of* V'. *Hence we obtain that* $\operatorname{rank}(V') \leq 3$. *This finishes the proof.*

Bibliographical Notes

The material presented in this chapter is part of the author's papers [16] and respectively [22].

Chapter 3

Lie algebras

Lie algebras are studied in various fields such as differential geometry, classical/quantum mechanics or the theory of particle physics. In differential geometry, Lie algebras arise naturally on the tangent space of symmetry (Lie) groups on manifolds. In Hamiltonian mechanics the phase space is an example of a Lie algebra while in quantum mechanics Heisenberg postulated the existence of an infinite-dimensional Lie algebra of operators: the theory of quantum mechanics follows more or less from properties of Lie algebras. In the theory of particle physics Lie algebras play a key role. For instance, bosonic string theory uses a Lie algebra to formulate operators and the state space. Beyond the remarkable applications in the above-mentioned fields, Lie algebras are objects of study in their own right. In this context, it is natural to consider the extending structures problem for Lie algebras. Certain results in this chapter will need no proof as they can be easily derived as special cases of the corresponding results concerning Leibniz algebras. The extending structures (ES) problem for Lie algebras comes down to the following question:

Let $\mathfrak{g}$ be a Lie algebra and E a vector space containing $\mathfrak{g}$ as a subspace. Describe and classify the set of all Lie algebra structures $[-,-]$ that can be defined on E such that $\mathfrak{g}$ becomes a Lie subalgebra of $(E, [-,-])$.

The chapter is organized as follows. In Section 3.1 we will perform the abstract construction of the *unified product* $\mathfrak{g} \natural V$: it is associated to a Lie algebra $\mathfrak{g}$, a vector space V and a system of data $\Omega(\mathfrak{g}, V) = \big(\triangleleft, \triangleright, f, \{-, -\}\big)$ called an extending datum of $\mathfrak{g}$ through V. Theorem 3.1.2 establishes the set of axioms that has to be satisfied by $\Omega(\mathfrak{g}, V)$ such that $\mathfrak{g} \natural V$ with a given canonical bracket becomes a Lie algebra, i.e., is a unified product. In this case, $\Omega(\mathfrak{g}, V) = \big(\triangleleft, \triangleright, f, \{-, -\}\big)$ will be called a *Lie extending structure* of $\mathfrak{g}$ through V. Now let $\mathfrak{g}$ be a Lie algebra, E a vector space containing $\mathfrak{g}$ as a subspace and V a given complement of $\mathfrak{g}$ in E. Theorem 3.1.4 provides the answer to the description part of the ES problem: there exists a Lie algebra structure $[-,-]$ on E such that $\mathfrak{g}$ is a subalgebra of $(E, [-,-])$ if and only if there exists an isomorphism of Lie algebras $(E, [-,-]) \cong \mathfrak{g} \natural V$, for some Lie extending structure $\Omega(\mathfrak{g}, V) = \big(\triangleleft, \triangleright, f, \{-, -\}\big)$ of $\mathfrak{g}$ through V. The answer to the classification part of the ES problem is given in Theorem 3.1.7 by constructing two classifying objects as in the case of Leibniz algebras. Theorem 3.1.7 offers the theoretical answer to the extending structures problem. The challenge we are left to deal with is a purely computational one: for a given Lie algebra $\mathfrak{g}$ that is a subspace in a vector space E with a given complement V, we have to compute explicitly the classifying object $\mathcal{H}^2_{\mathfrak{g}}(V, \mathfrak{g})$ and then list the set of all types of Lie algebra structures on E which extend the Lie algebra structure on $\mathfrak{g}$. This is highly nontrivial considering that the construction of $\mathcal{H}^2_{\mathfrak{g}}(V, \mathfrak{g})$ is very laborious. In Section 3.2 we shall identify a way of computing $\mathcal{H}^2_{\mathfrak{g}}(V, \mathfrak{g})$ for the case when the complement V is finite-dimensional, namely for what we call *flag extending structures* of $\mathfrak{g}$ to E in the sense of Definition 3.2.1. All flag extending structures of $\mathfrak{g}$ to E can be completely described by a recursive reasoning where the key step is the case when $\mathfrak{g}$ has codimension 1 as a subspace of E. This case is solved in Theorem 3.2.8 where $\mathcal{H}^2_{\mathfrak{g}}(V, \mathfrak{g})$ and $\mathcal{H}^2(V, \mathfrak{g})$ are completely described: both objects are quotient pointed sets of the set $\mathrm{TwDer}(\mathfrak{g})$ of all twisted derivations of $\mathfrak{g}$ introduced in Definition 3.2.2. The set $\mathrm{TwDer}(\mathfrak{g})$ contains the usual space of derivations $\mathrm{Der}(\mathfrak{g})$ via the canonical embedding which is an isomorphism in the case when $\mathfrak{g}$ is a perfect Lie algebra. Finally, two explicit examples are given in Example 3.2.12 and Example 3.2.13: in the first case all extending structures of a 5-dimensional perfect Lie algebra to a space of dimension 6 are classified while in the second one we list all types of extending structures of the non-perfect Lie algebra $\mathfrak{gl}(2, k)$ to a space of dimension 5.

Section 3.3 contains some special cases of the unified product such as: the crossed product, the bicrossed product and the skew crossed product which plays a key role in developing a Galois theory for Lie algebras.

The next two sections of this chapter contain some explicit computations concerning the factorization problem and its converse, the classifying complements problem. The approach used relies on Proposition 3.2.4 and Remark 3.2.5 which can be rephrased as follows: the set of all matched pairs of Lie algebras $(k_0, \mathfrak{h}, \triangleleft, \triangleright)$ (by k_0 we will denote the abelian Lie algebra of dimension 1) and the space $\mathrm{TwDer}(\mathfrak{h})$ of all twisted derivations of $\mathfrak{h}$ are in one-to-one correspondence; moreover, any Lie algebra $\mathfrak{L}$ containing $\mathfrak{h}$ as a subalgebra of codimension 1 is isomorphic to a bicrossed product $k_0 \bowtie \mathfrak{h} = \mathfrak{h}_{(\lambda, \Delta)}$, for some $(\lambda, \Delta) \in \mathrm{TwDer}(\mathfrak{h})$. The classification up to an isomorphism of all bicrossed products $\mathfrak{h}_{(\lambda, \Delta)}$ is given in the case when $\mathfrak{h}$ is perfect. As an application of our approach, the group $\mathrm{Aut}_{\mathrm{Lie}}(\mathfrak{h}_{(\lambda, \Delta)})$ of all automorphisms of such Lie algebras is fully described in Corollary 3.4.3: it appears as a subgroup of a certain semidirect product $\mathfrak{h} \ltimes \big(k^* \times \mathrm{Aut}_{\mathrm{Lie}}(\mathfrak{h})\big)$ of groups. At this point we mention that the classification of automorphisms groups of all indecomposable real Lie algebras of dimension up to five was obtained recently in [108] where the importance of this subject in mathematical physics is highlighted. For the special case of sympathetic Lie algebras $\mathfrak{h}$, Corollary 3.4.5 proves that, up to an isomorphism, there exists only one Lie algebra that contains $\mathfrak{h}$ as a Lie subalgebra of codimension one, namely the direct product $k_0 \times \mathfrak{h}$ and $\mathrm{Aut}_{\mathrm{Lie}}(k_0 \times \mathfrak{h}) \cong k^* \times \mathrm{Aut}_{\mathrm{Lie}}(\mathfrak{h})$. Now, k_0 is a subalgebra of $k_0 \bowtie \mathfrak{h} = \mathfrak{h}_{(\lambda, \Delta)}$ having $\mathfrak{h}$ as a complement: for a 5-dimensional perfect Lie algebra, all complements of k_0 in $\mathfrak{h}_{(\lambda, \Delta)}$ are described in Example 3.4.7 as matched pair deformations of $\mathfrak{h}$. Section 3.5 treats the same problem for a given $(2n+1)$-dimensional non-perfect Lie algebra $\mathfrak{h} := \mathfrak{l}(2n+1, k)$. Theorem 3.5.2 describes explicitly all Lie algebras containing $\mathfrak{l}(2n+1,k)$ as a subalgebra of codimension 1. They are parameterized by a set $\mathcal{T}(n)$ of matrices $(A, B, C, D, \lambda_0, \delta) \in \mathrm{M}_n(k)^4 \times k \times k^{2n+1}$: there are four such families of Lie algebras if the characteristic of k is $\neq 2$ and two families in characteristic 2. All complements of k_0 in two such bicrossed products $k_0 \bowtie \mathfrak{l}(2n+1,k)$ are described by computing all matched pair deformations of the Lie algebra $\mathfrak{l}(2n+1,k)$ in Proposition 3.5.4 and Proposition 3.5.8. In particular, in Example 3.5.9 we construct an example where the factorization index of k_0 in the 4-dimensional Lie algebra $\mathfrak{m}(4,k)$ is infinite: that is k_0 has an infinite family of non-isomorphic complements in $\mathfrak{m}(4,k)$. To conclude, there are three reasons for which we considered the Lie algebra $\mathfrak{l}(2n+1,k)$ in Section 3.5: on the one hand it provided us with an example of a finite-dimensional Lie algebra extension $\mathfrak{g} \subset \mathfrak{L}$ such that $\mathfrak{g}$ has infinitely many non-isomorphic complements as a Lie subalgebra in $\mathfrak{L}$. On the other hand, the Lie algebra $\mathfrak{l}(2n+1,k)$ serves for constructing two counterexamples in Remark 3.5.10 which show that some properties of Lie algebras are not preserved by the matched pair deformation.

The last part of this chapter aims at developing a Galois theory for Lie algebras. If $\mathfrak{g} \subseteq \mathfrak{h}$ is an extension of Lie algebras, Theorem 3.6.5 provides the explicit description of the Galois group $\mathrm{Gal}(\mathfrak{h}/\mathfrak{g})$ as a subgroup of the canonical semidirect product $\mathrm{GL}_k(V) \rtimes \mathrm{Hom}_k(V, \mathfrak{g})$ of groups, where V is a vector space that measures the codimension of $\mathfrak{g}$ in $\mathfrak{h}$. i.e., the 'degree' of the extension $\mathfrak{h}/\mathfrak{g}$. We point out that the group $\mathrm{GL}_k(V) \rtimes \mathrm{Hom}_k(V, \mathfrak{g})$ is a lot more complex than the classical general affine group $\mathrm{GL}_k(V) \rtimes V$ of an affine space V. Theorem 3.6.12 is the counterpart of Artin's Theorem for Lie algebras: if G is a finite group of invertible order in k acting on a Lie algebra $\mathfrak{h}$, then the Lie algebra $\mathfrak{h}$ is reconstructed as a skew crossed product $\mathfrak{h} \cong \mathfrak{h}^G \#^{\bullet} V$ between the Lie subalgebra of invariants $\mathfrak{h}^G$ and the kernel V of the Reynolds operator $t : \mathfrak{h} \to \mathfrak{h}^G$. The Galois group $\mathrm{Gal}(\mathfrak{h}/\mathfrak{h}^G)$ is also described and Example 3.6.13 shows that even in the case of faithful actions, the group $\mathrm{Gal}(\mathfrak{h}/\mathfrak{h}^G)$ is different from G, as opposed to the classical Galois theory of fields where the two groups coincide. Theorem 3.6.14 is Hilbert's 90 Theorem for Lie algebras: if G is a cyclic group then the kernel of the Reynolds operator $t : \mathfrak{h} \to \mathfrak{h}^G$ is determined. The structure of Lie algebras $\mathfrak{h}$ endowed with a certain type of action of a finite cyclic group is then described in Corollary 3.6.16: $\mathfrak{h}$ is isomorphic to a semidirect product between $\mathfrak{h}^G$ and an ideal of $\mathfrak{h}$. This is the Lie algebra counterpart of the structure theorem for cyclic Galois extensions of fields [158, Theorem 6.2]: if $G \leq \mathrm{Aut}(K)$ is a cyclic subgroup of order n of the automorphism group of a field K of characteristic zero and $k := K^G$, then K is isomorphic to the splitting field over k of a polynomial of the form $X^n - a \in k[X]$. Corollary 3.6.18 shows that if $\mathfrak{g} \subseteq \mathfrak{h}$ is a Lie subalgebra of codimension 1 in $\mathfrak{h}$, then the Galois group $\mathrm{Gal}(\mathfrak{h}/\mathfrak{g})$ is metabelian (in particular, solvable). Based on this, the Lie algebra counterpart of the concept of a radical extension of fields is proposed in Definition 3.6.20. As in the classical Galois theory, Theorem 3.6.21 proves that the Galois group $\mathrm{Gal}(\mathfrak{h}/\mathfrak{g})$ of a radical extension $\mathfrak{g} \subseteq \mathfrak{h}$ of finite-dimensional Lie algebras is a solvable group. Several other applications and concrete examples of Galois groups are presented. For instance, in Example 3.6.25 we present a Lie algebra extension $\mathfrak{g} \subseteq \mathfrak{h}$ whose Galois group is trivial i.e., $\mathrm{Gal}\,(\mathfrak{h}/\mathfrak{g}) = \{\mathrm{Id}_{\mathfrak{h}}\}$. However, we point out that in general the Galois group of a Lie algebra extension is far from being trivial.

3.1 Unified products for Lie algebras

In this section we deal with the ES problem for Lie algebras in its full generality. We start by introducing the unified product which provides the answer to the ES problem.

Definition 3.1.1 *Let $\mathfrak{g}$ be a Lie algebra and V a vector space. An extending datum of $\mathfrak{g}$ through V is a system $\Omega(\mathfrak{g}, V) = \big(\triangleleft, \triangleright, f, \{-, -\}\big)$ consisting of four bilinear maps*

$$\triangleleft : V \times \mathfrak{g} \to V, \quad \triangleright : V \times \mathfrak{g} \to \mathfrak{g}, \quad f : V \times V \to \mathfrak{g}, \quad \{-, -\} : V \times V \to V$$

Let $\Omega(\mathfrak{g}, V) = \big(\triangleleft, \triangleright, f, \{-, -\}\big)$ be an extending datum. We denote by $\mathfrak{g} \natural_{\Omega(\mathfrak{g},V)} V = \mathfrak{g} \natural V$ the vector space $\mathfrak{g} \times V$ with the bilinear map $[-, -] : (\mathfrak{g} \times V) \times (\mathfrak{g} \times V) \to \mathfrak{g} \times V$ defined by:

$$[(g, x), (h, y)] := \big([g, h] + x \triangleright h - y \triangleright g + f(x, y), \ \{x, y\} + x \triangleleft h - y \triangleleft g\big) \quad (3.1)$$

for all $g, h \in \mathfrak{g}$ and $x, y \in V$. The object $\mathfrak{g} \natural V$ is called the unified product of $\mathfrak{g}$ and $\Omega(\mathfrak{g}, V)$ if it is a Lie algebra with the bracket given by (3.1). In this case the extending datum $\Omega(\mathfrak{g}, V) = \big(\triangleleft, \triangleright, f, \{-, -\}\big)$ is called a Lie extending structure of $\mathfrak{g}$ through V. The maps $\triangleleft$ and $\triangleright$ are called the actions of $\Omega(\mathfrak{g}, V)$ and f is called the cocycle of $\Omega(\mathfrak{g}, V)$.

The extending datum $\Omega(\mathfrak{g}, V) = \big(\triangleleft, \triangleright, f, \{-, -\}\big)$, for which $(\triangleleft, \triangleright, f, \{-, -\})$ are all the trivial maps is an example of a Lie extending structure, called the *trivial extending structure* of $\mathfrak{g}$ through V. Let $\Omega(\mathfrak{g}, V)$ be an extending datum of $\mathfrak{g}$ through V. Then, the following relations, very useful in computations, hold in $\mathfrak{g} \natural V$:

$$[(g, 0), (h, y)] = \big([g, h] - y \triangleright g, \ -y \triangleleft g\big) \quad (3.2)$$
$$[(0, x), (h, y)] = \big(x \triangleright h + f(x, y), \ x \triangleleft h + \{x, y\}\big) \quad (3.3)$$

for all $g, h \in \mathfrak{g}$ and $x, y \in V$.

Theorem 3.1.2 *Let $\mathfrak{g}$ be a Lie algebra, V a k-vector space and $\Omega(\mathfrak{g}, V)$ an extending datum of $\mathfrak{g}$ by V. The following statements are equivalent:*

(1) *$\mathfrak{g} \natural V$ is a unified product.*

(2) *The following compatibilities hold for any $g, h \in \mathfrak{g}$, $x, y, z \in V$:*

(LE1) $f(x, x) = 0, \qquad \{x, x\} = 0;$

(LE2) $(V, \triangleleft)$ *is a right* $\mathfrak{g}$*-module;*

(LE3) $x \triangleright [g, h] = [x \triangleright g, h] + [g, x \triangleright h] + (x \triangleleft g) \triangleright h - (x \triangleleft h) \triangleright g;$

(LE4) $\{x, y\} \triangleleft g = \{x, y \triangleleft g\} + \{x \triangleleft g, y\} + x \triangleleft (y \triangleright g) - y \triangleleft (x \triangleright g);$

(LE5) $\{x, y\} \triangleright g = x \triangleright (y \triangleright g) - y \triangleright (x \triangleright g) + [g, f(x, y)] + f(x, y \triangleleft g) + f(x \triangleleft g, y);$

(LE6) $f(x, \{y, z\}) + f(y, \{z, x\}) + f(z, \{x, y\}) + x \triangleright f(y, z) + y \triangleright f(z, x) + z \triangleright f(x, y) = 0;$

(LE7) $\{x, \{y, z\}\} + \{y, \{z, x\}\} + \{z, \{x, y\}\} + x \triangleleft f(y, z) + y \triangleleft f(z, x) + z \triangleleft f(x, y) = 0.$

Proof: Instead of including a rather lengthy computation as a proof, we will only point out that this result can be obtained as a special case of Theorem 2.1.3. Indeed, using the anti-symmetry of a Lie algebra bracket together with (2.2) and (2.3) we obtain:

$$g \leftharpoonup y = -y \triangleright g, \quad g \rightharpoonup y = -y \triangleleft g$$

for all $g \in \mathfrak{g}$ and $y \in V$. Now it is straightforward to check that the compatibility conditions $(L1)$–$(L14)$ in Theorem 2.1.3 amount to the compatibilities $(LE1)$–$(LE7)$. □

We make a few remarks on the compatibilities in Theorem 3.1.2. Aside from the fact that V is not a Lie algebra, $(LE3)$ and $(LE4)$ are exactly the compatibilities defining a matched pair of Lie algebras [174, Definition 8.3.1]. The compatibility condition $(LE5)$ is called the *twisted module condition* for the action $\triangleright$; in the case when V is a Lie algebra it measures how far $(\mathfrak{g}, \triangleright)$ is from being a left V-module. $(LE6)$ is called the *twisted cocycle condition*: if $\triangleright$ is the trivial action and $(V, \{-, -\})$ is a Lie algebra then the compatibility condition $(LE6)$ is exactly the classical 2-cocycle condition for Lie algebras. $(LE7)$ is called the *twisted Jacobi condition*: it measures how far $\{-, -\}$ is from being a Lie structure on V. If either $\triangleleft$ or f is the trivial map, then $(LE7)$ is equivalent to $\{-, -\}$ being a Lie bracket on V.

From now on, in light of Theorem 3.1.2, a Lie extending structure of $\mathfrak{g}$ through V will be viewed as a system $\Omega(\mathfrak{g}, V) = \big(\triangleleft, \triangleright, f, \{-, -\}\big)$ satisfying the compatibility conditions $(LE1) - (LE7)$. We denote by $\mathcal{L}(\mathfrak{g}, V)$ the set of all Lie extending structures of $\mathfrak{g}$ through V.

Example 3.1.3 *We provide the first example of a Lie extending structure and the corresponding unified product. More examples will be given in Section 3.3 and Section 3.2.*

Let $\Omega(\mathfrak{g}, V) = \big(\triangleleft, \triangleright, f, \{-, -\}\big)$ *be an extending datum of a Lie algebra* $\mathfrak{g}$ *through a vector space* V *such that* $\triangleleft$ *and* $\triangleright$ *are both trivial maps, i.e.,* $x \triangleleft g = x \triangleright g = 0$, *for all* $x, y \in V$ *and* $g \in \mathfrak{g}$. *Then,* $\Omega(\mathfrak{g}, V) = \big(f, \{-, -\}\big)$ *is a Lie extending structure of* $\mathfrak{g}$ *through* V *if and only if* $(V, \{-, -\})$ *is a Lie algebra and* $f : V \times V \to \mathfrak{g}$ *is a classical* 2*-cocycle, that is:*

$$f(x, x) = 0, \quad [g, f(x, y)] = 0, \quad f\big(x, \{y, z\}\big) + f\big(y, \{z, x\}\big) + f\big(z, \{x, y\}\big) = 0$$

for all $g \in \mathfrak{g}$, $x, y, z \in V$. *In this case, the associated unified product* $\mathfrak{g} \natural_{\Omega(\mathfrak{g}, V)} V$ *will be denoted by* $\mathfrak{g} \#^f V$ *and we shall call it the* twisted product *of the Lie algebras* $\mathfrak{g}$ *and* V. *Hence, the twisted product associated to a given* 2*-cocycle* $f : V \times V \to \mathfrak{g}$ *between Lie algebras is the vector space* $\mathfrak{g} \times V$ *with the bracket given for any* $g, h \in \mathfrak{g}$ *and* $x, y \in V$ *by:*

$$[(g, x), (h, y)] := \big([g, h] + f(x, y), \{x, y\}\big) \tag{3.4}$$

The twisted product of two Lie algebras plays the crucial role in the classification of all 6*-dimensional nilpotent Lie algebras given in [89].*

Let $\Omega(\mathfrak{g}, V) = \big(\triangleleft, \triangleright, f, \{-, -\}\big) \in \mathcal{L}(\mathfrak{g}, V)$ be a Lie extending structure and $\mathfrak{g} \natural V$ the associated unified product. Then the canonical inclusion

$$i_{\mathfrak{g}} : \mathfrak{g} \to \mathfrak{g} \natural V, \qquad i_{\mathfrak{g}}(g) = (g, 0)$$

is an injective Lie algebra map. Therefore, we can see $\mathfrak{g}$ as a Lie subalgebra of $\mathfrak{g} \natural V$ through the identification $\mathfrak{g} \cong i_{\mathfrak{g}}(\mathfrak{g}) \cong \mathfrak{g} \times \{0\}$. Conversely, we will prove that any Lie algebra structure on a vector space E containing $\mathfrak{g}$ as a Lie subalgebra is isomorphic to a unified product. In this way, we obtain the answer to the description part of the extending structures problem. First we need to introduce the following concepts.

Let $\mathfrak{g}$ be a Lie algebra, E a vector space such that $\mathfrak{g}$ is a subspace of E and V a complement of $\mathfrak{g}$ in E. For a linear map $\varphi : E \to E$ we consider the diagram:

$$\begin{array}{ccccc} \mathfrak{g} & \xrightarrow{\ i\ } & E & \xrightarrow{\ \pi\ } & V \\ \downarrow{\scriptstyle Id} & & \downarrow{\scriptstyle \varphi} & & \downarrow{\scriptstyle Id} \\ \mathfrak{g} & \xrightarrow{\ i\ } & E & \xrightarrow{\ \pi\ } & V \end{array} \tag{3.5}$$

where $\pi : E \to V$ is the canonical projection of $E = \mathfrak{g} + V$ on V and $i : \mathfrak{g} \to E$ is the inclusion map. We say that $\varphi : E \to E$ *stabilizes* $\mathfrak{g}$ (resp. *co-stabilizes* V) if the left square (resp. the right square) of the diagram (3.5) is commutative.

Let $\{-, -\}$ and $\{-, -\}'$ be two Lie algebra structures on E both containing $\mathfrak{g}$ as a Lie subalgebra. $\{-, -\}$ and $\{-, -\}'$ are called *equivalent*, and we denote this by $(E, \{-, -\}) \equiv (E, \{-, -\}')$, if there exists a Lie algebra isomorphism $\varphi : (E, \{-, -\}) \to (E, \{-, -\}')$ which stabilizes $\mathfrak{g}$. $\{-, -\}$ and $\{-, -\}'$ are called *cohomologous*, and we denote this by $(E, \{-, -\}) \approx (E, \{-, -\}')$, if there exists a Lie algebra isomorphism $\varphi : (E, \{-, -\}) \to (E, \{-, -\}')$ which stabilizes $\mathfrak{g}$ and co-stabilizes V, i.e., the diagram (3.5) is commutative.

$\equiv$ and $\approx$ are both equivalence relations on the set of all Lie algebra structures on E containing $\mathfrak{g}$ as a Lie subalgebra and we denote by $\text{Extd}\,(E, \mathfrak{g})$ (resp. $\text{Extd}'\,(E, \mathfrak{g})$) the set of all equivalence classes via $\equiv$ (resp. $\approx$). Thus, $\text{Extd}\,(E, \mathfrak{g})$ is the classifying object of the extending structures problem: by explicitly computing $\text{Extd}\,(E, \mathfrak{g})$ we obtain a parameterization of the set of all isomorphism classes of Lie algebra structures on E that stabilizes $\mathfrak{g}$. $\text{Extd}'\,(E, \mathfrak{g})$ gives a classification of the ES problem from the point of view of the extension problem. Any two cohomologous brackets on E are of course equivalent, hence there exists a canonical projection

$$\text{Extd}'\,(E, \mathfrak{g}) \twoheadrightarrow \text{Extd}\,(E, \mathfrak{g})$$

The classification part of the extending structures problem will be solved by computing explicitly both classifying objects. Borrowing the terminology from Lie algebra cohomology, we will see that $\text{Extd}'\,(E, \mathfrak{g})$ is parameterized by a cohomological object denoted by $\mathcal{H}^2\,(V, \mathfrak{g})$, which will be explicitly constructed and which generalizes the classical second cohomology group for Lie algebras [77], while $\text{Extd}\,(E, \mathfrak{g})$ will be parameterized by a relative cohomological object, denoted by $\mathcal{H}^2_{\mathfrak{g}}\,(V, \mathfrak{g})$ which turns out to be a quotient of $\mathcal{H}^2\,(V, \mathfrak{g})$.

Theorem 3.1.4 *Let $\mathfrak{g}$ be a Lie algebra, E a vector space containing $\mathfrak{g}$ as a subspace and $[-, -]$ a Lie algebra structure on E such that $\mathfrak{g}$ is a Lie subalgebra in $(E, [-, -])$. Then there exists a Lie extending structure $\Omega(\mathfrak{g}, V) = (\triangleleft, \triangleright, f, \{-, -\})$ of $\mathfrak{g}$ through a subspace V of E and an isomorphism of Lie algebras $(E, [-, -]) \cong \mathfrak{g} \natural V$ that stabilizes $\mathfrak{g}$ and co-stabilizes V.*

Proof: The proof follows in a straightforward manner from Theorem 2.1.5. However, we will write down, for further use, the Lie extending structure. First we fix a linear map $p : E \to \mathfrak{g}$ such that $p(g) = g$, for all $g \in \mathfrak{g}$—such a map always exists as k is a field. Then $V := \text{Ker}(p)$ is a subspace of $\mathfrak{h}$ and a complement of $\mathfrak{g}$ in $\mathfrak{h}$, that is $\mathfrak{h} = \mathfrak{g} + V$ and $\mathfrak{g} \cap V = \{0\}$. Using p we define a Lie extending system of $\mathfrak{g}$ through V, called the canonical extending system associated to p, where the bilinear maps $\rightharpoonup : V \times \mathfrak{g} \to \mathfrak{g}$, $\leftharpoonup : V \times \mathfrak{g} \to V$, $\theta : V \times V \to \mathfrak{g}$ and $\{\, ,\, \} : V \times V \to V$ are given by the following formulas for any $g \in \mathfrak{g}$ and $x,\, y \in V$:

$$x \rightharpoonup g \;:=\; p\big([x,\, g]\big), \qquad x \leftharpoonup g := [x,\, g] - p\big([x,\, g]\big) \tag{3.6}$$

$$\theta(x, y) \;:=\; p\big([x,\, y]\big), \qquad \{x, y\} := [x,\, y] - p\big([x,\, y]\big) \tag{3.7}$$

□

Using Theorem 3.1.4, the classification of all Lie algebra structures on E that contains $\mathfrak{g}$ as a Lie subalgebra, reduces to the classification of all unified products $\mathfrak{g} \natural V$, associated to all Lie extending structures $\Omega(\mathfrak{g}, V) = (\triangleleft, \triangleright, f, \{-, -\})$, for a given complement V of $\mathfrak{g}$ in E. In order to construct the cohomological objects $\mathcal{H}^2_{\mathfrak{g}}(V, \mathfrak{g})$ and $\mathcal{H}^2(V, \mathfrak{g})$ which will parameterize the classifying sets $\mathrm{Extd}(E, \mathfrak{g})$ and respectively $\mathrm{Extd}'(E, \mathfrak{g})$ defined above, we need the following technical lemma:

Lemma 3.1.5 *Let $\Omega(\mathfrak{g}, V) = (\triangleleft, \triangleright, f, \{-, -\})$ and $\Omega'(\mathfrak{g}, V) = (\triangleleft', \triangleright', f', \{-, -\}')$ be two Lie algebra extending structures of $\mathfrak{g}$ through V and $\mathfrak{g} \natural V$, $\mathfrak{g} \natural' V$ the associated unified products. Then there exists a bijection between the set of all morphisms of Lie algebras $\psi : \mathfrak{g} \natural V \to \mathfrak{g} \natural' V$ which stabilizes $\mathfrak{g}$ and the set of pairs (r, v), where $r : V \to \mathfrak{g}$, $v : V \to V$ are two linear maps satisfying the following compatibility conditions for any $g \in \mathfrak{g}$, x, $y \in V$:*

(ML1) $v(x) \triangleleft' g = v(x \triangleleft g)$;

(ML2) $r(x \triangleleft g) = [r(x), g] - x \triangleright g + v(x) \triangleright' g$;

(ML3) $v(\{x, y\}) = \{v(x), v(y)\}' + v(x) \triangleleft' r(y) - v(y) \triangleleft' r(x)$;

(ML4) $r(\{x, y\}) = [r(x), r(y)] + v(x) \triangleright' r(y) - v(y) \triangleright' r(x) + f'\big(v(x), v(y)\big) - f(x, y)$

Under the above bijection the morphism of Lie algebras $\psi = \psi_{(r,v)} : \mathfrak{g} \natural V \to \mathfrak{g} \natural' V$ corresponding to (r, v) is given for any $g \in \mathfrak{g}$ and $x \in V$ by:

$$\psi(g, x) = (g + r(x), v(x))$$

Moreover, $\psi = \psi_{(r,v)}$ is an isomorphism if and only if $v : V \to V$ is an isomorphism and $\psi = \psi_{(r,v)}$ co-stabilizes V if and only if $v = \mathrm{Id}_V$.

Definition 3.1.6 *Let $\mathfrak{g}$ be a Lie algebra and V a k-vector space. Two Lie algebra extending structures of $\mathfrak{g}$ by V, $\Omega(\mathfrak{g}, V) = (\triangleleft, \triangleright, f, \{-, -\})$ and $\Omega'(\mathfrak{g}, V) = (\triangleleft', \triangleright', f', \{-, -\}')$ are called* equivalent, *and we denote this by $\Omega(\mathfrak{g}, V) \equiv \Omega'(\mathfrak{g}, V)$, if there exists a pair (r, v) of linear maps, where $r : V \to \mathfrak{g}$ and $v \in \mathrm{Aut}_k(V)$ such that $(\triangleleft', \triangleright', f', \{-, -\}')$ is implemented from $(\triangleleft, \triangleright, f, \{-, -\})$ using (r, v) via:*

$$\begin{aligned}
x \triangleleft' g &= v\big(v^{-1}(x) \triangleleft g\big)\\
x \triangleright' g &= r\big(v^{-1}(x) \triangleleft g\big) + v^{-1}(x) \triangleright g + [g, r\big(v^{-1}(x)\big)]\\
f'(x, y) &= f\big(v^{-1}(x), v^{-1}(y)\big) + r\big(\{v^{-1}(x), v^{-1}(y)\}\big) + [r\big(v^{-1}(x)\big), r\big(v^{-1}(y)\big)]\\
&\quad - r\big(v^{-1}(x) \triangleleft r\big(v^{-1}(y)\big)\big) - v^{-1}(x) \triangleright r\big(v^{-1}(y)\big)\big)\\
&\quad + r\big(v^{-1}(y) \triangleleft r\big(v^{-1}(x)\big)\big) + v^{-1}(y) \triangleright r\big(v^{-1}(x)\big)\big)\\
\{x, y\}' &= v\big(\{v^{-1}(x), v^{-1}(y)\}\big) - v\big(v^{-1}(x) \triangleleft r\big(v^{-1}(y)\big)\big)\\
&\quad + v\big(v^{-1}(y) \triangleleft r\big(v^{-1}(x)\big)\big)
\end{aligned}$$

for all $g \in \mathfrak{g}$, x, $y \in V$.

As a conclusion, the main result of this section, which gives the theoretical answer to the extending structure problem, follows:

Theorem 3.1.7 *Let $\mathfrak{g}$ be a Lie algebra, E a vector space that contains $\mathfrak{g}$ as a subspace and V a complement of $\mathfrak{g}$ in E. Then:*

(1) *$\equiv$ is an equivalence relation on the set $\mathcal{L}(\mathfrak{g}, V)$ of all Lie extending structures of $\mathfrak{g}$ through V. We denote by $\mathcal{H}^2_{\mathfrak{g}}(V, \mathfrak{g}) := \mathcal{L}(\mathfrak{g}, V)/\equiv$, the pointed quotient set.*

(2) *The map*

$$\mathcal{H}^2_{\mathfrak{g}}(V, \mathfrak{g}) \to \mathrm{Extd}\,(E, \mathfrak{g}), \qquad \overline{(\triangleleft, \triangleright, f, \{-, -\})} \to (\mathfrak{g} \natural V, [-, -])$$

is bijective, where $\overline{(\triangleleft, \triangleright, f, \{-, -\})}$ is the equivalence class of $(\triangleleft, \triangleright, f, \{-, -\})$ via $\equiv$.

Remark 3.1.8 *The second cohomological object $\mathcal{H}^2(V, \mathfrak{g})$ that parameterizes* Extd$'(E, \mathfrak{g})$ *is constructed in a simple manner as follows: two Lie algebra extending structures $\Omega(\mathfrak{g}, V) = (\triangleleft, \triangleright, f, \{-, -\})$ and $\Omega'(\mathfrak{g}, V) = (\triangleleft', \triangleright', f', \{-, -\}')$ are called* cohomologous, *and we denote this by $\Omega(\mathfrak{g}, V) \approx \Omega'(\mathfrak{g}, V)$ if and only if $\triangleleft' = \triangleleft$ and there exists a linear map $r : V \to \mathfrak{g}$ such that*

$$\begin{aligned}
x \triangleright' g &= x \triangleright g + r(x \triangleleft g) - [r(x),\, g] \\
f'(x, y) &= f(x, y) + r(\{x,\, y\}) + [r(x),\, r(y)] + \\
&\quad + y \triangleright r(x) - x \triangleright r(y) + r(y \triangleleft r(x)) - r(x \triangleleft r(y)) \\
\{x, y\}' &= \{x, y\} - x \triangleleft r(y) + y \triangleleft r(x)
\end{aligned}$$

for all $g \in \mathfrak{g}$, x, $y \in V$.

Similar to the proof of Theorem 3.1.7 we can easily see that $\Omega(\mathfrak{g}, V) \approx \Omega'(\mathfrak{g}, V)$ if and only if there exists an isomorphism of Lie algebras $\psi : \mathfrak{g} \natural V \to \mathfrak{g} \natural' V$ which stabilizes $\mathfrak{g}$ and co-stabilizes V. Thus, $\approx$ is an equivalence relation on the set $\mathcal{L}(\mathfrak{g}, V)$ of all Lie extending structures of $\mathfrak{g}$ through V. If we denote $\mathcal{H}^2(V, \mathfrak{g}) := \mathcal{L}(\mathfrak{g}, V)/\approx$, the map

$$\mathcal{H}^2(V, \mathfrak{g}) \to \mathrm{Extd}'\,(E, \mathfrak{g}), \qquad \overline{(\triangleleft, \triangleright, f, \{-, -\})} \to (\mathfrak{g} \natural V, [-, -])$$

is a bijection between $\mathcal{H}^2(V, \mathfrak{g})$ and the isomorphism classes of all Lie algebra structures on E which stabilizes $\mathfrak{g}$ and co-stabilizes V.

3.2 Flag extending structures: Examples

Theorem 3.1.7 offers the theoretical answer to the extending structures problem. The next challenge is a computational one: for a given Lie algebra

$\mathfrak{g}$ that is a subspace in a vector space E with a given complement V, we aim to compute explicitly the classifying object $\mathcal{H}^2_{\mathfrak{g}}(V, \mathfrak{g})$ and then to list all Lie algebra structures on E which extend the Lie algebra structure on $\mathfrak{g}$. In what follows we provide a way of answering this problem for a large class of such structures.

Definition 3.2.1 *Let $\mathfrak{g}$ be a Lie algebra and E a vector space containing $\mathfrak{g}$ as a subspace. A Lie algebra structure on E such that $\mathfrak{g}$ is a Lie subalgebra is called a* flag extending structure *of $\mathfrak{g}$ to E if there exists a finite chain of Lie subalgebras of E*

$$\mathfrak{g} = E_0 \subset E_1 \subset \cdots \subset E_m = E \tag{3.8}$$

such that E_i has codimension 1 in E_{i+1}, for all $i = 0, \cdots, m-1$.

In the context of Definition 3.2.1 we have that $\dim_k(V) = m$, where V is the complement of $\mathfrak{g}$ in E. The existence of such a chain of Lie subalgebras is quite common in the theory of solvable Lie algebras [55, Proposition 2], [138, Lie Theorem]. All flag extending structures of $\mathfrak{g}$ to E can be completely described by a recursive reasoning where the key step is $m = 1$. More precisely, this key step describes and classifies all unified products $\mathfrak{g} \natural V_1$, for a 1-dimensional vector space V_1. We will prove that they are parameterized by the space $\mathrm{TwDer}(\mathfrak{g})$ of all twisted derivations of $\mathfrak{g}$. Then, by replacing the initial Lie algebra $\mathfrak{g}$ with such a unified product $\mathfrak{g} \natural V$ which can be described in terms of $\mathfrak{g}$ only, we can iterate the process: in this way, on our second step we describe and classify all unified products of the form $(\mathfrak{g} \natural V_1) \natural V_2$, where V_1 and V_2 are vector spaces of dimension 1. Of course, after $m = \dim_k(V)$ steps, we obtain the description of all flag extending structures of $\mathfrak{g}$ to E. We start by introducing the following concept which generalizes the notion of derivation of a Lie algebra:

Definition 3.2.2 *A* twisted derivation *of a Lie algebra $\mathfrak{g}$ is a pair (λ, D) consisting of two linear maps $\lambda : \mathfrak{g} \to k$ and $D : \mathfrak{g} \to \mathfrak{g}$ such that for any g, $h \in \mathfrak{g}$:*

$$\lambda([g, h]) = 0 \tag{3.9}$$

$$D([g, h]) = [D(g), h] + [g, D(h)] + \lambda(g)D(h) - \lambda(h)D(g) \tag{3.10}$$

The set of all twisted derivations of $\mathfrak{g}$ will be denoted by $\mathrm{TwDer}(\mathfrak{g})$. The compatibility condition (3.9) is equivalent to $\mathfrak{g}' \subseteq \mathrm{Ker}(\lambda)$, where $\mathfrak{g}'$ is the derived algebra of $\mathfrak{g}$.

Examples 3.2.3 *1.* $\mathrm{TwDer}(\mathfrak{g})$ *contains the usual space of derivations* $\mathrm{Der}(\mathfrak{g})$ *via the canonical embedding*

$$\mathrm{Der}(\mathfrak{g}) \hookrightarrow \mathrm{TwDer}(\mathfrak{g}), \qquad D \mapsto (0, D)$$

which is an isomorphism if $\mathfrak{g}$ is a perfect Lie algebra.

2. Let $g_0 \in \mathfrak{g}$ and $\lambda : \mathfrak{g} \to k$ be a k-linear map such that $\mathfrak{g}' \subseteq \mathrm{Ker}(\lambda)$. We define the map

$$D_{g_0,\lambda} : \mathfrak{g} \to \mathfrak{g}, \quad D_{g_0,\lambda}(h) := [g_0,\, h] - \lambda(h) g_0$$

for all $h \in \mathfrak{g}$. Then $(\lambda, D_{g_0,\lambda})$ is a twisted derivation called an inner twisted derivation.

We shall prove now that the set of all Lie extending structures $\mathcal{L}(\mathfrak{g}, V)$ of a Lie algebra $\mathfrak{g}$ through a 1-dimensional vector space V is parameterized by $\mathrm{TwDer}(\mathfrak{g})$.

Proposition 3.2.4 *Let $\mathfrak{g}$ be a Lie algebra and V a vector space of dimension 1 with a basis $\{x\}$. Then there exists a bijection between the set $\mathcal{L}(\mathfrak{g}, V)$ of all Lie extending structures of $\mathfrak{g}$ through V and the space $\mathrm{TwDer}(\mathfrak{g})$ of all twisted derivations of $\mathfrak{g}$. Through the above bijection, the Lie extending structure $\Omega(\mathfrak{g}, V) = \big(\triangleleft, \triangleright, f, \{-,-\}\big)$ corresponding to $(\lambda, D) \in \mathrm{TwDer}(\mathfrak{g})$ is given by:*

$$x \triangleleft g = \lambda(g) x, \quad x \triangleright g = D(g), \quad f = 0, \quad \{-,-\} = 0 \tag{3.11}$$

for all $g \in \mathfrak{g}$. The unified product associated to the Lie extending structure (3.11) will be denoted by $\mathfrak{g} \natural_{(\lambda,\, D)} V$ and has the bracket given for any g, $h \in \mathfrak{g}$ by:[1]

$$[(g, 0),\, (h, 0)] = ([g, h],\, 0), \quad [(g, 0),\, (0, x)] = -(D(g),\, \lambda(g)x) \tag{3.12}$$

Remark 3.2.5 *Since the cocycle f in (3.11) is trivial we obtain that $\mathfrak{g} \natural_{(\lambda,\, D)} V = \mathfrak{g} \bowtie V$, where $\mathfrak{g} \bowtie V$ is a bicrossed product between $\mathfrak{g}$ and an abelian Lie algebra of dimension 1. Hence, Proposition 3.2.4 shows in fact that any unified product $\mathfrak{g} \natural_{(\lambda,\, D)} V$ between an arbitrary Lie algebra $\mathfrak{g}$ and a 1-dimensional vector space V is isomorphic to a bicrossed product $\mathfrak{g} \bowtie V$ between $\mathfrak{g}$ and the abelian Lie algebra of dimension 1.*

Example 3.2.6 *Let $A = (A, [-,\, -])$ be a Lie algebra. There is a bijection between the set of all Lie extending system of A through k and the set of all pairs $(\lambda,\, D) \in A^* \times \mathrm{End}_k(A)$ satisfying the following compatibilities for any a, $b \in A$:*

(FL1) $\lambda([a,\, b]) = 0$

(FL2) $D([a,\, b]) = [D(a),\, b] + [a,\, D(b)] + \lambda(a)D(b) - \lambda(b)D(a)$

The bijection is given such that the Lie extending system $\Lambda(A,\, k) = \big(\leftharpoonup, \rightharpoonup, \theta, \{-,\, -\}\big)$ associated to a twisted derivation $(\lambda,\, D)$ is defined for any x, $y \in k$ and $a \in A$ by:

$$x \leftharpoonup a := x\,\lambda(a), \quad x \rightharpoonup a := x\,D(a), \quad \theta(x,\, y) := 0, \quad \{x,\, y\} := 0 \tag{3.13}$$

[1] As usual, we define the bracket only in the points where the values are non-zero.

TwDer(A) *contains the usual space of derivations* Der(A) *via the canonical embedding* Der$(A) \hookrightarrow$ TwDer(A), $D \mapsto (0, D)$. *We point out that the above canonical embedding is bijective if A is a perfect Lie algebra. The unified product* $A \ltimes_{\Lambda(A,k)} k$ *associated to* $(\lambda,\, D) \in \mathrm{TwDer}(A)$ *is denoted by* $A_{(\lambda,\, D)}$ *and it is the vector space $A \times k$ with the bracket defined for any a, $b \in A$ and x, $y \in k$ by:*

$$[(a,x),\, (b,y)] := \big([a,\, b] + xD(b) - yD(a),\ x\lambda(b) - y\lambda(a)\big)$$

A Lie algebra $\mathfrak{g}$ contains A as a Lie subalgebra of codimension 1 *if and only if* $\mathfrak{g} \cong A_{(\lambda,\, D)}$, *for some* $(\lambda,\, D) \in \mathrm{TwDer}(A)$.

Next, we classify all Lie algebras $\mathfrak{g} \natural_{(\lambda,\, D)} V$ by computing the cohomological objects $\mathcal{H}^2_{\mathfrak{g}}(V, \mathfrak{g})$ and $\mathcal{H}^2\,(V, \mathfrak{g})$. First we need the following:

Definition 3.2.7 *Two twisted derivations* (λ, D) *and* $(\lambda', D') \in \mathrm{TwDer}(\mathfrak{g})$ *are called* equivalent *and we denote this by* $(\lambda, D) \equiv (\lambda', D')$ *if* $\lambda = \lambda'$ *and there exists a pair* $(g_0, q) \in \mathfrak{g} \times k^*$ *such that for any* $h \in \mathfrak{g}$ *we have:*

$$D(h) = qD'(h) + [g_0,\, h] - \lambda(h)g_0 \tag{3.14}$$

Hence, $(\lambda, D) \equiv (\lambda', D')$ if and only if $\lambda = \lambda'$ and there exists a non-zero scalar $q \in k$ such that $D - qD'$ is a inner twisted derivation. We provide below the first explicit classification result of the extending structures problem for Lie algebras. This is also the key step in the classification of all flag extending structures.

Theorem 3.2.8 *Let $\mathfrak{g}$ be a Lie algebra of codimension* 1 *in the vector space E and V a complement of $\mathfrak{g}$ in E. Then:*

(1) $\equiv$ *is an equivalence relation on the set* TwDer$(\mathfrak{g})$ *of all twisted derivations of* $\mathfrak{g}$.

(2) Extd$\,(E, \mathfrak{g}) \cong \mathcal{H}^2_{\mathfrak{g}}(V, \mathfrak{g}) \cong \mathrm{TwDer}(\mathfrak{g})/\equiv$. *The bijection between* TwDer$(\mathfrak{g})/\equiv$ *and* Extd$\,(E, \mathfrak{g})$, *the isomorphism classes of all Lie algebra structures on E that stabilize $\mathfrak{g}$, is given by:*

$$\overline{(\lambda,\, D)} \mapsto \mathfrak{g} \natural_{(\lambda,\, D)} V$$

where $\overline{(\lambda,\, D)}$ *is the equivalence class of* $(\lambda,\, D)$ *via the relation* $\equiv$ *and* $\mathfrak{g} \natural_{(\lambda,\, D)} V$ *is the Lie algebra constructed in (3.12).*

(3) $\mathcal{H}^2\,(V, \mathfrak{g}) \cong \mathrm{TwDer}(\mathfrak{g})/\approx$, *where* $\approx$ *is the following relation:* $(\lambda, D) \approx (\lambda', D')$ *if and only if* $\lambda = \lambda'$ *and* $D - D'$ *is a inner twisted derivation of* $\mathfrak{g}$.

Proof: Let (λ, D), $(\lambda', D') \in \mathrm{TwDer}(\mathfrak{g})$ be two twisted derivations of $\mathfrak{g}$ and $\Omega(\mathfrak{g}, V) = (\triangleleft, \triangleright, f, \{-,-\})$ respectively $\Omega'(\mathfrak{g}, V) = (\triangleleft', \triangleright', f', \{-,-\}')$ the corresponding Lie extending structure constructed in (3.11). We will prove that $(\lambda, D) \equiv (\lambda', D')$ if and only if there exists an isomorphism of Lie algebras $\mathfrak{g} \natural_{(\lambda,\, D)} V \cong \mathfrak{g} \natural_{(\lambda',\, D')} V$ that stabilizes $\mathfrak{g}$. This observation together with Proposition 3.2.4 and Theorem 3.1.7 will finish the proof.

Indeed, using Lemma 3.1.5 we obtain that there exists an isomorphism $\mathfrak{g} \natural_{(\lambda, D)} V \cong \mathfrak{g} \natural_{(\lambda', D')} V$ of Lie algebras which stabilizes $\mathfrak{g}$ if and only if there exists a pair (r, v), where $r : V \to \mathfrak{g}$ and $v : V \to V$ are linear maps satisfying the compatibility conditions $(ML1)$—$(ML4)$ and v is bijective. Since $\dim_k(V) = 1$, any linear map $r : V \to \mathfrak{g}$ is uniquely determined by an element $g_0 \in \mathfrak{g}$ such that $r(x) = g_0$, where $\{x\}$ is a basis in V. On the other hand, any automorphism v of V is uniquely determined by a non-zero scalar $q \in k^*$ such that $v(x) = qx$. It remains to check the compatibility conditions $(ML1)$–$(ML4)$, for this pair of maps $(r = r_{g_0}, v = v_q)$. Since $f = f' = 0$ and $\{-, -\} = \{-, -\}' = 0$ in the corresponding Lie extending structure, we obtain that the compatibility conditions $(ML3)$ and $(ML4)$ are trivially fulfilled. Now, the compatibility condition $(ML1)$ is equivalent to $q\lambda'(g)x = q\lambda(g)x$, for all $g \in \mathfrak{g}$, i.e., to the fact that $\lambda = \lambda'$, since $q \neq 0$. Finally, the compatibility condition $(ML1)$ takes the following equivalent form:

$$\lambda(g)g_0 = [g_0, g] - D(g) + qD'(g)$$

for all $g \in \mathfrak{g}$, which is precisely (3.14) from Definition 3.2.7. Thus, using Lemma 3.1.5 we have proved that $\mathfrak{g} \natural_{(\lambda, D)} V \cong \mathfrak{g} \natural_{(\lambda', D')} V$ (an isomorphism of Lie algebras that stabilizes $\mathfrak{g}$) if and only if $(\lambda, D) \equiv (\lambda', D')$ and the proof is finished. □

Theorem 3.2.8 takes the following simplified form in the case of perfect Lie algebras.

Corollary 3.2.9 *Let $\mathfrak{g}$ be a perfect Lie algebra of codimension* 1 *in the vector space E and V a complement of $\mathfrak{g}$ in E. Then:*

(1) $\mathrm{Extd}\,(E, \mathfrak{g}) \cong \mathcal{H}^2_{\mathfrak{g}}(V, \mathfrak{g}) \cong \mathrm{Der}(\mathfrak{g})/\approx$, *where $\approx$ is the equivalence relation on* $\mathrm{Der}(\mathfrak{g})$ *defined by: $D \approx D'$ if and only if there exists $q \in k^*$ such that $D - qD'$ is an inner derivation of $\mathfrak{g}$. The bijection between* $\mathrm{Der}(\mathfrak{g})/\approx$ *and* $\mathrm{Extd}\,(E, \mathfrak{g})$ *is given by*

$$\overline{D} \mapsto \mathfrak{g} \natural_D V$$

where $\overline{D}$ is the equivalence class of D via the relation $\approx$ and $\mathfrak{g} \natural_D V$ is the Lie algebra constructed in (3.12) for $\lambda = 0$.

(2) $\mathcal{H}^2\,(V, \mathfrak{g}) \cong \mathrm{Out}(\mathfrak{g})$.

Proof: Follows directly from Theorem 3.2.8 using Example 3.2.3: for a perfect Lie algebra $\mathfrak{g}$, we have that $\mathrm{TwDer}(\mathfrak{g}) = \{0\} \times \mathrm{Der}(\mathfrak{g})$. □

Remark 3.2.10 *We recall that the space of outer derivations of a Lie algebra $\mathfrak{g}$ is the quotient vector space* $\mathrm{Out}(\mathfrak{g}) := \mathrm{Der}(\mathfrak{g})/\mathrm{Inn}(\mathfrak{g})$. *Thus, by definition, two derivations D, $D' \in$* $\mathrm{Der}(\mathfrak{g})$ *are congruent and we denote this by $D \sim D'$ if $D - D' \in$* $\mathrm{Inn}(\mathfrak{g})$ *and then the space* $\mathrm{Out}(\mathfrak{g})$ *is defined as the quotient via this congruence relation, i.e.,* $\mathrm{Der}(\mathfrak{g})/\sim$. *Now, two congruent derivations are equivalent in the sense of Corollary 3.2.9 but the converse does not hold. This means that there exists a canonical surjection*

$$\mathrm{Out}(\mathfrak{g}) \twoheadrightarrow \mathrm{Der}(\mathfrak{g})/\approx$$

We also note that the classifying object $\mathrm{Der}(\mathfrak{g})/\approx$ *is just a pointed set as it does not carry a group structure: it is straightforward to see that the following possible group structure* $\overline{D}+\overline{D'}=\overline{D+D'}$ *is not well defined on* $\mathrm{Der}(\mathfrak{g})/\approx$. *However, this is not at all surprising from the point of view of non-abelian extension theory: the classifying object is not a group anymore but a pointed set.*

Now, we indicate a class of Lie algebras $\mathfrak{g}$ such that the classifying objects of Corollary 3.2.9 are both singletons. This class contains the semisimple Lie algebras or, more generally, the sympathetic Lie algebras [48]. In particular, it shows that for a semisimple Lie algebra $\mathfrak{g}$ there exists, up to an isomorphism that stabilizes $\mathfrak{g}$, a unique Lie algebra structure on a vector space of dimension $1+\dim_k(\mathfrak{g})$ which extends the one on $\mathfrak{g}$: more precisely, this unique Lie algebra structure is given by the direct product $\mathfrak{g}\times V$, between $\mathfrak{g}$ and an abelian Lie algebra of dimension 1.

Corollary 3.2.11 *Let* $\mathfrak{g}$ *be a Lie algebra of codimension* 1 *in the vector space* E *and* V *a complement of* $\mathfrak{g}$ *in* E. *Assume that* $\mathfrak{g}$ *is perfect and* $\mathrm{Der}(\mathfrak{g}) = \mathrm{Inn}(\mathfrak{g})$. *Then* $\mathcal{H}^2_{\mathfrak{g}}(V,\mathfrak{g}) = \mathcal{H}^2(V,\mathfrak{g}) = 0$.

Proof: We apply Corollary 3.2.9: since $\mathrm{Der}(\mathfrak{g}) = \mathrm{Inn}(\mathfrak{g})$ we obtain that the space of outer derivations $\mathrm{Out}(\mathfrak{g}) = 0$, hence, so is $\mathrm{Der}(\mathfrak{g})/\approx$, being a quotient of a null space. □

Next we provide two explicit examples for the above results by computing $\mathcal{H}^2_{\mathfrak{g}}(V,\mathfrak{g})$ and then describing all Lie algebra structures which extend the Lie algebra structure from $\mathfrak{g}$ to a vector space of dimension $1+\dim_k(\mathfrak{g})$. The detailed computations are rather long but straightforward and can be provided upon request. We start with the case when $\mathfrak{g}$ is a perfect Lie algebra which is not semisimple.

Example 3.2.12 *Let* k *be a field of characteristic zero and* $\mathfrak{g}$ *be the perfect* 5*-dimensional Lie algebra with a basis* $\{e_1, e_2, e_3, e_4, e_5\}$ *and bracket given by:*

$$[e_1,\, e_2] = e_3, \quad [e_1,\, e_3] = -2e_1, \quad [e_1,\, e_5] = [e_3,\, e_4] = e_4$$
$$[e_2,\, e_3] = 2e_2, \quad [e_2,\, e_4] = e_5, \quad [e_3,\, e_5] = -e_5$$

We shall compute the classifying object $\mathrm{Extd}\,(k^6,\mathfrak{g})$ *by proving that* $\mathrm{Extd}\,(k^6,\mathfrak{g}) \cong k^7/\equiv$, *where* $\equiv$ *is the equivalence relation on* k^7 *defined by:* $(a_1,\cdots,a_7)\equiv(a_1',\cdots,a_7')$ *if and only if there exists* $q\in k^*$ *such that* $a_2 = qa_2'$ *and* $2a_7 - a_1 = q(2a_7' - a_1')$.

Indeed, by a rather long but straightforward computation it can be proved that the space of derivations $\mathrm{Der}(\mathfrak{g})$ *coincides with the space of all matrices* *from $\mathcal{M}_5(k)$ of the form:*

$$A = \begin{pmatrix} a_1 & 0 & a_6 & 0 & 0 \\ 0 & -a_1 & -2a_2 & 0 & 0 \\ a_2 & a_4 & 0 & 0 & 0 \\ a_3 & 0 & a_5 & a_7 & a_4 \\ 0 & a_5 & -a_3 & a_2 & (a_7 - a_1) \end{pmatrix}$$

for all $a_1, \cdots, a_7 \in k$. Thus, any 6-dimensional Lie algebra that contains $\mathfrak{g}$ as a Lie subalgebra is isomorphic to one of the following seven-parameter Lie algebra denoted by $\mathfrak{g}_{(a_1,\cdots,a_7)}(x) := \mathfrak{g} \natural V$, which has the basis $\{e_1, e_2, e_3, e_4, e_5, x\}$ and bracket given by:

$$[e_1,\, x] = -a_1 e_1 - a_2 e_3 - a_3 e_4, \quad [e_2,\, x] = a_1 e_2 - a_4 e_3 - a_5 e_5$$

$$[e_3,\, x] = -a_6 e_1 + 2a_2 e_2 - a_5 e_4 + a_3 e_5, \; [e_4,\, x] = -a_7 e_4 - a_2 e_5,$$

$$[e_5,\, x] = -a_4 e_4 + (a_1 - a_7) e_5$$

for some scalars $a_1, \cdots, a_7 \in k$. Two such Lie algebras $\mathfrak{g}_{(a_1,\cdots,a_7)}(x)$ and $\mathfrak{g}_{(a_1',\cdots,a_7')}(x)$ are equivalent if and only if there exists $q \in k^*$ such that $a_2 = q a_2'$ and $2a_7 - a_1 = q(2a_7' - a_1')$, as needed.

Finally, we give an example in the case when $\mathfrak{g}$ is not perfect.

Example 3.2.13 *Let k be a field of characteristic zero and $gl(2, k)$ the Lie algebra of all 2×2 matrices over k with the usual Lie bracket defined by:*

$$[e_{ij},\, e_{kl}] = \delta_{jk} e_{il} - \delta_{il} e_{kj} \tag{3.15}$$

for all $1 \leq i, j \leq 2$, where δ is the Kronecker delta and e_{ij} the matrix units. We will compute the twisted derivations for $gl(2, k)$. The k-linear maps $\lambda : gl(2, k) \to k$ satisfying (3.9) are given as follows:

$$\lambda(e_{11}) = \lambda(e_{22}) := q \in k, \qquad \lambda(e_{12}) = \lambda(e_{21}) = 0$$

Now depending on the values of q we obtain, by a rather long but straightforward computation, the following k-linear maps $D : gl(2, k) \to gl(2, k)$ satisfying (3.10):

Case 1: *Suppose first that $q \notin \{0,\, 1,\, -1,\, 2\}$. Then the space of k-linear maps $D : gl(2, k) \to gl(2, k)$ satisfying (3.10) coincide with the space of all matrices from $\mathcal{M}_4(k)$ of the form:*

$$A = \begin{pmatrix} a_1 & -(1-q)^{-1} a_3 & -(1+q) a_2 & a_1 \\ a_2 & q^{-1}(a_4 - a_1) & 0 & (q-1)(q+1)^{-1} a_2 \\ a_3 & 0 & q^{-1}(a_1 - a_4) & (q+1)(q-1)^{-1} a_3 \\ a_4 & (1-q)^{-1} a_3 & (1+q)^{-1} a_2 & a_4 \end{pmatrix}$$

for all $a_1, \cdots, a_4 \in k$. Thus, in this case any 5-dimensional Lie algebra that contains $\mathfrak{gl}(2,k)$ as a Lie subalgebra is isomorphic to one of the following five-parameter Lie algebra denoted by $\mathfrak{gl}(2,k)_{(q,a_1,\cdots,a_4)}(x)$, which has the basis $\{e_{11}, e_{12}, e_{21}, e_{22}, x\}$ and bracket given by:

$$\begin{aligned} [e_{11}, x] &= -a_1e_{11} - a_2e_{12} - a_3e_{21} - a_4e_{22} - qx \\ [e_{12}, x] &= (1-q)^{-1}a_3e_{11} + q^{-1}(a_1 - a_4)e_{12} - (1-q)^{-1}a_3e_{22} \\ [e_{21}, x] &= (1+q)^{-1}a_2e_{11} + q^{-1}(a_4 - a_1)e_{21} - (1+q)^{-1}a_2e_{22} \\ [e_{22}, x] &= -a_1e_{11} - (q-1)(q+1)^{-1}a_2e_{12} - (q+1)(q-1)^{-1}a_3e_{21} \\ &\quad -a_4e_{22} - qx \end{aligned}$$

Two such Lie algebras $\mathfrak{gl}(2,k)_{(q,a_1,\cdots,a_4)}(x)$ and $\mathfrak{gl}(2,k)_{(q,a'_1,\cdots,a'_4)}(x)$ are equivalent if and only if there exists $p \in k^*$ such that $a_2 = pa'_2$, $a_3 = pa'_3$ and $a_1 - a_4 = p(a'_1 - a'_4)$.

Case 2: *Assume that $q = 0$. Then the space of k-linear maps $D : \mathfrak{gl}(2,k) \to \mathfrak{gl}(2,k)$ satisfying (3.10) coincide with the space of all matrices from $\mathcal{M}_4(k)$ of the form:*

$$A = \begin{pmatrix} a_1 & -a_3 & -a_2 & a_1 \\ a_2 & a_4 & 0 & -a_2 \\ a_3 & 0 & -a_4 & -a_3 \\ a_1 & a_3 & a_2 & a_1 \end{pmatrix}$$

for all $a_1, \cdots, a_5 \in k$. Thus, in this case any 5-dimensional Lie algebra that contains $\mathfrak{gl}(2,k)$ as a Lie subalgebra has the basis $\{e_{11}, e_{12}, e_{21}, e_{22}, x\}$ and bracket given by:

$$[e_{11}, x] = -a_1e_{11} - a_2e_{12} - a_3e_{21} - a_1e_{22}, \quad [e_{12}, x] = a_3e_{11} - a_4e_{12} - a_3e_{22}$$
$$[e_{21}, x] = a_2e_{11} + a_4e_{21} - a_2e_{22}, \quad [e_{22}, x] = -a_1e_{11} + a_2e_{12} + a_3e_{21} - a_1e_{22}$$

Two such Lie algebras are equivalent if and only if there exists $p \in k^*$ such that $a_1 = pa'_1$.

Case 3: *Assume that $q = 1$. Then the space of k-linear maps $D : \mathfrak{gl}(2,k) \to \mathfrak{gl}(2,k)$ satisfying (3.10) coincide with the space of all matrices from $\mathcal{M}_4(k)$ of the form:*

$$A = \begin{pmatrix} a_1 & a_3 & -2^{-1}a_2 & a_1 \\ a_2 & a_4 - a_1 & 0 & 0 \\ 0 & 0 & a_1 - a_4 & 2a_3 \\ a_4 & -a_3 & 2^{-1}a_2 & a_4 \end{pmatrix}$$

for all $a_1, \cdots, a_4 \in k$. Thus, in this case any 5-dimensional Lie algebra that contains $\mathfrak{gl}(2,k)$ as a Lie subalgebra has the basis $\{e_{11}, e_{12}, e_{21}, e_{22}, x\}$ and bracket given by:

$$\begin{aligned} [e_{11}, x] &= -a_1e_{11} - a_2e_{12} - a_4e_{22} - x, \\ [e_{12}, x] &= -a_3e_{11} + (a_1 - a_4)e_{12} + a_3e_{22} \\ [e_{21}, x] &= 2^{-1}a_2e_{11} + (a_2 - a_1)e_{21} - 2^{-1}a_2e_{22}, \\ [e_{22}, x] &= -a_1e_{11} - 2a_3e_{21} - a_4e_{22} - x \end{aligned}$$

Two such Lie algebras are equivalent if and only if there exists $p \in k^*$ such that $a_2 = pa_2^{'}$, $a_3 = pa_3^{'}$ and $a_1 - a_4 = p(a_1^{'} - a_4^{'})$.

Case 4: *Assume that $q = -1$. Then the space of k-linear maps $D : gl(2,k) \to gl(2,k)$ satisfying (3.10) coincide with the space of all matrices from $\mathcal{M}_4(k)$ of the form:*

$$A = \begin{pmatrix} a_1 & -2^{-1}a_3 & a_2 & a_1 \\ 0 & a_1 - a_4 & 0 & 2a_2 \\ a_3 & 0 & a_4 - a_1 & 0 \\ a_4 & 2^{-1}a_3 & -a_2 & a_4 \end{pmatrix}$$

for all $a_1, \cdots, a_4 \in k$. Thus, in this case any 5-dimensional Lie algebra that contains $gl(2,k)$ as a Lie subalgebra has the basis $\{e_{11}, e_{12}, e_{21}, e_{22}, x\}$ and bracket given by:

$$\begin{aligned} [e_{11},\, x] &= -a_1e_{11} - a_3e_{21} - a_4e_{22} + x, \\ [e_{12},\, x] &= 2^{-1}a_3e_{11} + (a_4 - a_1)e_{12} - 2^{-1}a_3e_{22} \\ [e_{21},\, x] &= -a_2e_{11} + (a_1 - a_4)e_{21} + a_2e_{22}, \\ [e_{22},\, x] &= -a_1e_{11} - 2a_2e_{12} - a_4e_{22} + x \end{aligned}$$

Two such Lie algebras are equivalent if and only if there exists $p \in k^*$ such that $a_3 = pa_3^{'}$, $a_2 = pa_2^{'}$ and $a_1 - a_4 = p(a_1^{'} - a_4^{'})$.

Case 5: *Assume that $q = 2$. Then the space of k-linear maps $D : gl(2,k) \to gl(2,k)$ satisfying (3.10) coincide with the space of all matrices from $\mathcal{M}_4(k)$ of the form:*

$$A = \begin{pmatrix} a_1 & a_3 & -3^{-1}a_2 & a_1 \\ a_2 & 2^{-1}(a_4 - a_1) & a_5 & 3^{-1}a_2 \\ a_3 & 0 & 2^{-1}(a_1 - a_4) & 3a_3 \\ a_4 & -a_3 & 3^{-1}a_2 & a_4 \end{pmatrix}$$

for all $a_1, \cdots, a_5 \in k$. Thus, in this case any 5-dimensional Lie algebra that contains $gl(2,k)$ as a Lie subalgebra has the basis $\{e_{11}, e_{12}, e_{21}, e_{22}, x\}$ and bracket given by:

$$\begin{aligned} [e_{11},\, x] &= -a_1e_{11} - a_2e_{12} - a_3e_{21} - a_4e_{22} - 2x \\ [e_{12},\, x] &= -a_3e_{11} + 2^{-1}(a_1 - a_4)e_{12} + a_3e_{22} \\ [e_{21},\, x] &= 3^{-1}a_2e_{11} - a_5e_{12} + 2^{-1}(a_4 - a_1)e_{21} - 3^{-1}a_2e_{22} \\ [e_{22},\, x] &= -a_1e_{11} - 3^{-1}a_2e_{12} - 3a_3e_{21} - a_4e_{22} - 2x \end{aligned}$$

Two such Lie algebras are equivalent if and only if there exists $p \in k^*$ such that $a_2 = pa_2^{'}$, $a_3 = pa_3^{'}$, $a_5 = pa_5^{'}$ and $a_1 - a_4 = p(a_1^{'} - a_4^{'})$.

Thus we have described the classifying object $\text{Extd}\,(k^5,\, gl(2,k)) \cong \mathcal{H}^2_{gl(2,k)}\,(V,\, gl(2,k))$*: it is equal to the disjoint union of the five quotient spaces described above. This is easy to see having in mind that if two twisted derivations (λ_1, D_1) are equivalent in the sense of Corollary 3.2.9 then $\lambda_1 = \lambda_2$.*

3.3 Special cases of unified products for Lie algebras

In this section we show that crossed products and bicrossed products of two Lie algebras are both special cases of the unified product. We make the following convention: if one of the maps $\triangleleft$, $\triangleright$, f or $\{-, -\}$ of an extending datum $\Omega(\mathfrak{g}, V) = \big(\triangleleft, \triangleright, f, \{-, -\}\big)$ is trivial then we will omit it from the quadruple $\big(\triangleleft, \triangleright, f, \{-, -\}\big)$.

Crossed products and the extension problem

Let $\Omega(\mathfrak{g}, V) = \big(\triangleleft, \triangleright, f, \{-, -\}\big)$ be an extending datum of $\mathfrak{g}$ through V such that $\triangleleft$ is the trivial map, i.e., $x \triangleleft g = 0$, for all x, $y \in V$ and $g \in \mathfrak{g}$. Then, $\Omega(\mathfrak{g}, V) = \big(\triangleleft, \triangleright, f, \{-, -\}\big) = \big(\triangleright, f, \{-, -\}\big)$ is a Lie extending structure of $\mathfrak{g}$ through V if and only if $(V, \{-,-\})$ is a Lie algebra and the following compatibilities hold for any g, $h \in \mathfrak{g}$ and x, y, $z \in V$:

- $f(x, x) = 0$
- $x \triangleright [g, h] = [x \triangleright g, h] + [g, x \triangleright h]$
- $\{x, y\} \triangleright g = x \triangleright (y \triangleright g) - y \triangleright (x \triangleright g) + [g, f(x, y)]$
- $f(x, \{y, z\}) + f(y, \{z, x\}) + f(z, \{x, y\}) + x \triangleright f(y, z) + y \triangleright f(z, x)$
  $+ z \triangleright f(x, y) = 0$

In this case, the associated unified product $\mathfrak{g} \natural_{\Omega(\mathfrak{g},V)} V = \mathfrak{g} \#_{\triangleleft}^{f} V$ is the *crossed product* of the Lie algebras $\mathfrak{g}$ and V. A system $(\mathfrak{g}, V, \triangleright, f)$ consisting of two Lie algebras $\mathfrak{g}$, V and two bilinear maps $\triangleright : V \times \mathfrak{g} \to \mathfrak{g}$, $f : V \times V \to \mathfrak{g}$ satisfying the above four compatibility conditions will be called a *crossed system of Lie algebras*. The crossed product associated to the crossed system $(\mathfrak{g}, V, \triangleright, f)$ is the Lie algebra defined as follow: $\mathfrak{g} \#_{\triangleleft}^{f} V = \mathfrak{g} \times V$ with the bracket given for any g, $h \in \mathfrak{g}$ and x, $y \in V$ by:

$$[(g, x), (h, y)] := \big([g, h] + x \triangleright h - y \triangleright g + f(x, y), \{x, y\}\big) \tag{3.16}$$

The crossed product of Lie algebras provides the answer to the following restricted version of the extending structures problem: *Let $\mathfrak{g}$ be a Lie algebra, and E a vector space containing $\mathfrak{g}$ as a subspace. Describe and classify all Lie algebra structures on E such that $\mathfrak{g}$ is an ideal of E.*

Indeed, let $(\mathfrak{g}, V, \triangleright, f)$ be a crossed system of two Lie algebras. Then, $\mathfrak{g} \cong \mathfrak{g} \times \{0\}$ is an ideal in the Lie algebra $\mathfrak{g} \#_{\triangleleft}^{f} V$ since $[(g, 0), (h, y)] := \big([g, h] - y \triangleright g, 0\big)$. Conversely, crossed products describe all Lie algebra structures on a vector space E such that a given Lie algebra $\mathfrak{g}$ is an ideal of E.

Corollary 3.3.1 *Let $\mathfrak{g}$ be a Lie algebra, E a vector space containing $\mathfrak{g}$ as a subspace. Then any Lie algebra structure on E that contains $\mathfrak{g}$ as an ideal is isomorphic to a crossed product of Lie algebras $\mathfrak{g}\#_{\triangleleft}^{f} V$.*

Remark 3.3.2 *We have proved Corollary 3.3.1 based on Theorem 3.1.4 by taking a linear retraction p of the canonical inclusion $i : \mathfrak{g} \hookrightarrow E$ and the crossed system arising from p. The classical proof of Corollary 3.3.1, given in the extension theory of Lie algebras, is completely different by the way the crossed system is constructed: since $\mathfrak{g}$ is an ideal of the Lie algebra E we can consider the quotient Lie algebra $\mathfrak{h} := E/\mathfrak{g}$. Let $\pi : E \to \mathfrak{h}$ be the canonical projection and $s : \mathfrak{h} \to E$ be a linear section of π. We define the action $\triangleright = \triangleright_s$ and the cocycle $f = f_s$ associated to s by the formulas:*

$$\triangleright : \mathfrak{h} \times \mathfrak{g} \to \mathfrak{g}, \qquad x \triangleright g := [s(x), g] \tag{3.17}$$

$$f : \mathfrak{h} \times \mathfrak{h} \to \mathfrak{g}, \qquad f(x, y) := [s(x), s(y)] - s([x, y]) \tag{3.18}$$

Then, $(\mathfrak{g}, \mathfrak{h}, \triangleright = \triangleright_s, f = f_s)$ is a crossed system of Lie algebras and the map $\psi : \mathfrak{g}\#_{\triangleleft}^{f} \mathfrak{h} \to E$, $\psi(g, x) := g + s(x)$ is an isomorphism of Lie algebras with the inverse $\psi^{-1}(z) := (z - s(\pi(z)), \pi(z))$, for all $z \in E$.

The restricted version of the extending structures problem is in fact an equivalent reformulation of the (non-abelian) extension problem. Indeed, first of all we remark that any crossed product $\mathfrak{g}\#_{\triangleleft}^{f} V$ is an extension of $\mathfrak{g}$ by V via the following sequence:

$$0 \longrightarrow \mathfrak{g} \xrightarrow{\ i\ } \mathfrak{g}\#_{\triangleleft}^{f} V \xrightarrow{\ \pi\ } V \longrightarrow 0$$

where $i : \mathfrak{g} \to \mathfrak{g}\#_{\triangleleft}^{f} V$, $i(g) := (g, 0)$ and $\pi : \mathfrak{g}\#_{\triangleleft}^{f} V \to V$, $\pi(g, x) := x$. Conversely, let $\mathfrak{E}$ be an extension of $\mathfrak{g}$ by $\mathfrak{h}$, that is, there exists an exact sequence of Lie algebras of the form:

$$0 \longrightarrow \mathfrak{g} \xrightarrow{\ i\ } \mathfrak{E} \xrightarrow{\ \pi\ } \mathfrak{h} \longrightarrow 0 \tag{3.19}$$

By identifying $\mathfrak{g} \cong \mathrm{Im(i)} = \mathrm{Ker}(\pi)$ we view $\mathfrak{g}$ as an ideal of $\mathfrak{E}$. Then, it follows from Corollary 3.3.1 that there exists a crossed system $(\mathfrak{g}, \mathfrak{h}, \triangleright, f)$ of Lie algebras such that $\mathfrak{g}\#_{\triangleleft}^{f} \mathfrak{h} \cong \mathfrak{E}$, an isomorphism of Lie algebras. Furthermore, using Theorem 3.1.4, the isomorphism can be chosen such that it stabilizes $\mathfrak{g}$ and co-stabilizes $\mathfrak{h}$.

The abelian case: Cotangent extending structures

Let E be a vector space, and $\mathfrak{g}$ a subspace of E with the abelian Lie algebra structure. A Lie algebra structure on E containing $\mathfrak{g}$ as an ideal is called a *cotangent extending structure*. The terminology is a generalization of the concept introduced in [193, Section 3.1]. Let V be a given complement of $\mathfrak{g}$ in E: it follows from Corollary 3.3.1 and Theorem 3.1.4 that the set of all cotangent extending structures of the abelian Lie algebra $\mathfrak{g}$ to the vector space E is parameterized by the set of all triples $(\triangleright, f, \{-, -\})$, such that $(V, \{-, -\})$ is a Lie algebra, $(\mathfrak{g}, \triangleright)$ is a left V-module and $f : V \times V \to \mathfrak{g}$ is a bilinear map such that $f(x, x) = 0$ and

$$f(x, \{y, z\}) + f(y, \{z, x\}) + f(z, \{x, y\}) + x \triangleright f(y, z) + y \triangleright f(z, x) + z \triangleright f(x, y) = 0$$

for all $x, y, z \in V$. For such a triple $(\triangleright, f, \{-, -\})$, the bracket of the cotangent extending structure on $E \cong \mathfrak{g} \times V$ is given by:

$$[(g, x), (h, y)] := (x \triangleright h - y \triangleright g + f(x, y), \{x, y\}) \tag{3.20}$$

for all $g, h \in \mathfrak{g}$ and $x, y \in V$. Moreover, any cotangent bracket on E has the form (3.20).

Skew crossed products of Lie algebras

Consider the bilinear map $\rightharpoonup: V \times \mathfrak{g} \to \mathfrak{g}$ to be trivial, i.e., $x \rightharpoonup g = 0$, for all $x \in V$ and $g \in \mathfrak{g}$. Then $\Lambda(\mathfrak{g}, V) = \big(\leftharpoonup, \rightharpoonup := 0, \theta, \{-, -\}\big)$ is a Lie extending system of $\mathfrak{g}$ through V if and only if the following compatibility conditions hold for any $a \in \mathfrak{g}$, $x, y, z \in V$:

(T1) $(V, \leftharpoonup)$ is a right Lie $\mathfrak{g}$-module, $\theta(x, x) = 0$ and $\{x, x\} = 0$

(T2) $\{x, y\} \leftharpoonup a = \{x, y \leftharpoonup a\} + \{x \leftharpoonup a, y\}$

(T3) $[\theta(x, y), a] = \theta(x, y \leftharpoonup a) + \theta(x \leftharpoonup a, y)$

(T4) $\sum_{(c)} \theta(x, \{y, z\}) = 0$

(T5) $\sum_{(c)} \{x, \{y, z\}\} + \sum_{(c)} x \leftharpoonup \theta(y, z) = 0$

In this case the trivial map $\rightharpoonup$ will be omitted when writing down the Lie extending system $\Lambda(\mathfrak{g}, V)$. The associated unified product $\mathfrak{g} \natural V$ will be denoted by $\mathfrak{g} \#^{\bullet} V$ and we will call it the skew crossed product associated to the system $\Lambda(\mathfrak{g}, V) = \big(\leftharpoonup, \theta, \{-, -\}\big)$ satisfying (T1)–(T5). Thus, $\mathfrak{g} \#^{\bullet} V$ is the vector space $\mathfrak{g} \times V$ with the Lie bracket $[-, -]$ defined for any $a, b \in \mathfrak{g}$ and $x, y \in V$ by:

$$[(a, x), (b, y)] := \big([a, b] + \theta(x, y), \{x, y\} + x \leftharpoonup b - y \leftharpoonup a\big) \tag{3.21}$$

An explicit example of a skew crossed product is given in Example 3.6.7 where we write $\mathfrak{sl}(2, k)$ as a skew crossed product $k \#^{\bullet} k^2$ between the abelian Lie algebras of dimensions one and two, associated to a certain right action $\leftharpoonup$ and a cocycle θ.

Moreover, if the cocycle θ of a Lie extending structure $\Lambda(\mathfrak{g}, V) = \big(\leftharpoonup, \theta, \{-, -\}\big)$ is also the trivial map, then the skew crossed product $\mathfrak{g} \#^{\bullet} V$ is just the usual semidirect product $\mathfrak{g} \rtimes V$ of two Lie algebras written in the right side convention. We point out that in our notational convention the Lie algebra $\mathfrak{g} \rtimes V$ contains $V \cong \{0\} \times V$ as an ideal.

Another important special case of the unified product, namely the bicrossed product of Lie algebras, was first introduced in [172, Theorem 4.1] and independently in [170, Theorem 3.9] and it answers the factorization problem.

Bicrossed products of Lie algebras and the factorization problem

As mentioned in the previous chapter, the bicrossed product arises as a special case of the unified product. Indeed, let $\Omega(\mathfrak{g}, V) = \big(\triangleleft, \triangleright, f, \{-, -\}\big)$ be an extending datum of $\mathfrak{g}$ through V such that f is the trivial map, i.e., $f(x, y) = 0$, for all $x, y \in V$. Then, $\Omega(\mathfrak{g}, V) = \big(\triangleleft, \triangleright, \{-, -\}\big)$ is a Lie extending structure of $\mathfrak{g}$ through V if and only if $(V, \{-,-\})$ is a Lie algebra and $(\mathfrak{g}, V, \triangleleft, \triangleright)$ is a *matched pair of Lie algebras* as defined in [172, Theorem 4.1] and independently in [170, Theorem 3.9]: i.e., $\mathfrak{g}$ is a left V-module under $\triangleright : V \otimes \mathfrak{g} \to \mathfrak{g}$, V is a right $\mathfrak{g}$-module under $\triangleleft : V \otimes \mathfrak{g} \to V$ and the following compatibilities hold for all $g, h \in \mathfrak{g}$, $x, y \in V$:

$$x \triangleright [g, h] = [x \triangleright g, h] + [g, x \triangleright h] + (x \triangleleft g) \triangleright h - (x \triangleleft h) \triangleright g \quad (3.22)$$

$$\{x, y\} \triangleleft g = \{x, y \triangleleft g\} + \{x \triangleleft g, y\} + x \triangleleft (y \triangleright g) - y \triangleleft (x \triangleright g) \quad (3.23)$$

In this case, the associated unified product $\mathfrak{g} \natural_{\Omega(\mathfrak{g},V)} V = \mathfrak{g} \bowtie V$ is precisely the *bicrossed product* of the matched pair $(\mathfrak{g}, V, \triangleleft, \triangleright)$ of Lie algebras; more precisely, $\mathfrak{g} \bowtie \mathfrak{h} := \mathfrak{g} \times \mathfrak{h}$, as a vector space, is a Lie algebra with the bracket defined by

$$\{(g, x), (h, y)\} := \Big([g, h] + x \triangleright h - y \triangleright g, [x, y] + x \triangleleft h - y \triangleleft g\Big) \quad (3.24)$$

for all $g, h \in \mathfrak{g}$ and $x, y \in \mathfrak{h}$. Throughout we adopt the name bicrossed product established in group theory [224] and Hopf algebra theory [146]. Other names used in the literature for the above product are: *bicrossproduct* in [172, Theorem 4.1], *double cross sum* in [174, Proposition 8.3.2], *double Lie algebra* [170, Definition 3.3], or *knit product* in [184]. Important examples of bicrossed products of Lie algebras are Manin's triples [170, Definition 1.13 and Theorem 1.12].

Any bicrossed product $\mathfrak{g} \bowtie \mathfrak{h}$ factorizes through $\mathfrak{g} = \mathfrak{g} \times \{0\}$ and $\mathfrak{h} = \{0\} \times \mathfrak{h}$; the converse also holds [174, Proposition 8.3.2]: if a Lie algebra $\mathfrak{L}$ factorizes through $\mathfrak{g}$ and $\mathfrak{h}$, then there exists an isomorphism of Lie algebras $\mathfrak{L} \cong \mathfrak{g} \bowtie \mathfrak{h}$, where $\mathfrak{g} \bowtie \mathfrak{h}$ is the bicrossed product associated to the matched pair $(\mathfrak{g}, \mathfrak{h}, \triangleleft, \triangleright)$ whose actions are constructed from the unique decomposition:

$$[x, g] = x \triangleright g + x \triangleleft g \in \mathfrak{g} + \mathfrak{h} \quad (3.25)$$

for all $x \in \mathfrak{h}$ and $g \in \mathfrak{g}$. The matched pair $(\mathfrak{g}, \mathfrak{h}, \triangleleft, \triangleright)$ defined by (3.25) is called the *canonical matched pair* associated to the factorization $\mathfrak{L} = \mathfrak{g} + \mathfrak{h}$. Thus, the factorization problem can be restated in a purely computational manner: *Let $\mathfrak{g}$ and $\mathfrak{h}$ be two given Lie algebras. Describe the set of all matched pairs $(\mathfrak{g}, \mathfrak{h}, \triangleleft, \triangleright)$ and classify up to an isomorphism all bicrossed products $\mathfrak{g} \bowtie \mathfrak{h}$.*

In the case that k is algebraically closed of characteristic zero and Ξ is a finite-dimensional Lie algebra then the famous Levi-Malcev theorem [55, Theorem 5] proves that there exists a Lie subalgebra $\mathfrak{h}$ of Ξ, called a Levi subalgebra, such that Ξ factorizes through $\mathrm{Rad}(\Xi)$ and $\mathfrak{h}$, where $\mathrm{Rad}(\Xi)$ is the radical of Ξ. Thus, any finite-dimensional Lie algebra Ξ is isomorphic to a bicrossed product between $\mathrm{Rad}(\Xi)$ and a semi-simple Lie algebra $\mathfrak{h} \cong \Xi/\mathrm{Rad}(\Xi)$.

Remark 3.3.3 *An interesting equivalent description for the factorization of a Lie algebra Ξ through two Lie subalgebras is proved in [34, Proposition 2.2]. A linear map $f : \Xi \to \Xi$ is called a* complex product structure *on Ξ [35, Definition 2.1] if $f \neq \pm\mathrm{Id}$, $f^2 = f$ and f is* integrable, *that is for any x, $y \in \Xi$ we have:*

$$f([x, y]) = [f(x), y] + [x, f(y)] - f\big([f(x), f(y)]\big)$$

If the characteristic of k is $\neq 2$, then the linear map $f : \mathfrak{g} \bowtie \mathfrak{h} \to \mathfrak{g} \bowtie \mathfrak{h}$, $f(g,h) := (g, -h)$, for all $g \in \mathfrak{g}$ and $h \in \mathfrak{h}$ is a complex product structure on any bicrossed product $\mathfrak{g} \bowtie \mathfrak{h}$ of Lie algebras. Conversely, if f is a complex product structure on Ξ, then Ξ factorizes through two Lie subalgebras $\Xi = \Xi_+ + \Xi_+$, where $\Xi_\pm$ denote the eigenspaces corresponding to the eigenvalue ± 1 of f [34, Proposition 2.2].

If a Lie algebra $\mathfrak{L}$ *factorizes* through $\mathfrak{g}$ and $\mathfrak{h}$, $\mathfrak{h}$ is called a *complement* of $\mathfrak{g}$ in $\mathfrak{L}$; if $\mathfrak{g}$ is an ideal of $\mathfrak{L}$, then a complement $\mathfrak{h}$, if it exists, is unique being isomorphic to the quotient Lie algebra $\mathfrak{L}/\mathfrak{g}$. In general, if $\mathfrak{g}$ is only a subalgebra of $\mathfrak{L}$, then we are very far from having unique complements; for a given extension $\mathfrak{g} \subset \mathfrak{L}$ of Lie algebras, the number of types of isomorphisms of all complements of $\mathfrak{g}$ in $\mathfrak{L}$ is called the *factorization index* of $\mathfrak{g}$ in $\mathfrak{L}$ and is denoted by $[\mathfrak{L} : \mathfrak{g}]^f$—a theoretical formula for computing $[\mathfrak{L} : \mathfrak{g}]^f$ is given in [15, Theorem 4.5].

Let $(\mathfrak{g}, \mathfrak{h}, \triangleleft, \triangleright)$ be a matched pair of Lie algebras. A linear map $r : \mathfrak{h} \to \mathfrak{g}$ is called a *deformation map* [15, Definition 4.1] of the matched pair $(\mathfrak{g}, \mathfrak{h}, \triangleright, \triangleleft)$ if the following compatibility holds for any x, $y \in \mathfrak{h}$:

$$r\big([x, y]\big) - \big[r(x), r(y)\big] = r\big(y \triangleleft r(x) - x \triangleleft r(y)\big) + x \triangleright r(y) - y \triangleright r(x) \quad (3.26)$$

We denote by $\mathcal{DM}(\mathfrak{h}, \mathfrak{g} \,|\, (\triangleright, \triangleleft))$ the set of all deformation maps of the matched pair $(\mathfrak{g}, \mathfrak{h}, \triangleright, \triangleleft)$. If $r \in \mathcal{DM}(\mathfrak{h}, \mathfrak{g} \,|\, (\triangleright, \triangleleft))$ then $\mathfrak{h}_r := \mathfrak{h}$, as a vector space, with the new bracket defined for any x, $y \in \mathfrak{h}$ by:

$$[x, y]_r := [x, y] + x \triangleleft r(y) - y \triangleleft r(x) \quad (3.27)$$

is a Lie algebra called the *r-deformation* of $\mathfrak{h}$. A Lie algebra $\overline{\mathfrak{h}}$ is a complement of $\mathfrak{g} \cong \mathfrak{g} \times \{0\}$ in the bicrossed product $\mathfrak{g} \bowtie \mathfrak{h}$ if and only if $\overline{\mathfrak{h}} \cong \mathfrak{h}_r$, for some deformation map $r \in \mathcal{DM}(\mathfrak{h}, \mathfrak{g} \,|\, (\triangleright, \triangleleft))$ ([15, Theorem 4.3]).

3.4 Matched pair deformations and the factorization index for Lie algebras: The case of perfect Lie algebras

Computing all matched pairs between two given Lie algebras $\mathfrak{g}$ and $\mathfrak{h}$ and classifying all associated bicrossed products $\mathfrak{g} \bowtie \mathfrak{h}$ is a challenging problem. In the case when $\mathfrak{g} := k = k_0$, the abelian Lie algebra of dimension 1, they are parameterized by the set $\mathrm{TwDer}(\mathfrak{h})$ of all twisted derivations of the Lie algebra $\mathfrak{h}$ as defined in Definition 3.2.2. As a special case of Proposition 3.2.4 and Remark 3.2.5 we have:

Proposition 3.4.1 *Let $\mathfrak{h}$ be a Lie algebra. Then there exists a bijection between the set of all matched pairs $(k_0, \mathfrak{h}, \triangleleft, \triangleright)$ and the space $\mathrm{TwDer}(\mathfrak{h})$ of all twisted derivations of $\mathfrak{h}$ given such that the matched pair $(k_0, \mathfrak{h}, \triangleleft, \triangleright)$ corresponding to $(\lambda, \Delta) \in \mathrm{TwDer}(\mathfrak{h})$ is defined by:*

$$h \triangleright a = a\, \lambda(h), \quad h \triangleleft a = a\, \Delta(h) \tag{3.28}$$

for all $h \in \mathfrak{h}$ and $a \in k = k_0$. The bicrossed product $k_0 \bowtie \mathfrak{h}$ associated to the matched pair (3.28) is denoted by $\mathfrak{h}_{(\lambda,\, \Delta)}$ and has the bracket given for any a, $b \in k$ and x, $y \in \mathfrak{h}$ by:

$$\{(a, x),\, (b, y)\} := \Big(b\, \lambda(x) - a\, \lambda(y),\, [x,\, y] + b\, \Delta(x) - a\, \Delta(y)\Big) \tag{3.29}$$

A Lie algebra $\mathfrak{L}$ contains $\mathfrak{h}$ as a subalgebra of codimension 1 if and only if $\mathfrak{L}$ is isomorphic to $\mathfrak{h}_{(\lambda,\, \Delta)}$, for some $(\lambda, \Delta) \in \mathrm{TwDer}(\mathfrak{h})$.

Suppose $\{e_i \,|\, i \in I\}$ is a basis for the Lie algebra $\mathfrak{h}$. Then, $\mathfrak{h}_{(\lambda,\, \Delta)}$ has $\{F,\, e_i \,|\, i \in I\}$ as a basis and the bracket given for any $i \in I$ by

$$[e_i,\, F] = \lambda(e_i)\, F + \Delta(e_i), \quad [e_i,\, e_j] = [e_i,\, e_j]_{\mathfrak{h}} \tag{3.30}$$

where $[-,\, -]_{\mathfrak{h}}$ is the bracket on $\mathfrak{h}$. Above we identify $e_i = (0, e_i)$ and denote $F = (1, 0)$ in the bicrossed product $k_0 \bowtie \mathfrak{h}$. Classifying the Lie algebras $\mathfrak{h}_{(\lambda,\, \Delta)}$ is a difficult task. In what follows we deal with this problem for a perfect Lie algebra $\mathfrak{h}$: in this case $\mathrm{TwDer}(\mathfrak{h}) = \{0\} \times \mathrm{Der}(\mathfrak{h})$ and we denote by $\mathfrak{h}_{(\Delta)} = \mathfrak{h}_{(0,\, \Delta)}$, for any $\Delta \in \mathrm{Der}(\mathfrak{h})$.

Theorem 3.4.2 *Let $\mathfrak{h}$ be a perfect Lie algebra and Δ, $\Delta' \in \mathrm{Der}(\mathfrak{h})$. Then there exists a bijection between the set of all morphisms of Lie algebras $\varphi : \mathfrak{h}_{(\Delta)} \to \mathfrak{h}_{(\Delta')}$ and the set of all triples $(\alpha,\, h,\, v) \in k \times \mathfrak{h} \times \mathrm{Hom}_{\mathrm{Lie}}(\mathfrak{h},\, \mathfrak{h})$ satisfying the following compatibility condition for all $x \in \mathfrak{h}$:*

$$v\big(\Delta(x)\big) - \alpha\, \Delta'\big(v(x)\big) = [v(x),\, h] \tag{3.31}$$

The bijection is given such that the Lie algebra map $\varphi = \varphi_{(\alpha,\, h,\, v)}$ corresponding to $(\alpha,\, h,\, v)$ is given by the formula

$$\varphi : \mathfrak{h}_{(\Delta)} \to \mathfrak{h}_{(\Delta')}, \quad \varphi(a, x) = (a\, \alpha,\, a\, h + v(x)) \tag{3.32}$$

for all $(a, x) \in \mathfrak{h}_{(\Delta)} = k_0 \bowtie \mathfrak{h}$. *Furthermore,* $\varphi = \varphi_{(\alpha, h, v)}$ *is an isomorphism of Lie algebras if and only if* $\alpha \neq 0$ *and* $v \in \mathrm{Aut}_{\mathrm{Lie}}(\mathfrak{h})$.

Proof: Any linear map $\varphi : k \times \mathfrak{h} \to k \times \mathfrak{h}$ is uniquely determined by a quadruple (α, h, β, v), where $\alpha \in k$, $h \in \mathfrak{h}$ and $\beta : \mathfrak{h} \to k$, $v : \mathfrak{h} \to \mathfrak{h}$ are k-linear maps such that:

$$\varphi(a, x) = \varphi_{(\alpha, h, \beta, v)} = (a\,\alpha + \beta(x),\ a\,h + v(x))$$

We will prove that φ defined above is a Lie algebra map if and only if β is the trivial map, v is a Lie algebra map and (3.31) holds. It is enough to test the compatibility:

$$\varphi\big([(a, x),\, (b, y)]\big) = [\varphi(a, x),\, \varphi(b, y)] \tag{3.33}$$

for all generators of $\mathfrak{h}_{(\Delta)} = k \times \mathfrak{h}$, i.e., elements of the form $(1, 0)$ and $(0, x)$, for all $x \in \mathfrak{h}$. Moreover, since $\mathfrak{h}$ is perfect (i.e., $\lambda = 0$) the bracket on $\mathfrak{h}_{(\Delta)}$ given by (3.29) takes the form: $\{(a, x),\, (b, y)\} = \Big(0,\ [x,\, y] + b\,\Delta(x) - a\,\Delta(y)\Big)$. Using this formula we obtain that (3.33) holds for $(0, x)$ and $(0, y)$ if and only if

$$\beta\big([x,\, y]\big) = 0, \quad v\big([x,\, y]\big) = [v(x),\, v(y)] + \beta(y)\Delta(v(x)) - \beta(x)\Delta(v(y))$$

As $\mathfrak{h}$ is perfect, these two conditions are equivalent to the fact that $\beta = 0$ and v is a Lie algebra map. Finally, as $\beta = 0$, we can easily show that (3.33) holds in $(1, 0)$ and $(0, x)$ if and only if (3.31) holds. Thus, we have obtained that φ is a Lie algebra map if and only if v is a Lie algebra map, $\beta = 0$ and (3.31) holds. In what follows we denote by $\varphi_{(\alpha, h, v)}$ the Lie algebra map corresponding to a quadruple (α, h, β, v) with $\beta = 0$. Suppose first that $\varphi := \varphi_{(\alpha, h, v)}$ is a Lie algebra isomorphism. Then, there exists a Lie algebra map $\overline{\varphi} := \varphi_{(\gamma, g, w)} : \mathfrak{h}_{(\Delta')} \to \mathfrak{h}_{(\Delta)}$ such that $\varphi \circ \overline{\varphi}(a, x) = \overline{\varphi} \circ \varphi(a, x) = (a, x)$ for all $a \in k$, $x \in \mathfrak{h}$. Thus, for all $a \in k$ and $x \in \mathfrak{h}$, we have:

$$a\alpha\gamma = a, \qquad a\gamma + v(ag) + v\big(w(x)\big) = x = a\alpha g + w(ah) + w\big(v(x)\big) \tag{3.34}$$

By the first part of (3.34) for $a = 1$ we obtain $\alpha\gamma = 1$ and thus $\alpha \neq 0$ while the second part of (3.34) for $a = 0$ implies v bijective. To end with, assume that $\alpha \neq 0$ and $v \in \mathrm{Aut}_{\mathrm{Lie}}(\mathfrak{h})$. Then, it is straightforward to see that $\varphi = \varphi_{(\alpha, h, v)}$ is an isomorphism with the inverse given by $\varphi^{-1} := \varphi_{(\alpha^{-1},\, -\alpha^{-1}v^{-1}(h),\, v^{-1})}$. □

Let k^* be the units group of k and $(\mathfrak{h}, +)$ the underlying abelian group of the Lie algebra $\mathfrak{h}$. Then the map given for any $\alpha \in k^*$, $v \in \mathrm{Aut}_{\mathrm{Lie}}(\mathfrak{h})$ and $h \in \mathfrak{h}$ by:

$$\varphi : k^* \times \mathrm{Aut}_{\mathrm{Lie}}(\mathfrak{h}) \to \mathrm{Aut}_{\mathrm{Gr}}\,(\mathfrak{h}, +), \qquad \varphi(\alpha,\, v)\,(h) := \alpha^{-1}\, v(h)$$

is a morphism of groups. Thus, we can construct the semidirect product of groups $\mathfrak{h} \ltimes_{\varphi} \big(k^* \times \mathrm{Aut}_{\mathrm{Lie}}(\mathfrak{h})\big)$ associated to φ. The next result shows that $\mathrm{Aut}_{\mathrm{Lie}}(\mathfrak{h}_{(\Delta)})$ is isomorphic to a certain subgroup of the semidirect product of groups $\mathfrak{h} \ltimes_{\varphi} \big(k^* \times \mathrm{Aut}_{\mathrm{Lie}}(\mathfrak{h})\big)$.

Corollary 3.4.3 *Let $\mathfrak{h}$ be a perfect Lie algebra and Δ, $\Delta' \in \mathrm{Der}(\mathfrak{h})$. Then the Lie algebras $\mathfrak{h}_{(\Delta)}$ and $\mathfrak{h}_{(\Delta')}$ are isomorphic if and only if there exists a triple $(\alpha, h, v) \in k^* \times \mathfrak{h} \times \mathrm{Aut}_{\mathrm{Lie}}(\mathfrak{h})$ such that $v \circ \Delta - \alpha \Delta' \circ v = [v(-), h]$. Furthermore, there exists an isomorphism of groups*

$$\mathrm{Aut}_{\mathrm{Lie}}(\mathfrak{h}_{(\Delta)}) \cong \mathcal{G}(\mathfrak{h}, \Delta) := \{ (\alpha, h, v) \in k^* \times \mathfrak{h} \times \mathrm{Aut}_{\mathrm{Lie}}(\mathfrak{h}) \mid v \circ \Delta - \alpha \Delta \circ v = [v(-), h] \}$$

where $\mathcal{G}(\mathfrak{h}, \Delta)$ is a group with respect to the following multiplication:

$$(\alpha, h, v) \cdot (\beta, g, w) := (\alpha\beta, \beta h + v(g), v \circ w) \tag{3.35}$$

for all (α, h, v), $(\beta, g, w) \in\in \mathcal{G}(\mathfrak{h}, \Delta)$. Moreover, the canonical map

$$\mathcal{G}(\mathfrak{h}, \Delta) \longrightarrow \mathfrak{h} \ltimes_{\varphi} (k^* \times \mathrm{Aut}_{\mathrm{Lie}}(\mathfrak{h})), \qquad (\alpha, h, v) \mapsto (\alpha^{-1} h, (\alpha, v))$$

in an injective morphism of groups.

Proof: The first part follows trivially from Theorem 3.4.2. Consider now $\gamma, \psi \in \mathrm{Aut}_{\mathrm{Lie}}(\mathfrak{h}_{(\Delta)})$. Using again Theorem 3.4.2, we can find (α, h, v), $(\beta, g, w) \in k^* \times \mathfrak{h} \times \mathrm{Aut}_{\mathrm{Lie}}(\mathfrak{h})$ such that $\gamma = \varphi_{(\alpha, h, v)}$ and $\psi = \varphi_{(\beta, g, w)}$. Then, for all $a \in k$, $x \in \mathfrak{h}$ we have:

$$\begin{aligned} \varphi_{(\alpha, h, v)} \circ \varphi_{(\beta, g, w)}(a, x) &= \varphi_{(\alpha, h, v)}\big(a\,\beta,\, ag + w(x)\big) \\ &= \big(\alpha\,\beta\, a,\, a\beta\, h + av(g) + v \circ w(x)\big) \\ &= \varphi_{(\alpha\,\beta,\, \beta\, h + v(g),\, v \circ w)}(a, x) \end{aligned}$$

Thus, $\mathrm{Aut}_{\mathrm{Lie}}(\mathfrak{h}_{(\Delta)})$ is isomorphic to $\mathcal{G}(\mathfrak{h}, \Delta)$ with the multiplication given by (3.35). The last assertion follows by a routine computation. □

Remark 3.4.4 *Let $\Delta = [x_0, -]$ be an inner derivation of a perfect Lie algebra $\mathfrak{h}$. Then the group $\mathrm{Aut}_{\mathrm{Lie}}(\mathfrak{h}_{([x_0, -])})$ admits a simpler description as follows:*

$$\mathcal{G}(\mathfrak{h}, [x_0, -]) = \{ (\alpha, h, v) \in k^* \times \mathfrak{h} \times \mathrm{Aut}_{\mathrm{Lie}}(\mathfrak{h}) \mid v(x_0) - \alpha\, x_0 + h \in \mathrm{Z}(\mathfrak{h}) \}$$

where $\mathrm{Z}(\mathfrak{h})$ is the center of $\mathfrak{h}$. Assume in addition that $\mathfrak{h}$ has trivial center, i.e., $\mathrm{Z}(\mathfrak{h}) = \{0\}$; it follows that there exists an isomorphism of groups

$$\mathrm{Aut}_{\mathrm{Lie}}(\mathfrak{h}_{([x_0, -])}) \cong k^* \times \mathrm{Aut}_{\mathrm{Lie}}(\mathfrak{h})$$

since in this case any element h from a triple $(\alpha, h, v) \in \mathcal{G}(\mathfrak{h}, [x_0, -])$ must be equal to $\alpha\, x_0 - v(x_0)$. Moreover, in this context, the multiplication given by (3.35) is precisely that of a direct product of groups.

For sympathetic Lie algebras, Theorem 3.4.2 takes the following form which considerably improves [19, Corollary 4.10], where the classification is made only up to an isomorphism of Lie algebras which acts as identity on $\mathfrak{h}$.

Corollary 3.4.5 *Let $\mathfrak{h}$ be a sympathetic Lie algebra. Then up to an isomorphism of Lie algebras there exists only one Lie algebra that contains $\mathfrak{h}$ as a Lie subalgebra of codimension one, namely the direct product $k_0 \times \mathfrak{h}$ of Lie algebras. Furthermore, there exists an isomorphism of groups $\mathrm{Aut}_{\mathrm{Lie}}(k_0 \times \mathfrak{h}) \cong k^* \times \mathrm{Aut}_{\mathrm{Lie}}(\mathfrak{h})$.*

Proof: Since $\mathfrak{h}$ is perfect, any Lie algebra that contains $\mathfrak{h}$ as a Lie subalgebra of codimension 1 is isomorphic to $\mathfrak{h}_{(D)}$, for some $D \in \mathrm{Der}(\mathfrak{h})$. As $\mathfrak{h}$ is also complete, any derivation is inner. For an arbitrary derivation $D = [d, -]$ we can prove that $\mathfrak{h}_{(D)} \cong \mathfrak{h}_{(0)}$, where $0 = [0, -]$ is the trivial derivation and moreover $\mathfrak{h}_{(0)}$ is just the direct product of Lie algebras $k_0 \times \mathfrak{h}$. Indeed, by taking $(\alpha, h, v) := (1, -d, \mathrm{Id}_{\mathfrak{h}})$ one can see that relation (3.31) holds for $D = [d, -]$ and $D' = [0, -]$, that is $\mathfrak{h}_{(D)} \cong \mathfrak{h}_{(0)}$. The final part follows from Remark 3.4.4. □

Remark 3.4.6 *Let $\mathfrak{h}$ be a perfect Lie algebra with a basis $\{e_i \,|\, i \in I\}$, $\Delta \in \mathrm{Der}(\mathfrak{h})$ a given derivation and consider the extension $k_0 \subseteq \mathfrak{h}_{(\Delta)} = k_0 \bowtie \mathfrak{h}_{(\Delta)}$. In order to determine all complements of k_0 in $\mathfrak{h}_{(\Delta)}$ we have to describe the set of all deformation maps $r : \mathfrak{h} \to k_0$ of the matched pair (3.28). A deformation map is completely determined by a family of scalars $(a)_{i\in I}$ satisfying the following compatibility condition for any $i, j \in I$:*

$$r([e_i, e_j]_{\mathfrak{h}}) = r\big(a_i\, \Delta(e_j) - a_j\, \Delta(e_i)\big)$$

via the relation $r(e_i) = a_i$. For such an $r = (a_i)_{i\in I}$, the r-deformation of $\mathfrak{h}$ is the Lie algebra $\mathfrak{h}_r$ having $\{e_i \,|\, i \in I\}$ as a basis and the bracket defined for any $i, j \in I$ by:

$$[e_i, e_j]_r = [e_i, e_j]_{\mathfrak{h}} + a_j\, \Delta(e_i) - a_i\, \Delta(e_j)$$

Any complement of k_0 in $\mathfrak{h}_{(\Delta)}$ is isomorphic to such an $\mathfrak{h}_r$. An explicit example in dimension 5 is given below.

Example 3.4.7 *Let k be a field of characteristic $\neq 2$ and $\mathfrak{h}$ the perfect 5-dimensional Lie algebra with a basis $\{e_1, e_2, e_3, e_4, e_5\}$ and bracket given by:*

$$[e_1, e_2] = e_3, \quad [e_1, e_3] = -2e_1, \quad [e_1, e_5] = [e_3, e_4] = e_4$$
$$[e_2, e_3] = 2e_2, \quad [e_2, e_4] = e_5, \quad [e_3, e_5] = -e_5$$

By a straightforward computation it can be proved that the space of derivations $\mathrm{Der}(\mathfrak{h})$ coincides with the space of all matrices from $\mathcal{M}_5(k)$ of the form:

$$A = \begin{pmatrix} a_1 & 0 & -2a_4 & 0 & 0 \\ 0 & -a_1 & -2a_2 & 0 & 0 \\ a_2 & a_4 & 0 & 0 & 0 \\ a_3 & 0 & a_5 & a_6 & a_4 \\ 0 & a_5 & -a_3 & -a_2 & (a_6 - a_1) \end{pmatrix}$$

for all $a_1, \cdots, a_6 \in k$. *Thus* $\mathfrak{h}$ *is not complete since* $\mathrm{Der}(\mathfrak{h})$ *has dimension* 6. *One can show easily that the derivation* $\Delta := e_{11} - e_{41} - e_{22} + e_{53} - e_{44} - 2\, e_{55}$ *is not inner, where* $e_{i\,j} \in \mathcal{M}_n(k)$ *is the matrix having* 1 *in the* $(i, j)^{th}$ *position and zeros elsewhere. For the derivation* Δ *we consider the extension* $k_0 \subseteq k_0 \bowtie \mathfrak{h} = \mathfrak{h}_{(\Delta)}$ *and we will describe all the complements of* k_0 *in* $\mathfrak{h}_{(\Delta)}$. *By a routine computation it can be seen that* $r : \mathfrak{h} \to k_0$ *is a deformation map of the matched pair (3.28) if and only if* $r := 0$ *(the trivial map) or* r *is given by*

$$r(e_1) := a, \quad r(e_2) := -a^{-1}, \quad r(e_3) = 2, \quad r(e_4) = r(e_5) = 0$$

for some $a \in k^*$. *Thus a Lie algebra* $\mathfrak{C}$ *is a complement of* k_0 *in* $\mathfrak{h}_{(\Delta)}$ *if and only if* $\mathfrak{C} \cong \mathfrak{h}$ *or* $\mathfrak{C} \cong \mathfrak{h}_a$, *where* $\mathfrak{h}_a$ *is the* 5*-dimensional Lie algebra with basis* $\{e_1, e_2, e_3, e_4, e_5\}$ *and bracket given by:*

$$\begin{aligned}
&[e_1,\, e_2]_a := -a^{-1}\, e_1 + a\, e_2 + e_3 + a^{-1}\, e_4, \quad [e_1,\, e_3]_a := -2\, e_4 - a\, e_5,\\
&[e_1,\, e_4]_a := a\, e_4\\
&[e_1,\, e_5]_a := e_4 + 2a\, e_5, \quad [e_2,\, e_3]_a := a^{-1}\, e_5, \quad [e_2,\, e_4]_a := e_5 - a^{-1}\, e_4\\
&[e_2,\, e_5]_a := -2a^{-1}\, e_5, \quad [e_3,\, e_4]_a := 3\, e_4, \quad [e_3,\, e_5]_a := 3\, e_5
\end{aligned}$$

for any $a \in k^*$. *Remark that none of the matched pair deformations* $\mathfrak{h}_a$ *of the Lie algebra* $\mathfrak{h}$ *is perfect since the dimension of the derived algebra* $[\mathfrak{h}_a,\, \mathfrak{h}_a]$ *is equal to* 3.

3.5 Matched pair deformations and the factorization index for Lie algebras: The case of non-perfect Lie algebras

In Section 3.4 we have described and classified all bicrossed products $k_0 \bowtie \mathfrak{h}$ for a perfect Lie algebra $\mathfrak{h}$; furthermore, Remark 3.4.6 and Example 3.4.7 describe all complements of k_0 in a given bicrossed product $k_0 \bowtie \mathfrak{h}$. In this section we approach the same questions for a given non-perfect Lie algebra $\mathfrak{h} := \mathfrak{l}\,(2n+1, k)$, where $\mathfrak{l}\,(2n+1, k)$ is the $(2n+1)$-dimensional Lie algebra with basis $\{E_i,\, F_i,\, G \,|\, i = 1, \cdots, n\}$ and bracket given for any $i = 1, \cdots, n$ by:

$$[E_i,\, G] := E_i, \quad [G,\, F_i] := F_i$$

First, we shall describe all bicrossed products $k_0 \bowtie \mathfrak{l}\,(2n+1, k)$: they will explicitly describe all Lie algebras which contain $\mathfrak{l}\,(2n+1, k)$ as a subalgebra of codimension 1. Then, as the second step, we shall find all r-deformations of the Lie algebra $\mathfrak{l}\,(2n+1, k)$, for two given extensions $k_0 \subseteq k_0 \bowtie \mathfrak{l}\,(2n+1, k)$. Based on Proposition 3.4.1 we have to compute first the space $\mathrm{TwDer}\big(\mathfrak{l}\,(2n+1, k)\big)$ of all twisted derivations.

Proposition 3.5.1 *There exists a bijection between* $\mathrm{TwDer}\big(\mathfrak{l}\,(2n+1,k)\big)$ *and the set of all matrices* $(A,\, B,\, C,\, D,\, \lambda_0,\, \delta) \in \mathrm{M}_n(k)^4 \,\times k \times k^{2n+1}$ *satisfying the following conditions:*

$$\lambda_0\, A \;=\; -\,\delta_{2n+1}\, I_n, \quad (2+\lambda_0)\, B = 0, \quad (2-\lambda_0)\, C = 0, \quad \lambda_0 D = \delta_{2n+1} I_n \tag{3.36}$$

where $\delta = (\delta_1, \cdots, \delta_{2n+1}) \in k^{2n+1}$. *The bijection is given such that the twisted derivation* $(\lambda, \Delta) \in \mathrm{TwDer}\big(\mathfrak{l}\,(2n+1,k)\big)$ *associated to* $(A,\, B,\, C,\, D,\, \lambda_0,\, \delta)$ *is given by:*

$$\lambda(E_i) = \lambda(F_i) := 0, \quad \lambda(G) := \lambda_0 \tag{3.37}$$

$$\Delta := \begin{pmatrix} A & B & \delta_1 \\ C & D & : \\ 0 & 0 & \delta_{2n+1} \end{pmatrix} \tag{3.38}$$

$\mathcal{T}(n)$ *denotes the set of all* $(A,\, B,\, C,\, D,\, \lambda_0,\, \delta) \in \mathrm{M}_n(k)^4 \times k \times k^{2n+1}$ *satisfying (3.36).*

Proof: The first compatibility condition (3.10) shows that a linear map $\lambda : \mathfrak{l}\,(2n+1,k) \to k$ of a twisted derivation (λ, D) must have the form given by (3.37), for some $\lambda_0 \in k$. We shall fix such a map for a given $\lambda_0 \in k$. We write down the linear map $\Delta : \mathfrak{l}\,(2n+1,k) \to \mathfrak{l}\,(2n+1,k)$ as a matrix associated to the basis $\{E_1, \cdots, E_n, F_1, \cdots, F_n, G\}$ of $\mathfrak{l}\,(2n+1,k)$, as follows:

$$\Delta = \begin{pmatrix} A & B & d_{1,2n+1} \\ C & D & : \\ d_{2n+1,1} & .. & d_{2n+1,2n+1} \end{pmatrix}$$

for some matrices $A,\, B,\, C,\, D \in \mathrm{M}_n(k)$ and some scalars $d_{i,j} \in k$, for all i, $j = 1, \cdots, 2n+1$. We denote $A = (a_{ij})$, $B = (b_{ij})$, $C = (c_{ij})$, $D = (d_{ij})$. It remains to check the compatibility condition (3.10) for Δ, i.e., the following compatibility

$$\Delta([g,\, h]) = [\Delta(g),\, h] + [g,\, \Delta(h)] + \lambda(g)\Delta(h) - \lambda(h)\Delta(g)$$

for all $g \neq h \in \{E_1, \cdots, E_n, F_1, \cdots, F_n, G\}$. As this is a routine, straightforward computation we will only indicate the main steps of the proof. We can easily see that the compatibility condition (3.10) holds for $(g,\, h) = (E_i,\, E_j)$ if and only if $d_{2n+1,i} = 0$, for all $i = 1, \cdots n$. In the same way, (3.10) holds for $(g,\, h) = (F_i,\, F_j)$ if and only if $d_{2n+1,n+i} = 0$, for all $i = 1, \cdots n$. This shows that Δ has the form (3.38), that is the first $2n$ entries from the last row of the matrix Δ are all zeros and we will denote the last column of D by $(d_{1,2n+1}, \cdots, d_{2n+1,2n+1}) = \delta = (\delta_1, \cdots, \delta_{2n+1})$. It follows from here that (3.10) holds trivially for the pair $(g,\, h) = (E_i,\, F_j)$. An easy computation shows that (3.10) holds for $(g, h) = (E_i, G)$ if and only if the following equation holds

$$(1-\lambda_0)\Big(\sum_{j=1}^{n} a_{j,i}E_j + \sum_{j=1}^{n} c_{j,i}F_j\Big) = \sum_{j=1}^{n} a_{j,i}E_j - \sum_{j=1}^{n} c_{j,i}F_j + \delta_{2n+1}E_i$$

which is equivalent to $-\lambda_0 A = \delta_{2n+1} I_n$ and $(2-\lambda_0)C = 0$, i.e., the first and the third equations from (3.36). A similar computation shows that (3.10) holds for $(g, h) = (G, F_i)$ if and only if $(2+\lambda_0)B = 0$ and $\lambda_0 D = \delta_{2n+1} I_n$ and the proof is finished. □

Let $\mathfrak{l}(2n+1,k)_{(A,\,B,\,C,\,D,\,\lambda_0,\,\delta)}$ be the bicrossed product $k_0 \bowtie \mathfrak{l}(2n+1,k)$ associated to the matched pair given by the twisted derivation $\big(A = (a_{ji}),\, B = (b_{ji}),\, C = (c_{ji}),\, D = (d_{ji}),\, \lambda_0,\, \delta = (\delta_j)\big) \in \mathcal{T}(n)$. From now on we will use the following convention: if one of the elements of the 6-tuple $(A, B, C, D, \lambda_0, \delta)$ is equal to 0 then we will omit it when writing down the Lie algebra $\mathfrak{l}(2n+1,k)_{(A,\,B,\,C,\,D,\,\lambda_0,\,\delta)}$. A basis of $\mathfrak{l}(2n+1,k)_{(A,\,B,\,C,\,D,\,\lambda_0,\,\delta)}$ will be denoted by $\{E_i, F_i, G, H \mid i = 1, \cdots, n\}$: these Lie algebras can be explicitly described by first computing the set $\mathcal{T}(n)$ and then using Proposition 3.4.1. Considering the equations (3.36) which define $\mathcal{T}(n)$, a discussion involving the field k and the scalar λ_0 is mandatory. For two sets X and Y we shall denote by $X \sqcup Y$ the disjoint union of X and Y. As a conclusion of the above results we obtain:

Theorem 3.5.2 (1) *If k is a field such that* $\mathrm{char}(k) \neq 2$ *then*

$$\mathcal{T}(n) \cong \big((k \setminus \{0, \pm 2\}) \times k^{2n+1}\big) \sqcup \big(\mathrm{M}_n(k)^2 \times k^{2n}\big) \sqcup \big(\mathrm{M}_n(k) \times k^{2n+1}\big) \sqcup \big(\mathrm{M}_n(k) \times k^{2n+1}\big)$$

and the four families of Lie algebras containing $\mathfrak{l}(2n+1,k)$ as a subalgebra of codimension 1 are the following:

- *the Lie algebra $\mathfrak{l}^1(2n+1,k)_{(\lambda_0,\,\delta)}$ with the bracket given for any $i = 1, \cdots, n$ by:*

$$[E_i, G] = E_i, \quad [G, F_i] = F_i, \quad [E_i, H] = -\lambda_0^{-1}\,\delta_{2n+1}\,E_i,$$
$$[F_i, H] = \lambda_0^{-1}\,\delta_{2n+1}\,F_i,$$
$$[G, H] = \lambda_0\,H + \sum_{j=1}^{n} \delta_j\,E_j + \sum_{j=1}^{n} \delta_{n+j}\,F_j + \delta_{2n+1} G$$

for all $(\lambda_0, \delta) \in (k \setminus \{0, \pm 2\}) \times k^{2n+1}$.

- *the Lie algebra $\mathfrak{l}^2(2n+1,k)_{(A,\,D,\,\delta)}$ with the bracket given for any $i = 1, \cdots, n$ by:*

$$[E_i, G] = E_i, \quad [G, F_i] = F_i, \quad [E_i, H] = \sum_{j=1}^{n} a_{ji}\,E_j$$

$$[F_i, H] = \sum_{j=1}^{n} d_{ji}\,F_j, \quad [G, H] = \sum_{j=1}^{n} \delta_j\,E_j + \sum_{j=1}^{n} \delta_{n+j}\,F_j$$

for all $(A = (a_{ij}),\, D = (d_{ij}),\, \delta) \in \mathrm{M}_n(k) \times \mathrm{M}_n(k) \times k^{2n}$.

• *the Lie algebra* $\mathfrak{l}^3\,(2n+1, k)_{(C,\,\delta)}$ *with the bracket given for any* $i = 1, \cdots, n$ *by:*

$$[E_i,\, G] = E_i, \quad [G,\, F_i] = F_i, \quad [E_i,\, H] = -2^{-1}\delta_{2n+1}E_i + \sum_{j=1}^{n} c_{ji}\, F_j$$

$$[F_i,\, H] = 2^{-1}\delta_{2n+1}F_i, \quad [G,\, H] = 2\, H + \sum_{j=1}^{n} \delta_j\, E_j + \sum_{j=1}^{n} \delta_{n+j}\, F_j + \delta_{2n+1}G$$

for all $(C = (c_{ij}),\, \delta) \in \mathrm{M}_n(k) \times k^{2n+1}$.

• *the Lie algebra* $\mathfrak{l}^4\,(2n+1, k)_{(B,\,\delta)}$ *with the bracket given for any* $i = 1, \cdots, n$ *by:*

$$[E_i,\, G] = E_i, \quad [G,\, F_i] = F_i, \quad [F_i,\, H] = \sum_{j=1}^{n} b_{ji}\, E_j - 2^{-1}\delta_{2n+1}F_i$$

$$[E_i,\, H] = 2^{-1}\delta_{2n+1}E_i, \quad [G,\, H] = -2\, H + \sum_{j=1}^{n} \delta_j\, E_j + \sum_{j=1}^{n} \delta_{n+j}\, F_j + \delta_{2n+1}G$$

for all $(B = (b_{ij}),\, \delta) \in \mathrm{M}_n(k) \times k^{2n+1}$.

(2) *If* $\mathrm{char}(k) = 2$ *then*

$$\mathcal{T}(n) \cong \left(\mathrm{M}_n(k)^4 \times k^{2n}\right) \sqcup \left(k^* \times k^{2n+1}\right)$$

and the two families of Lie algebras containing $\mathfrak{l}\,(2n+1, k)$ *as a subalgebra of codimension* 1 *are the following:*

• *the Lie algebra* $\mathfrak{l}_1\,(2n+1, k)_{(A,\, B,\, C,\, D,\, \delta)}$ *with the bracket given for any* $i = 1, \cdots, n$ *by:*

$$[E_i,\, G] = E_i, \quad [G,\, F_i] = F_i, \quad [E_i,\, H] = \sum_{j=1}^{n} \left(a_{ji}\, E_j + c_{ji}\, F_j\right)$$

$$[F_i,\, H] = \sum_{j=1}^{n} \left(b_{ji}\, E_j + d_{ji}\, F_j\right), \quad [G,\, H] = \sum_{j=1}^{n} \delta_j\, E_j + \sum_{j=1}^{n} \delta_{n+j}\, F_j$$

for all $(A,\, B,\, C,\, D,\, \delta) \in \mathrm{M}_n(k)^4 \times k^{2n}$.

• *the Lie algebra* $\mathfrak{l}_2\,(2n+1, k)_{(\lambda_0,\, \delta)}$ *with the bracket given for any* $i = 1, \cdots, n$ *by:*

$$[E_i,\, G] = E_i, \quad [G,\, F_i] = F_i, \quad [E_i,\, H] = -\lambda_0^{-1}\delta_{2n+1}E_i$$

$$[F_i,\, H] = \lambda_0^{-1}\delta_{2n+1}F_i, \quad [G,\, H] = \lambda_0\, H + \sum_{j=1}^{n} \delta_j\, E_j + \sum_{j=1}^{n} \delta_{n+j}\, F_j + \delta_{2n+1}G$$

for all $(\lambda_0,\, \delta) \in k^* \times k^{2n+1}$.

Proof: The proof relies on the use of Proposition 3.4.1 and Proposition 3.5.1 as well as the equations (3.36) defining $\mathcal{T}(n)$. Besides the discussion on the characteristic of k, it is also necessary to consider whether λ_0 belongs to the set $\{0, 2, -2\}$. In the case that $\mathrm{char}(k) \neq 2$, the first Lie algebra listed is the bicrossed product which corresponds to the case when $\lambda_0 \notin \{0, 2, -2\}$. In this case, we can easily see that $\big(A, B, C, D, \lambda_0, \delta = (\delta_j)\big) \in \mathcal{T}(n)$ if and only if $B = C = 0$, $A = -\lambda_0^{-1}\delta_{2n+1}I_n$ and $D = \lambda_0^{-1}\delta_{2n+1}I_n$. The Lie algebra $\mathfrak{l}^1\,(2n+1, k)_{(\lambda_0,\,\delta)}$ is exactly the bicrossed product $k_0 \bowtie \mathfrak{l}\,(2n+1, k)$ corresponding to this twisted derivation. The Lie algebra $\mathfrak{l}^2\,(2n+1, k)_{(A,\,D,\,\delta)}$ is the bicrossed product $k_0 \bowtie \mathfrak{l}\,(2n+1, k)$ corresponding to the case $\lambda_0 = 0$ while the last two Lie algebras are the bicrossed products $k_0 \bowtie \mathfrak{l}\,(2n+1, k)$ associated to the case when $\lambda_0 = 2$ and respectively $\lambda_0 = -2$.

If the characteristic of k is equal to 2 we distinguish the following two possibilities: the Lie algebra $\mathfrak{l}_1\,(2n+1, k)_{(A,\,B,\,C,\,D,\,\delta)}$ is the bicrossed product $k_0 \bowtie \mathfrak{l}\,(2n+1, k)$ associated to $\lambda_0 = 0$ while the Lie algebra $\mathfrak{l}_2\,(2n+1, k)_{(\lambda_0,\,\delta)}$ is the same bicrossed product but associated to $\lambda_0 \neq 0$. □

Let k be a field of characteristic $\neq 2$ and $\mathfrak{l}^1\,(2n+1, k)_{(\lambda_0,\,\delta)}$ the Lie algebra of Theorem 3.5.2. In order to keep the computations efficient we will consider $\lambda_0 := 1$ and $\delta := (0, \cdots, 0, 1)$ and we denote by $L\,(2n+2, k) := \mathfrak{l}^1\,(2n+1, k)_{(1,\,(0,\cdots,0,1))}$, the $(2n+2)$-dimensional Lie algebra having a basis $\{E_i,\, F_i,\, G,\, H \,|\, i = 1, \cdots, n\}$ and the bracket defined for any $i = 1, \cdots, n$ by:

$$\begin{aligned} &[E_i,\, G] = E_i, \quad [G,\, F_i] = F_i, \quad [E_i,\, H] = -E_i, \quad [F_i,\, H] = F_i,\\ &[G,\, H] = H + G \end{aligned} \tag{3.39}$$

We consider the Lie algebra extension $kH \subset L\,(2n+2, k)$, where $kH \cong k_0$ is the abelian Lie algebra of dimension 1. Of course, $L\,(2n+2, k)$ factorizes through kH and $\mathfrak{l}\,(2n+1, k)$, i.e., $L\,(2n+2, k) = kH \bowtie \mathfrak{l}\,(2n+1, k)$—the actions $\triangleleft : \mathfrak{l}\,(2n+1, k) \times kH \to \mathfrak{l}\,(2n+1, k)$ and $\triangleright : \mathfrak{l}\,(2n+1, k) \times kH \to kH$ of the canonical matched pair are given by:

$$E_i \triangleleft H := -E_i, \quad F_i \triangleleft H := F_i, \quad G \triangleleft H := G, \quad G \triangleright H := H \tag{3.40}$$

and all undefined actions are zero. Next we compute the set $\mathcal{DM}\big(\mathfrak{l}\,(2n+1, k),\, kH \,|\, (\triangleright, \triangleleft)\big)$ of all deformation maps of the matched pair $(kH, \mathfrak{l}\,(2n+1, k), \triangleright, \triangleleft)$ given by (3.40).

Lemma 3.5.3 *Let k be a field of characteristic $\neq 2$. Then there exists a bijection*

$$\mathcal{DM}\big(\mathfrak{l}\,(2n+1, k),\, kH \,|\, (\triangleright, \triangleleft)\big) \cong \big(k^n \setminus \{0\}\big) \sqcup \big(k^n \times k\big)$$

The bijection is given such that the deformation map $r = r_a : \mathfrak{l}\,(2n+1, k) \to kH$ associated to $a = (a_i) \in k^n \setminus \{0\}$ is given by

$$r(E_i) := a_i\, H, \quad r(F_i) := 0, \quad r(G) := H \tag{3.41}$$

while the deformation map $r = r_{(b,\,c)} : \mathfrak{l}(2n+1, k) \to kH$ *associated to* $(b = (b_i), c) \in k^n \times k$ *is given as follows*

$$r(E_i) := 0, \quad r(F_i) := b_i\, H, \quad r(G) := c\, H \tag{3.42}$$

for all $i = 1, \cdots, n$.

Proof: Any linear map $r : \mathfrak{l}(2n+1, k) \to kH$ is uniquely determined by a triple $(a = (a_i),\, b = (b_i),\, c) \in k^n \times k^n \times k$ via: $r(E_i) := a_i\, H$, $r(F_i) := b_i\, H$ and $r(G) := c\, H$, for all $i = 1, \cdots, n$. We need to check under what conditions such a map $r = r_{(a,\,b,\,c)}$ is a deformation map. Since kH is abelian, equation (3.26) comes down to

$$r([x,\, y]) = r\big(y \triangleleft r(x) - x \triangleleft r(y)\big) + x \triangleright r(y) - y \triangleright r(x) \tag{3.43}$$

which needs to be checked for all $x,\, y \in \{E_i,\, F_i,\, G \,|\, i = 1, \cdots, n\}$. Notice that (3.43) is symmetrical i.e., if (3.43) is fulfilled for (x, y) then (3.43) is also fulfilled for (y, x). By a routine computation, it can be seen that $r = r_{(a,\,b,\,c)}$ is a deformation map if and only if

$$a_i\, b_j = 0, \qquad (1-c)a_i = 0 \tag{3.44}$$

for all $i,\, j = 1, \cdots, n$. Indeed, (3.43) holds for $(x, y) = (E_i, F_j)$ if and only if $a_i\, b_j = 0$ and it holds for $(x, y) = (E_i, G)$ if and only if $a_i = a_i\, c$. The other cases left to study are either automatically fulfilled or equivalent to one of the two conditions above. The first condition of (3.44) divides the description of deformation maps into two cases: the first one corresponds to $a = (a_i) \neq 0$ and we automatically have $b = 0$ and $c = 1$. The second case corresponds to $a := 0$ which implies that (3.44) holds for any $(b, c) \in k^n \times k$. □

The next result describes all deformations of $\mathfrak{l}(2n+1, k)$ associated to the canonical matched pair $(kH, \mathfrak{l}(2n+1, k), \triangleright, \triangleleft)$ given by (3.40).

Proposition 3.5.4 *Let* k *be a field of characteristic* $\neq 2$ *and the extension of Lie algebras* $kH \subset L(2n+2, k)$. *Then a Lie algebra* $\mathfrak{C}$ *is a complement of* kH *in* $L(2n+2, k)$ *if and only if* $\mathfrak{C}$ *is isomorphic to one of the Lie algebras from the three families defined below:*

- *the Lie algebra* $\mathfrak{l}_{(a)}(2n+1, k)$ *having the bracket defined for any* $i = 1, \cdots, n$ *by:*

$$[E_i,\, E_j]_a := a_i\, E_j - a_j\, E_i, \quad [E_i,\, F_j]_a := -a_i\, F_j, \quad [E_i,\, G]_a := -a_i\, G \tag{3.45}$$

for all $a = (a_i) \in k^n \setminus \{0\}$.

- *the Lie algebra* $\mathfrak{l}'_{(b)}(2n+1, k)$ *having the bracket defined for any* $i = 1, \cdots, n$ *by:*

$$[E_i,\, F_j]_b := -b_j\, E_i, \qquad [E_i,\, G]_b := -E_i \tag{3.46}$$

$$[F_i,\, F_j]_b := b_j\, F_i - b_i\, F_j, \quad [F_i,\, G]_b := F_i - b_i\, G \tag{3.47}$$

for all $b = (b_i) \in k^n$.

• *the Lie algebra* $\mathfrak{l}''_{(b)}\,(2n+1, k)$ *having the bracket defined for any* $i = 1, \cdots, n$ *by:*

$$[E_i,\, F_j]_b := -b_j\, E_i, \quad [F_i,\, F_j]_b := b_j\, F_i - b_i\, F_j, \quad [F_i,\, G]_b := -b_i\, G \quad (3.48)$$

for all $b = (b_i) \in k^n$.

Thus the factorization index $[L\,(2n+2, k) : kH]^f$ *is equal to the number of types of isomorphisms of Lie algebras of the set*

$$\{\mathfrak{l}_{(a)}\,(2n+1, k),\ \mathfrak{l}'_{(b)}\,(2n+1, k),\ \mathfrak{l}''_{(b)}\,(2n+1, k) \mid a \in k^n \setminus \{0\},\ b \in k^n\}$$

Proof: $\mathfrak{l}\,(2n+1, k)$ is a complement of kH in $L\,(2n+2, k)$ and we can write $L\,(2n+2, k) = kH \bowtie \mathfrak{l}\,(2n+1, k)$, where the bicrossed product is associated to the matched pair given in (3.40). Hence, by [15, Theorem 4.3], any other complement $\mathfrak{C}$ of kH in $L\,(2n+2, k)$ is isomorphic to an r-deformation of $\mathfrak{l}\,(2n+1, k)$, for some deformation map $r : \mathfrak{l}\,(2n+1, k) \to kH$ of the matched pair (3.40). These are described in Lemma 3.5.3. The Lie algebra $\mathfrak{l}_{(a)}\,(2n+1, k)$ is precisely the r_a-deformation of $\mathfrak{l}\,(2n+1, k)$, where r_a is given by (3.41). On the other hand the $r_{(b,c)}$-deformation of $\mathfrak{l}\,(2n+1, k)$, where $r_{(b,\,c)}$ is given by (3.42) for some $(b = (b_i),\, c) \in k^n \times k$, is the Lie algebra denoted by $\mathfrak{l}_{(b,\,c)}\,(2n+1, k)$ having the bracket given for any $i = 1, \cdots, n$ by:

$$[E_i,\, F_j]_{(b,c)} := -b_j\, E_i, \qquad [E_i,\, G]_{(b,c)} := (1-c)\, E_i \qquad (3.49)$$

$$[F_i,\, F_j]_{(b,c)} := b_j\, F_i - b_i\, F_j, \quad [F_i,\, G]_{(b,c)} := (c-1)\, F_i - b_i\, G \quad (3.50)$$

for all $(b = (b_i),\, c) \in k^n \times k$. Now, for $c \neq 1$ we can see that $\mathfrak{l}_{(b,\,c)}\,(2n+1, k) \cong \mathfrak{l}'_{(b)}\,(2n+1, k)$ (by sending G to $(c-1)^{-1}G$) while $\mathfrak{l}_{(b,\,1)}\,(2n+1, k) = \mathfrak{l}''_{(b)}\,(2n+1, k)$ and we are done. □

Remark 3.5.5 *An attempt to compute* $[L\,(2n+2, k) : kH]^f$ *for an arbitrary integer* n *is hopeless. However, one can easily see that* $\mathfrak{l}'_{(0)}\,(2n+1, k) = \mathfrak{l}\,(2n+1, k)$ *and* $\mathfrak{l}''_{(0)}\,(2n+1, k) = k_0^{2n+1}$, *the abelian Lie algebra of dimension* $2n+1$. *Thus,* $[L\,(2n+2, k) : kH]^f \geq 2$. *The case* $n = 1$ *is presented below:*

Example 3.5.6 *Let* k *be a field of characteristic* $\neq 2$ *and consider* $\{E,\, F,\, G\}$ *the basis of* $\mathfrak{l}\,(3, k)$ *with the bracket given by* $[E,\, G] = E$ *and* $[G,\, F] = F$. *Then, the factorization index* $[L\,(4, k) : kH]^f = 3$. *More precisely, the isomorphism classes of all complements of* kH *in* $L\,(4, k)$ *are represented by the following three Lie algebras:* $\mathfrak{l}\,(3, k)$, k_0^3 *and the Lie algebra* L_{-1} *having* $\{E,\, F,\, G\}$ *as a basis and the bracket given by*

$$[F,\, E] = F, \quad [E,\, G] = -G$$

Since char$(k) \neq 2$ *the Lie algebras* $\mathfrak{l}\,(3, k)$ *and* L_{-1} *are not isomorphic [99, Exercise 3.2]. For* $a \in k^*$ *the Lie algebra* $\mathfrak{l}_{(a)}\,(3, k)$ *has the bracket given by:*

$[E, F] = -a\,F$ and $[E, G] = -a\,G$. Thus, $\mathfrak{l}_{(a)}\,(3, k) \cong \mathfrak{l}_{(1)}\,(3, k)$, and the latter is isomorphic to the Lie algebra L_{-1}. On the other hand we have: $\mathfrak{l}''_{(0)}\,(3, k) = k_0^3$ and for $b \neq 0$ we can easily see that $\mathfrak{l}''_{(b)}\,(3, k) \cong \mathfrak{l}''_{(1)}\,(3, k) \cong \mathfrak{l}\,(3, k)$. Finally, $\mathfrak{l}'_{(0)}\,(3, k) = \mathfrak{l}\,(3, k)$ and for $b \neq 0$ we have that $\mathfrak{l}'_{(b)}\,(3, k) \cong \mathfrak{l}'_{(1)}\,(3, k)$—the latter is the Lie algebra having $\{f_1, f_2, f_3\}$ as a basis and the bracket given by: $[f_1, f_2] = -f_1$, $[f_1, f_3] = f_1$ and $[f_3, f_2] = f_2 + f_3$. This Lie algebra is also isomorphic to $\mathfrak{l}\,(3, k)$, via the isomorphism which sends f_1 to E, f_3 to G and f_2 to $F - G$.

Let k be a field of characteristic $\neq 2$ and $\mathfrak{l}^2\,(2n+1, k)_{(A,\,D,\,\delta)}$ the Lie algebra of Theorem 3.5.2. In order to simplify computations we will assume $A = D := I_n$ and $\delta := (1, 0, \cdots, 0, 1)$. Let $\mathfrak{m}\,(2n+2, k) := \mathfrak{l}^2\,(2n+1, k)_{(I_n,\,I_n,\,(1,0,\cdots,0,1))}$ be the $(2n+2)$-dimensional Lie algebra having $\{E_i, F_i, G, H \,|\, i = 1, \cdots, n\}$ as a basis and the bracket defined for any $i = 1, \cdots, n$ by:

$$[E_i, G] = E_i, \quad [G, F_i] = F_i, \quad [E_i, H] = E_i, \quad [F_i, H] = F_i,$$
$$[G, H] = E_1 + F_n$$

We consider the Lie algebra extension $kH \subset \mathfrak{m}\,(2n+2, k)$, where $kH \cong k_0$ is the abelian Lie algebra of dimension 1. Of course, $\mathfrak{m}\,(2n+2, k)$ factorizes through kH and $\mathfrak{l}\,(2n+1, k)$, i.e., $\mathfrak{m}\,(2n+2, k) = kH \bowtie \mathfrak{l}\,(2n+1, k)$. Moreover, the canonical matched pair $\triangleleft : \mathfrak{l}\,(2n+1, k) \times kH \to \mathfrak{l}\,(2n+1, k)$ and $\triangleright : \mathfrak{l}\,(2n+1, k) \times kH \to kH$ associated to this factorization is given as follows:

$$E_i \triangleleft H := E_i, \quad F_i \triangleleft H := F_i, \quad G \triangleleft H := E_1 + F_n \tag{3.51}$$

and all undefined actions are zero. In particular, we should notice that the left action $\triangleright : \mathfrak{l}\,(2n+1, k) \times kH \to kH$ is trivial. Next, we describe the set $\mathcal{DM}\left(\mathfrak{l}\,(2n+1, k),\, kH \,|\, (\triangleright, \triangleleft)\right)$ of all deformation maps of the matched pair $(kH, \mathfrak{l}\,(2n+1, k), \triangleright, \triangleleft)$ given by (3.51).

Lemma 3.5.7 *Let k be a field of characteristic $\neq 2$. Then there exists a bijection*

$$\mathcal{DM}\left(\mathfrak{l}\,(2n+1, k),\, kH \,|\, (\triangleright, \triangleleft)\right) \cong \left(k^n \setminus \{0\}\right) \sqcup \left(k^n \setminus \{0\}\right) \sqcup k$$

The bijection is given such that the deformation map $r = r_a : \mathfrak{l}\,(2n+1, k) \to kH$ associated to $a = (a_i) \in k^n \setminus \{0\}$ is given by

$$r(E_i) := a_i\,H, \quad r(F_i) := 0, \quad r(G) := (a_1 - 1)H \tag{3.52}$$

the deformation map $r = r_b : \mathfrak{l}\,(2n+1, k) \to kH$ associated to another $b = (b_i) \in k^n \setminus \{0\}$ is given by

$$r(E_i) := 0, \quad r(F_i) := b_i\,H, \quad r(G) := (b_n + 1)H \tag{3.53}$$

while the deformation map $r = r_c : \mathfrak{l}\,(2n+1, k) \to kH$ associated to $c \in k$ is given by

$$r(E_i) := 0, \quad r(F_i) := 0, \quad r(G) := c\,H \tag{3.54}$$

for all $i = 1, \cdots, n$.

Proof: Any linear map $r : \mathfrak{l}(2n+1, k) \to kH$ is uniquely determined by a triple $(a = (a_i),\, b = (b_i),\, c) \in k^n \times k^n \times k$ via: $r(E_i) := a_i\, H$, $r(F_i) := b_i\, H$ and $r(G) := c\, H$, for all $i = 1, \cdots, n$. We only need to check when such a map $r = r_{(a,\, b,\, c)}$ is a deformation map. Since kH is the abelian Lie algebra and the left action $\triangleright : \mathfrak{l}(2n+1, k) \times kH \to kH$ is trivial, equation (3.26) comes down to:

$$r([x,\, y]) = r\big(y \triangleleft r(x) - x \triangleleft r(y)\big) \tag{3.55}$$

Since (3.55) is symmetrical it is enough to check it only for pairs of the form $(E_i,\, E_j)$, $(F_i,\, F_j)$, $(E_i,\, F_j)$, $(E_i,\, G)$, and $(F_i,\, G)$, for all $i,\, j = 1, \cdots, n$. It is straightforward to see that (3.55) is trivially fulfilled for the pairs $(E_i,\, E_j)$, $(F_i,\, F_j)$ and $(E_i,\, F_j)$. Moreover, (3.55) evaluated for $(E_i,\, G)$ and respectively $(F_i,\, G)$ yields $a_i(a_1 + b_n - c - 1) = 0$ and $b_i(a_1 + b_n - c + 1) = 0$ for all $i = 1, \cdots, n$. Therefore, keeping in mind that we work over a field of characteristic $\neq 2$, the triples $(a = (a_i),\, b = (b_i),\, c) \in k^n \times k^n \times k$ for which $r_{(a,\, b,\, c)}$ becomes a deformation map are given as follows: $(a = (a_i) \in k^n \setminus \{0\},\, b = 0,\, c = a_1 - 1)$, $(a = 0,\, b = (b_i) \in k^n \setminus \{0\},\, c = b_n + 1)$ and $(a = 0,\, b = 0,\, c \in k)$. The corresponding deformation maps are exactly those listed above. □

The next result describes all deformations of $\mathfrak{l}(2n+1, k)$ associated to the canonical matched pair $(kH, \mathfrak{l}(2n+1, k), \triangleright, \triangleleft)$ given by (3.51).

Proposition 3.5.8 *Let k be a field of characteristic $\neq 2$ and the extension of Lie algebras $kH \subset \mathfrak{m}(2n+2, k)$. Then a Lie algebra $\mathfrak{C}$ is a complement of kH in $\mathfrak{m}(2n+2, k)$ if and only if $\mathfrak{C}$ is isomorphic to one of the Lie algebras from the three families defined below:*

- *the Lie algebra $\overline{\mathfrak{l}}_{(a)}(2n+1, k)$ having the bracket defined for any $i = 1, \cdots, n$ by:*

$$[E_i,\, E_j]_a := a_j\, E_i - a_i\, E_j, \quad [E_i,\, F_j]_a := -a_i\, F_j \tag{3.56}$$
$$[E_i,\, G]_a := a_1\, E_i - a_i(E_1 + F_n), \quad [G,\, F_i]_a := (2 - a_1)F_i \tag{3.57}$$

for all $a = (a_i) \in k^n \setminus \{0\}$.

- *the Lie algebra $\overline{\mathfrak{l}'}_{(b)}(2n+1, k)$ having the bracket defined for any $i = 1, \cdots, n$ by:*

$$[F_i,\, F_j]_b := b_j\, F_i - b_i\, F_j, \quad [E_i,\, F_j]_b := b_j\, E_i \tag{3.58}$$
$$[E_i,\, G]_b := (2 + b_n)\, E_i, \quad [G,\, F_i]_b := b_i(E_1 + F_n) - b_n\, F_i \tag{3.59}$$

for all $b = (b_i) \in k^n \setminus \{0\}$.

- *the Lie algebra $\overline{\mathfrak{l}''}_{(c)}(2n+1, k)$ having the bracket defined for any $i = 1, \cdots, n$ by:*

$$[E_i,\, G]_c := (1 + c)\, E_i, \quad [G,\, F_i]_c := (1 - c)\, F_i \tag{3.60}$$

for all $c \in k$;

Thus the factorization index $[\mathfrak{m}(2n+2, k) : kH]^f$ is equal to the number of types of isomorphisms of Lie algebras of the set

$$\{\overline{\mathfrak{l}}_{(a)}(2n+1, k),\ \overline{\mathfrak{l}'}_{(b)}(2n+1, k),\ \overline{\mathfrak{l}''}_{(c)}(2n+1, k) \mid a,\, b \in k^n \setminus \{0\},\, c \in k\}$$

Proof: As in the proof of Proposition 3.5.4 we make use of [15, Theorem 4.3]. More precisely, this implies that all complements $\mathfrak{C}$ of kH in $\mathfrak{m}\,(2n+2,k)$ are isomorphic to an r-deformation of $\mathfrak{l}\,(2n+1,k)$, for some deformation map $r : \mathfrak{l}\,(2n+1,k) \to kH$ of the matched pair (3.51). These are described in Lemma 3.5.7. By a straightforward computation it can be seen that $\bar{\mathfrak{l}}_{(a)}\,(2n+1,k)$ is exactly the complement corresponding to the deformation map given by (3.52), $\overline{\mathfrak{l}'}_{(b)}\,(2n+1,k)$ corresponds to the deformation map given by (3.53), while $\overline{\mathfrak{l}''}_{(c)}\,(2n+1,k)$ is implemented by the deformation map given by (3.54). □

Example 3.5.9 *Let k be a field of characteristic $\neq 2$. Then, the factorization index $[\mathfrak{m}\,(4,k) : kH]^f$ depends essentially on the field k. We will prove that all complements of kH in $\mathfrak{m}\,(4,k)$ are isomorphic to a Lie algebra of the form:*

$$L_\alpha : [x,\, z] = x, \quad [y,\, z] = \alpha y, \quad \text{with} \quad \alpha \in k$$

Hence, $[\mathfrak{m}\,(4,k) : kH]^f = \infty$, if $|k| = \infty$ and $[\mathfrak{m}\,(4,k) : kH]^f = (1+p^n)/2$, if $|k| = p^n$, where $p \geq 3$ is a prime number. Indeed, for $n = 1$, the Lie algebras described in Proposition 3.5.8 become:

$$\begin{aligned}
\bar{\mathfrak{l}}_{(a)}\,(3,k): &\quad [E,\,F]_a := -a\,F,\ [E,\,G]_a := -a\,F,\ [G,\,F]_a := (2-a)F\\
\overline{\mathfrak{l}'}_{(b)}\,(3,k): &\quad [E,\,F]_b := b\,E,\ [E,\,G]_b := (2+b)\,E,\ [G,\,F]_b := b\,E\\
\overline{\mathfrak{l}''}_{(c)}\,(3,k): &\quad [E,\,G]_c := (1+c)\,E,\ [G,\,F]_c := (1-c)\,F
\end{aligned}$$

$a,\, b \in k^*$, $c \in k$. To start with, we should notice that the first two Lie algebras $\bar{\mathfrak{l}}_{(a)}$ and $\overline{\mathfrak{l}'}_{(b)}$ are isomorphic for all $a,\, b \in k^*$. The isomorphism $\gamma : \bar{\mathfrak{l}}_{(a)} \to \overline{\mathfrak{l}'}_{(b)}$ is given as follows:

$$\begin{aligned}
\gamma(E) &:= 2^{-1}(b-a)\,E + 2^{-1}(b-a+2)\,F + 2^{-1}(a-b)\,G, \quad \gamma(F) := E\\
\gamma(G) &:= 2^{-1}(b-a+4)\,E + 2^{-1}(b-a+4)\,F + 2^{-1}(a-b-2)\,G
\end{aligned}$$

Moreover, the map $\varphi : \bar{\mathfrak{l}}_{(a)} \to L_0$ given by:

$$\varphi(E) := y + a\,z, \qquad \varphi(F) := x, \qquad \varphi(G) := x + y + (a-2)\,z$$

is an isomorphism of Lie algebras for all $a \in k^*$. Therefore, the first two Lie algebras are both isomorphic to L_0 for all $a,\, b \in k^*$. We are left to study the family $\overline{\mathfrak{l}''}_{(c)}$. If $c = -1$ then $\overline{\mathfrak{l}''}_{(-1)}$ is again isomorphic to L_0. Suppose now that $c \neq -1$. Then the map $\psi : \overline{\mathfrak{l}''}_{(c)} \to L_{(c-1)(c+1)^{-1}}$ given by:

$$\psi(E) := x, \qquad \psi(F) := y, \qquad \psi(G) := (c+1)\,z$$

is an isomorphism of Lie algebras. Finally, we point out here that if $\alpha \notin \{\beta,\, \beta^{-1}\}$ then L_α is not isomorphic to L_β (see, for instance [99, Exercise 3.2]) and the conclusion follows.

We end this section with two remarks.

Remarks 3.5.10 1) *The deformation of a given Lie algebra* $\mathfrak{h}$ *associated to a matched pair* $(\mathfrak{g}, \mathfrak{h}, \triangleright, \triangleleft)$ *of Lie algebras and to a deformation map* r *as defined by (3.27) is a very general method of constructing new Lie algebras out of a given Lie algebra. It is therefore natural to ask if the properties of a Lie algebra are preserved by this new type of deformation. We will see that in general most properties are not preserved. First of all we remark that the Lie algebra* $\mathfrak{h} := \mathfrak{l}(2n+1, k)$ *is metabelian, that is* $[[\mathfrak{h}, \mathfrak{h}], [\mathfrak{h}, \mathfrak{h}]] = 0$. *Now, if we look at the matched pair deformation* $\mathfrak{h}_r = \mathfrak{l}_{(a)}(2n+1, k)$ *of* $\mathfrak{h}$ *given by (3.45) of Proposition 3.5.4, for* $a = (a_i) \in k^n \setminus \{0\}$ *we can easily see that* $\mathfrak{l}_{(a)}(2n+1, k)$ *is not a metabelian Lie algebra, but a* 3*-step solvable Lie algebra. Thus the property of being metabelian is not preserved by the* r*-deformation of a Lie algebra.*

2) *Next we consider an example of a somewhat different nature. First notice that* $\mathfrak{h} := \mathfrak{l}(2n+1, k)$ *is not a self-dual Lie algebra. Indeed, if* B : $\mathfrak{l}(2n+1, k) \times \mathfrak{l}(2n+1, k) \to k$ *is an arbitrary invariant bilinear form then we can easily prove that* $B(E_i, -) = 0$ *and thus any invariant form is degenerate. On the other hand, the* r*-deformation of* $\mathfrak{l}(2n+1, k)$ *denoted by* $\mathfrak{l}''_{(0)}(2n+1, k)$ *in Remark 3.5.5 is self-dual since it coincides with the* $(2n+1)$*-dimensional abelian Lie algebra.*

3.6 Application: Galois groups and group actions on Lie algebras

The complete description of the automorphism group $\mathrm{Aut}_{\mathrm{Lie}}(\mathfrak{h})$ of a given Lie algebra $\mathfrak{h}$ is an old and notoriously difficult problem intimately related to the structure of Lie algebras. One of the most important results dealing with this problem shows that the automorphism group of a finite-dimensional simple Lie algebra over an algebraically closed field of characteristic zero is generated, with few exceptions, by the invariant automorphisms [141, Theorem 4]. This allows for a full description of the automorphism group of any finite-dimensional reductive Lie algebra. Beyond the theoretical interest in this problem, the description of the automorphism group of an arbitrary Lie algebra turns out to be of crucial importance for the construction of solutions to Einstein's field equations for Bianchi geometries, in the study of $(4+1)$-dimensional spacetimes with applications in cosmology [78, 108, 125, 79] or for discrete symmetries of differential equations. The classification of automorphism groups for indecomposable real Lie algebras is known only up to dimension six and it has been only recently finished [108, 125]. For other contributions to the subject see [47, 128] and the references therein. Perhaps the strongest motivation for studying $\mathrm{Aut}_{\mathrm{Lie}}(\mathfrak{h})$ comes from Hilbert's invariant theory, whose foundation was set at the level of Lie algebras in the classical papers [53, 56, 225, 229]; for more recent work on the subject we refer the reader to [57, 94, 195]. An action as automorphisms of a group G on a Lie algebra $\mathfrak{h}$ is a morphism of groups $\varphi : G \to \mathrm{Aut}_{\mathrm{Lie}}(\mathfrak{h})$. Particular attention was given to the situation where G is a finite subgroup of $\mathrm{Aut}_{\mathrm{Lie}}(\mathfrak{h})$ with the canonical action on $\mathfrak{h}$; in this case achieving the description of the subgroups of $\mathrm{Aut}_{\mathrm{Lie}}(\mathfrak{h})$ is the key step. If $\varphi : G \to \mathrm{Aut}_{\mathrm{Lie}}(\mathfrak{h})$ is an action, then we can consider the subalgebra of invariants $\mathfrak{h}^G$ and we obtain an extension $\mathfrak{h}^G \subseteq \mathfrak{h}$ of Lie algebras. The fundamental problem of invariant theory [53, 130, 188, 155], in the setting of Lie algebras comes down to finding under which conditions on G and $\mathfrak{h}$ the algebraic/geometric properties can be transferred between the two Lie algebras $\mathfrak{h}^G$ and $\mathfrak{h}$. Turning to the problem we started with, since describing the automorphism group $\mathrm{Aut}_{\mathrm{Lie}}(\mathfrak{h})$ of a given Lie algebra $\mathfrak{h}$ is an extremely complicated task, it is natural to start by considering only those automorphisms of $\mathfrak{h}$ which fix a given subalgebra $\mathfrak{g} \neq 0$ of $\mathfrak{h}$. Thus, we can define the Galois group $\mathrm{Gal}\,(\mathfrak{h}/\mathfrak{g})$ of the extension $\mathfrak{g} \subseteq \mathfrak{h}$ as the subgroup of all Lie algebra automorphisms $\sigma : \mathfrak{h} \to \mathfrak{h}$ that fix $\mathfrak{g}$, i.e., $\sigma(g) = g$, for all $g \in \mathfrak{g}$. In an ideal situation, after computing $\mathrm{Gal}\,(\mathfrak{h}/\mathfrak{g})$ for as many subalgebras $\mathfrak{g}$ of $\mathfrak{h}$ as possible, we will have a complete picture of the entire group $\mathrm{Aut}_{\mathrm{Lie}}(\mathfrak{h})$ as well. Having defined the group $\mathrm{Gal}\,(\mathfrak{h}/\mathfrak{g})$, the following question arises naturally:

What is the counterpart in the context of Lie algebras of the classical Galois theory for fields?

At first sight, the chances of developing a promising Galois theory for Lie algebras are very low since even the basic concepts from field theory such as the algebraic/separable/normal extensions, the splitting fields of a polynomial, etc., are rather difficult to define in the context of Lie algebras. Moreover, it is unlikely to have a fundamental theorem establishing a bijective correspondence between the subgroups of $\mathrm{Gal}\,(\mathfrak{h}/\mathfrak{g})$ and the Lie subalgebras $\mathfrak{g}'$ of $\mathfrak{h}$ such that $\mathfrak{g} \subseteq \mathfrak{g}' \subseteq \mathfrak{h}$ as in the case of the classical Galois theory (Example 3.6.4). On the other hand, several counterparts of the classical Galois theory for fields were proved in the context of associative algebras [91], differential Galois theory [176], Hopf algebras [187], von Neumann algebras [190], structured ring spectra [209] or stable homotopy theory [181]. In this context, invariant theory seems to provide a better approach for our problem: if $G \leq \mathrm{Aut}(K)$ is a finite group of automorphisms of a field K then the famous Artin's theorem states that $k := K^G \subseteq K$ is a finite Galois extension of degree $[K : k] = |G|$ and $\mathrm{Gal}(K/k) = G$ [158, Theorem 1.8]. Furthermore, the extension $k \subseteq K$ has a normal basis as a consequence of being Galois; that is, there exists $x \in K$ such that $\{\sigma(x) \,|\, \sigma \in G\}$ is a k-basis of K [158, Theorem 1.8]. One of the many generalizations of Artin's theorem deals with arbitrary actions [187, Example 8.1.2]: if $\varphi : G \to \mathrm{Aut}(K)$ is an action as automorphisms of a finite group G on a field K, then K/K^G is a Galois extension in the classical sense with Galois group G if and only if G acts faithfully on K. A version of Artin's theorem for Hopf-Galois extensions (a concept which generalizes Galois extensions for fields [187, Example 8.1.2]) was obtained in [50, Theorem 1.18]. At this level, Hopf-Galois extensions satisfying the normal basis property coincide with crossed products [187, Corollary 8.2.5]. This last observation allows us to restate Artin's theorem in a more convenient but equivalent manner, as follows: if $G \leq \mathrm{Aut}(K)$ is a finite group of automorphisms of a field K, then K is isomorphic to a crossed product algebra $k \#_\sigma k[G]^*$ between the field of invariants $k = K^G$ and the dual algebra of the group algebra $k[G]$, associated to some cocycle $\sigma : k[G]^* \otimes k[G]^* \to k$. With this last conclusion in mind, we replace the category we work in: instead of fields we consider Lie algebras together with group actions as automorphisms. Now the question if an Artin type theorem holds for Lie algebras has a positive answer with a slight amending though: the role of the classical crossed product of Lie algebras, as it arises in the theory of Chevalley and Eilenberg [77] will be played by the skew crossed product of Lie algebras introduced in Section 3.3.

We start by introducing the following definition:

Definition 3.6.1 *We say that a group G acts as automorphisms on the Lie algebra $\mathfrak{h}$ if there exists a morphism of groups $\varphi : G \to \mathrm{Aut}_{\mathrm{Lie}}(\mathfrak{h})$ and we shall use the notation $\varphi(g)(x) = g \triangleright x$, for all $g \in G$ and $x \in \mathfrak{h}$. The action is called faithful if φ is injective.*

Since $\varphi(g)$ is a Lie algebra map we have that $g \triangleright [x, y] = [g \triangleright x, g \triangleright y]$, for all $g \in G$ and $x, y \in \mathfrak{h}$. The subalgebra of invariants $\mathfrak{h}^G$ of the action φ of G on $\mathfrak{h}$ is defined by:

$$\mathfrak{h}^G := \{x \in \mathfrak{h} \mid g \triangleright x = x, \forall\, g \in G\}$$

Then $\mathfrak{h}^G \subseteq \mathfrak{h}$ is a Lie subalgebra of $\mathfrak{h}$. If G is a finite group and $|G|$ is invertible in the base field k then the trace map or Reynolds operator (we borrowed the terminology from the classical invariant theory of groups acting on associative algebras [130]) defined for any $x \in \mathfrak{h}$ by:

$$t = t_\triangleright : \mathfrak{h} \to \mathfrak{h}^G, \qquad t(x) := |G|^{-1} \sum_{g \in G} g \triangleright x \tag{3.61}$$

is a linear retraction of the canonical inclusion $\mathfrak{h}^G \hookrightarrow \mathfrak{h}$. Furthermore, for any $a \in \mathfrak{h}^G$ and $x \in \mathfrak{h}$ we have $t([a, x]) = [a, t(x)]$.

Example 3.6.2 *(1) The basic example of a group acting on a Lie algebra is provided by any subgroup G of $\mathrm{Aut}_{\mathrm{Lie}}(\mathfrak{h})$ with the canonical action given by $\sigma \triangleright x := \sigma(x)$, for all $\sigma \in G$ and $x \in \mathfrak{h}$. Automorphic Lie algebras [168] introduced in the context of integrable systems are examples of Lie algebras of invariants; for further details we refer to [151].*

(2) The group $\mathrm{GL}(n, k)$ acts on $\mathfrak{gl}(n, k)$ (resp. $\mathfrak{sl}(n, k)$) by conjugation, i.e., $U \triangleright X := UXU^{-1}$, for all $U \in \mathrm{GL}(n, k)$ and $X \in \mathfrak{gl}(n, k)$ (resp. $X \in \mathfrak{sl}(n, k)$). Thus, any subgroup of $\mathrm{GL}(n, k)$ (such as $\mathrm{SL}(n, k)$, the permutation group S_n on n letters, the cyclic group C_n, or more generally any finite group of order n), acts on the Lie algebras $\mathfrak{gl}(n, k)$ and $\mathfrak{sl}(n, k)$ by the same action. The subalgebras of invariants for these actions are exactly the centralizers

in $\mathfrak{gl}(n,k)$ *(resp.* $\mathfrak{sl}(n,k)$*) of* $\mathrm{GL}(n,k)$ *(or its subgroups). For example, the symmetric group* S_n *acts as automorphisms on* $\mathfrak{gl}(n,k)$ *via the action:*

$$S_n \to \mathrm{Aut}_{\mathrm{Lie}}(\mathfrak{gl}(n,k)), \tag{3.62}$$
$$\tau \triangleright e_{ij} := \left(e_{1\tau(1)} + \cdots + e_{n\tau(n)}\right) e_{ij} \left(e_{1\tau(1)} + \cdots + e_{n\tau(n)}\right)^{-1}$$

for all $\tau \in S_n$ *and* $i,\, j = 1, \cdots, n$*, where* e_{ij} *is* $n \times n$ *matrix which has 1 in the* $(i,j)^{th}$*-position and zeros elsewhere. We will describe the subalgebras of invariants* $\mathfrak{gl}(n,k)^{C_n}$ *and respectively* $\mathfrak{gl}(n,k)^{S_n}$ *of the action defined by 3.62. We start by looking at* $\mathfrak{gl}(n,k)^{C_n}$*, where the cyclic group* C_n *is considered to be the subgroup of* S_n *generated by the cycle* $(12...n)$*. An easy computation gives:*

$$(1\,2\ldots n) \triangleright A = (e_{12} + e_{23} + e_{34} + \cdots + e_{n1}) A (e_{21} + e_{32} + e_{43} + \cdots + e_{1n}) \tag{3.63}$$

It follows that $A = \sum_{i,j=1}^{n} a_{ij}\, e_{ij} \in \mathfrak{gl}(n,k)^{C_n}$ *if and only if* $(12...n) \triangleright A = A$*. This yields:*

$$a_{11} = a_{nn}, \quad a_{1j} = a_{n,j-1}, \quad a_{j1} = a_{j-1,n}, \quad a_{ij} = a_{i-1,j-1}, \text{ for all } i,\, j \geqslant 2 \tag{3.64}$$

Therefore, by a careful analysis of the above compatibilities, we obtain that $\mathfrak{gl}(n,k)^{C_n}$ *is the* n*-dimensional subalgebra of* $\mathfrak{gl}(n,k)$ *consisting of all* $n \times n$*-matrices of the form:*

$$\begin{pmatrix} a_1 & a_2 & \cdots & a_n \\ a_n & a_1 & \cdots & a_{n-1} \\ a_{n-1} & a_n & \cdots & a_{n-2} \\ \cdot & \cdot & \cdots & \cdot \\ a_2 & a_3 & \cdots & a_1 \end{pmatrix}$$

for all $a_1, \cdots, a_n \in k$*. Next in line is* $\mathfrak{gl}(n,k)^{S_n}$*. As* S_n *is generated by the transposition* (12) *and the cycle* $(12...n)$*, it follows that a matrix* $A = \sum_{i,j=1}^{n} a_{ij}\, e_{ij} \in \mathfrak{gl}(n,k)^{S_n}$ *if and only if* $\tau \triangleright A = A$*, for* $\tau = (1\,2)$ *and* $\tau = (1\,2\ldots n)$*. Using again the formula given in (3.62) we obtain:*

$$(1\,2) \triangleright A = (e_{12} + e_{21} + e_{33} + \cdots + e_{nn}) A (e_{12} + e_{21} + e_{33} + \cdots + e_{nn}) \tag{3.65}$$

which yields:

$$a_{11} = a_{22}, \quad a_{12} = a_{21}, \quad a_{1j} = a_{2j}, \quad a_{j1} = a_{j2}, \text{ for all } j \geqslant 3.$$

The above compatibilities together with those in equation (3.64) come down to: $a_{ii} = \alpha \in k$ *and* $a_{ij} = \beta \in k$*, for all* $i,\, j = 1, \cdots, n,\, i \neq j$*. Thus,* $\mathfrak{gl}(n,k)^{S_n} = \left\{\alpha I_n + \beta \sum_{\substack{i,\,j=1 \\ i \neq j}}^{n} e_{ij} \mid \alpha,\, \beta \in k\right\}$*. Both subalgebras of invariants* $\mathfrak{gl}(n,k)^{C_n}$*, and respectively* $\mathfrak{gl}(n,k)^{S_n}$*, are abelian.*

(3) The actions as automorphisms of abelian groups on Lie algebras can be seen as the dual concept of the well-studied gradings on Lie algebras: for an overview and the importance of the problem introduced by Kac [144] we refer to [98, 153, 217] and the references therein.

Let $G = (G, +)$ *be an abelian group and* $\hat{G}$ *be the group of characters on* G*, i.e., all morphisms of groups* $G \to k^*$. *A* G*-graded Lie algebra is a Lie algebra* $\mathfrak{h}$ *such that* $\mathfrak{h} = \oplus_{g \in G} \mathfrak{h}_g$, *where any* $\mathfrak{h}_g$ *is a subspace of* $\mathfrak{h}$ *such that* $[\mathfrak{h}_g, \mathfrak{h}_{g'}] \subseteq \mathfrak{h}_{g+g'}$, *for all* g, $g' \in G$. *If* $\mathfrak{h} = \oplus_{g \in G} \mathfrak{h}_g$ *is a* G*-graded Lie algebra then the map*

$$\varphi : \hat{G} \to \mathrm{Aut}_{\mathrm{Lie}}(\mathfrak{h}), \qquad \varphi(\chi)(x_g) := \chi(g)\, x_g$$

for all $\chi \in \hat{G}$, $g \in G$ *and* $x_g \in \mathfrak{h}_g$ *is a faithful action of* $\hat{G}$ *on* $\mathfrak{h}$. *Conversely, if* $\varphi : \hat{G} \to \mathrm{Aut}_{\mathrm{Lie}}(\mathfrak{h})$ *is an injective morphism of groups, then* $\mathfrak{h} = \oplus_{g \in G} \mathfrak{h}_g$ *is a* G*-graded Lie algebra where* $\mathfrak{h}_g := \{y \in \mathfrak{h} \,|\, \chi \triangleright y = \chi(g) y, \ \forall \chi \in \hat{G}\}$. *In some special cases we can say more. For instance, if* k *is an algebraically closed field of characteristic zero and* G *is a finitely generated abelian group, then there exists a one-to-one correspondence between the set of all* G*-gradings on a given Lie algebra* $\mathfrak{h}$ *and the set of all faithful actions* $\hat{G} \to \mathrm{Aut}_{\mathrm{Lie}}(\mathfrak{h})$ *of* $\hat{G}$ *on* $\mathfrak{h}$ *[153, Proposition 4.1]. Working with actions instead of gradings comes with the advantage of not assuming the group* G *to be abelian nor the actions to be faithful.*

(4) As a special case of (3) let us take $\mathfrak{h} = \oplus_{i \in \mathbb{Z}} \mathfrak{h}_i$ *to be a* $\mathbb{Z}$*-graded Lie algebra. Then the multiplicative group of units* k^* *acts on* $\mathfrak{h}$ *via the following morphism of groups:*

$$\varphi : k^* \to \mathrm{Aut}_{\mathrm{Lie}}(\mathfrak{h}), \qquad \varphi(u)(y_i) := u^i\, y_i \tag{3.66}$$

for all $u \in k^*$, $i \in \mathbb{Z}$ *and* $y_i \in \mathfrak{h}_i$ *a homogeneous element of degree* i. *Moreover, the subalgebra of invariants* $\mathfrak{h}^{k^*} = \mathfrak{h}_0$, *the Lie subalgebra of all elements of degree zero. The typical example of a* $\mathbb{Z}$*-graded Lie algebra is the Witt algebra* $\mathfrak{W}$ *which is the vector space having* $\{e_i \,|\, i \in \mathbb{Z}\}$ *as a basis and the bracket* $[e_i, e_j] := (i-j)\, e_{i+j}$, *for all* i, $j \in \mathbb{Z}$. *Another example is given by* $\mathfrak{h} := \mathfrak{sl}(2, k)$, *the Lie algebra with basis* $\{e_1, e_2, e_3\}$ *and the usual bracket* $[e_1, e_2] = e_3$, $[e_1, e_3] = -2\, e_1$ *and* $[e_2, e_3] = 2\, e_2$ *viewed with the standard grading: namely* e_1 *has degree* -1, e_2 *has degree* 1 *and* e_3 *has degree* 0. *We obtain that the group* k^* *acts on* $\mathfrak{sl}(2, k)$ *via:*

$$\varphi : k^* \to \mathrm{Aut}_{\mathrm{Lie}}\big(\mathfrak{sl}(2, k)\big), \qquad u \triangleright (\alpha e_1 + \beta e_2 + \gamma e_3) := u^{-1} \alpha e_1 + u \beta e_2 + \gamma e_3 \tag{3.67}$$

for all $u \in k^*$ *and* α, β, $\gamma \in k$. *The algebra of invariants* $\mathfrak{sl}(2, k)^{k^*}$ *is the abelian Lie algebra having* e_3 *as a basis.*

We are now ready to introduce the Galois group of a given Lie algebra extension.

Definition 3.6.3 *Let $\mathfrak{g} \subseteq \mathfrak{h}$ be an extension of Lie algebras. We define the Galois group* $\mathrm{Gal}\,(\mathfrak{h}/\mathfrak{g})$ *as the subgroup of* $\mathrm{Aut}_{\mathrm{Lie}}(\mathfrak{h})$ *consisting of all Lie algebra automorphisms of* $\mathfrak{h}$ *that fix* $\mathfrak{g}$*, i.e.:*

$$\mathrm{Gal}\,(\mathfrak{h}/\mathfrak{g}) := \{\sigma \in \mathrm{Aut}_{\mathrm{Lie}}(\mathfrak{h}) \,|\, \sigma(g) = g, \,\forall\, g \in \mathfrak{g}\}$$

Since $\mathrm{Gal}\,(\mathfrak{h}/\mathfrak{g}) \leq \mathrm{Aut}_{\mathrm{Lie}}(\mathfrak{h})$ we can consider the subalgebra of invariants $\mathfrak{h}^{\mathrm{Gal}(\mathfrak{h}/\mathfrak{g})}$. Of course, we have that $\mathfrak{g} \subseteq \mathfrak{h}^{\mathrm{Gal}(\mathfrak{h}/\mathfrak{g})}$. As it can be seen from the example below, a fundamental theorem establishing a bijective correspondence between the subgroups of $\mathrm{Gal}\,(\mathfrak{h}/\mathfrak{g})$ and the Lie subalgebras $\mathfrak{g}'$ of $\mathfrak{h}$ such that $\mathfrak{g} \subseteq \mathfrak{g}' \subseteq \mathfrak{h}$ does not hold in the context of Lie algebras.

Example 3.6.4 *Let $\mathfrak{h} := \mathfrak{aff}(2, k)$ be the 2-dimensional affine Lie algebra with basis $\{e_1, e_2\}$ and bracket $[e_1, e_2] = e_2$ and $\mathfrak{g} := k e_1$ the abelian Lie subalgebra. Then* $\mathrm{Gal}\,(\mathfrak{aff}(2, k)/\mathfrak{g})$ *is isomorphic to k^* the multiplicative group of units of k while the subalgebra of invariants $\mathfrak{aff}(2, k)^{k^*} = \mathfrak{g}$. Of course, between $\mathfrak{g}$ and $\mathfrak{aff}(2, k)$ there are no proper intermediary subalgebras while k^* has many subgroups (such as the cyclic groups $U_n(k)$ of n-roots of unity) whose subalgebras of invariants coincide with $\mathfrak{g}$.*

In what follows we will describe the group $\mathrm{Gal}\,(\mathfrak{h}/\mathfrak{g})$. First we fix a linear map $p : \mathfrak{h} \to \mathfrak{g}$ such that $p(g) = g$, for all $g \in \mathfrak{g}$—such a map always exists as k is a field. Then $V := \mathrm{Ker}(p)$ is a subspace of $\mathfrak{h}$ and a complement of $\mathfrak{g}$ in $\mathfrak{h}$. As in (the proof of) Theorem 3.1.4, there exists a canonical extending system associated to p, where the bilinear maps $\rightharpoonup : V \times \mathfrak{g} \to \mathfrak{g}$, $\leftharpoonup : V \times \mathfrak{g} \to V$, $\theta : V \times V \to \mathfrak{g}$ and $\{\, ,\, \} : V \times V \to V$ are given by the formulas (3.6) and (3.7). Thus we can construct the unified product $\mathfrak{g} \natural V$ associated to the canonical extending structure, which is a Lie algebra with the bracket given by the formula (3.1). The map $\varphi : \mathfrak{g} \natural V \to \mathfrak{h}$, defined by $\varphi(g, x) := g + x$, is an isomorphism of Lie algebras with the inverse given by $\varphi^{-1}(y) := \big(p(y),\, y - p(y)\big)$, for all $y \in \mathfrak{h}$. Since φ fixes $\mathfrak{g} \cong \mathfrak{g} \times \{0\}$ we obtain that the map

$$\mathrm{Gal}\,(\mathfrak{h}/\mathfrak{g}) \to \mathrm{Gal}\,(\mathfrak{g} \natural V/\mathfrak{g}), \qquad \sigma \mapsto \varphi^{-1} \circ \sigma \circ \varphi \tag{3.68}$$

is an isomorphism of groups with the inverse given by $\psi \mapsto \varphi \circ \psi \circ \varphi^{-1}$. It follows from Lemma 3.1.5 that there exists a bijection between the set of all elements $\psi \in \mathrm{Gal}\,(\mathfrak{g} \natural V/\mathfrak{g})$ and the set of all pairs $(\sigma, r) \in \mathrm{GL}_k(V) \times \mathrm{Hom}_k(V, \mathfrak{g})$, satisfying the following four compatibility conditions for any $g \in \mathfrak{g}$, x, $y \in V$:

(G1) $\sigma(x \leftharpoonup g) = \sigma(x) \leftharpoonup g$, that is $\sigma : V \to V$ is a right Lie $\mathfrak{g}$-module map;

(G2) $r(x \leftharpoonup g) = [r(x),\, g] + \big(\sigma(x) - x\big) \rightharpoonup g$;

(G3) $\sigma(\{x, y\}) = \{\sigma(x),\, \sigma(y)\} + \sigma(x) \leftharpoonup r(y) - \sigma(y) \leftharpoonup r(x)$;

(G4) $r(\{x, y\}) = [r(x), r(y)] + \sigma(x) \rightharpoonup r(y) - \sigma(y) \rightharpoonup r(x) + \theta\big(\sigma(x), \sigma(y)\big) - \theta(x, y)$.

The bijection is such that $\psi = \psi_{(\sigma,r)} \in \mathrm{Gal}\,(\mathfrak{g} \natural V/\mathfrak{g})$ corresponding to $(\sigma, r) \in \mathrm{GL}_k(V) \times \mathrm{Hom}_k(V,\, \mathfrak{g})$ is given by $\psi(g,\, x) := (g + r(x),\, \sigma(x))$, for all $g \in \mathfrak{g}$ and $x \in V$. We point out that $\psi_{(\sigma,r)}$ is indeed an element of $\mathrm{Gal}\,(\mathfrak{g} \natural V/\mathfrak{g})$ with the inverse given by $\psi^{-1}_{(\sigma,r)}(g,\, x) = \big(g - r(\sigma^{-1}(x)),\, \sigma^{-1}(x)\big)$, for all $g \in \mathfrak{g}$ and $x \in V$. We denote by $\mathbb{G}^V_{\mathfrak{g}}\,\big(\leftharpoonup, \rightharpoonup, \theta, \{-,\, -\}\big)$ the set of all pairs $(\sigma,\, r) \in \mathrm{GL}_k(V) \times \mathrm{Hom}_k(V,\, \mathfrak{g})$ satisfying the compatibility conditions (G1)–(G4). It is straightforward to see that $\mathbb{G}^V_{\mathfrak{g}}\,\big(\leftharpoonup, \rightharpoonup, \theta, \{-,\, -\}\big)$ is a subgroup of the semidirect product of groups $\mathbb{G}^V_{\mathfrak{g}} := \mathrm{GL}_k(V) \rtimes \mathrm{Hom}_k(V,\, \mathfrak{g})$ with the group structure given by formula (0.2). Now, for any $(\sigma,\, r)$ and $(\sigma',\, r') \in \mathbb{G}^V_{\mathfrak{g}}\,\big(\leftharpoonup, \rightharpoonup, \theta, \{-,\, -\}\big)$, $g \in \mathfrak{g}$ and $x \in V$ we have:

$$\psi_{(\sigma,\, r)} \circ \psi_{(\sigma',\, r')}(g,\, x) = \big(g + r'(x) + r(\sigma'(x)),\, \sigma(\sigma'(x)\big) = \psi_{(\sigma\circ\sigma',\, r\circ\sigma' + r')}(g,\, x)$$

i.e., $\psi_{(\sigma,\, r)} \circ \psi_{(\sigma',\, r')} = \psi_{(\sigma\circ\sigma',\, r\circ\sigma'+r')}$. Finally, we recall that $\mathfrak{h} = \mathfrak{g} + V$ and $\mathfrak{g} \cap V = \{0\}$, i.e., any element $y \in \mathfrak{h}$ has a unique decomposition as $y = g + x$, for $g \in \mathfrak{g}$ and $x \in V = \mathrm{Ker}(p)$. Putting all together we proved the following:

Theorem 3.6.5 *Let $\mathfrak{g} \subseteq \mathfrak{h}$ be an extension of Lie algebras, $p : \mathfrak{h} \to \mathfrak{g}$ a linear retraction of the inclusion $\mathfrak{g} \subseteq \mathfrak{h}$, $V = \mathrm{Ker}(p)$ and consider $\Lambda(\mathfrak{g},\, V) = \big(\leftharpoonup, \rightharpoonup, \theta, \{-,\, -\}\big)$ to be the canonical Lie extending system associated to p. Then there exists an isomorphism of groups defined for any $(\sigma,\, r) \in \mathbb{G}^V_{\mathfrak{g}}\,\big(\leftharpoonup, \rightharpoonup, \theta, \{-,\, -\}\big)$, $g \in \mathfrak{g}$ and $x \in V$ by:*

$$\Omega : \mathbb{G}^V_{\mathfrak{g}}\,\big(\leftharpoonup, \rightharpoonup, \theta, \{-,\, -\}\big) \to \mathrm{Gal}\,(\mathfrak{h}/\mathfrak{g}), \quad \Omega(\sigma, r)(g + x) := g + r(x) + \sigma(x) \tag{3.69}$$

In particular, there exists an embedding $\mathrm{Gal}\,(\mathfrak{h}/\mathfrak{g}) \hookrightarrow \mathrm{GL}_k(V) \rtimes \mathrm{Hom}_k(V,\, \mathfrak{g})$, where the right-hand side is the semidirect product associated to the canonical right action of $\mathrm{GL}_k(V)$ on $\mathrm{Hom}_k(V,\, \mathfrak{g})$.

In the finite-dimensional case we obtain the Lie algebra counterpart of the fact that the Galois group of a Galois extension of fields of degree m embeds in the symmetric group S_m.

Corollary 3.6.6 *Let $\mathfrak{g} \subseteq \mathfrak{h}$ be an extension of Lie algebras such that $\dim_k(\mathfrak{g}) = n$ and $\dim_k(\mathfrak{h}) = n + m$. Then the Galois group $\mathrm{Gal}\,(\mathfrak{h}/\mathfrak{g})$ embeds in the canonical semidirect product of groups $\mathrm{GL}(m,\, k) \rtimes \mathrm{M}_{n\times m}(k)$.*

The first examples based on Theorem 3.6.5 are given below.

Example 3.6.7 *Consider the extension of Lie algebras $ke_3 \subseteq \mathfrak{sl}(2, k)$ with the notations of Example 3.6.2 and take $p : \mathfrak{sl}(2, k) \to ke_3$ given by $p(e_1) = p(e_2) := 0$ and $p(e_3) := e_3$. Then $V = \mathrm{Ker}(p) = ke_1 + ke_2$ and the canonical Lie extending system associated to p given by (3.6)–(3.7) takes the following form: $\rightharpoonup : V \times ke_3 \to V$ and $\{-, -\} : V \times V \to V$ are both trivial maps, while the action $\leftharpoonup : V \times ke_3 \to V$ and the cocycle $\theta : V \times V \to ke_3$ are given by*

$e_1 \leftharpoonup e_3 = -2e_1$, $e_2 \leftharpoonup e_3 = 2e_2$ and $\theta(e_1, e_1) = \theta(e_2, e_2) = 0$, $\theta(e_1, e_2) = -\theta(e_2, e_1) = e_3$. In particular, this shows that $\mathfrak{sl}(2, k)$ is isomorphic as a Lie algebra to the skew crossed product $ke_3 \#^{\bullet} V$ of the abelian Lie algebras of dimensions one and two. Now, an element $\sigma \in \mathrm{GL}_k(V)$ will be written in the matrix form $\sigma = (\sigma_{ij}) \in \mathrm{M}_2(k)$ and a linear map $r \in \mathrm{Hom}_k(V, ke_3)$ as a family of two scalars $r = (r_1, r_2) \in k^2$ given by $r(e_1) = r_1 e_3$ and $r(e_2) = r_2 e_3$. A straightforward computation proves that the pair $\big(\sigma = (\sigma_{ij}), r = (r_1, r_2)\big)$ satisfies (G1)–(G4) if and only if $r_1 = r_2 = 0$, $\sigma_{12} = \sigma_{21} = 0$ and $\sigma_{11}\sigma_{22} = 1$. This proves that the group $\mathbb{G}^V_{ke_3}(\leftharpoonup, \theta)$ identifies with the group of units k^* and hence $\mathrm{Gal}\,(\mathfrak{sl}(2, k)/ke_3) \cong k^*$. More precisely, $\tau \in \mathrm{Gal}\,(\mathfrak{sl}(2, k)/ke_3)$ if and only if there exists $u \in k^*$ such that $\tau(ae_1 + be_2 + ce_3) = uae_1 + u^{-1}be_2 + ce_3$, for all $a, b, c \in k$.

Our next example proves that the Galois group of the extension of two consecutive Lie Heisenberg algebras is the 2-dimensional special affine group $\mathrm{SL}_2(k) \rtimes k^2$.

Example 3.6.8 *Let $n \in \mathbb{N}^*$ and consider $\mathfrak{h}^{2n+1}$ to be the $(2n+1)$-dimensional Heisenberg Lie algebra having $\{x_1, \cdots, x_n, y_1, \cdots, y_n, w\}$ as a basis and the bracket given by $[x_i, y_i] = w$, for all $i = 1, \cdots, n$. If we consider the canonical Lie algebra extension $\mathfrak{h}^{2n+1} \subset \mathfrak{h}^{2n+3}$, then there exists an isomorphism of groups:*

$$\mathrm{Gal}\,(\mathfrak{h}^{2n+3}/\mathfrak{h}^{2n+1}) \cong \mathrm{SL}_2(k) \rtimes k^2$$

where $\mathrm{SL}_2(k) \rtimes k^2$ is the semidirect product of groups corresponding to the canonical right action $\triangleleft : k^2 \times \mathrm{SL}_2(k) \to k^2$ given by $(a, b) \triangleleft B = (a, b)B$, for all $(a, b) \in k^2$, $B \in \mathrm{SL}_2(k)$. To start with, we point out that $\mathfrak{h}^{2n+3}$ can be realized as a unified product between $\mathfrak{h}^{2n+1}$ and the vector space V with k basis $\{x_{n+1}, y_{n+1}\}$ corresponding to the Lie extending system with one non-trivial map, namely $\theta : V \times V \to \mathfrak{h}^{2n+1}$ given by $\theta(x_{n+1}, y_{n+1}) = w$. The conclusion now follows by applying Theorem 3.6.5. First notice that any pair $(\sigma, r) \in \mathrm{GL}_k(V) \times \mathrm{Hom}_k(V, \mathfrak{g})$ fulfills trivially the compatibility conditions (G1) and (G3). Furthermore, the compatibility condition (G2) yields $r(x_{n+1}) = \alpha w$ and $r(y_{n+1}) = \beta w$ for some $\alpha, \beta \in k$. Finally, if we denote $\sigma(x_{n+1}) = ax_{n+1} + bx_{n+1}$ and respectively $\sigma(y_{n+1}) = cx_{n+1} + dx_{n+1}$ for some $a, b, c, d \in k$, the compatibility condition (G4) gives $ad - bc = 1$. Therefore, the set of pairs $(\sigma, r) \in \mathrm{GL}_k(V) \times \mathrm{Hom}_k(V, \mathfrak{g})$ satisfying (G1)–(G4) is in fact equal to $\mathrm{SL}_2(k) \times k^2$. The proof is now finished by identifying the maps $\sigma \in \mathrm{GL}_k(V)$, $r \in \mathrm{Hom}_k(V, \mathfrak{g})$ with their corresponding matrices in $\mathrm{SL}_2(k)$ respectively k^2 and noticing that the multiplication on $\mathbb{G}^V_{\mathfrak{h}^{2n+1}}(\theta)$ comes down to that corresponding to the semidirect product induced by the action defined above.

Now we provide an example of a Lie algebra extension having a metabelian Galois group.

Example 3.6.9 *For a positive integer n, let $\mathfrak{l}(2n+1)$ be the metabelian Lie algebra with basis $\{E_i, F_i, G \mid i = 1, \cdots, n\}$ and bracket given by $[E_i, G] = E_i$, $[G, F_i] = F_i$, for all $i = 1, \cdots, n$. Then there exists an isomorphism of groups:*

$$\mathrm{Gal}\,(\mathfrak{l}(2n+3)/\mathfrak{l}(2n+1)) \cong (k^* \times k^*) \rtimes \mathcal{M}_{n\times 2}(k)$$

where $(k^* \times k^*) \rtimes \mathcal{M}_{n\times 2}(k)$ is the semidirect product corresponding to the right action $\triangleleft : \mathcal{M}_{n\times 2}(k) \times (k^* \times k^*) \to \mathcal{M}_{n\times 2}(k)$ given by $B \triangleleft (a\, b) = B \begin{pmatrix} a & 0 \\ 0 & b \end{pmatrix}$, for all a, $b \in k^*$ and $B \in \mathcal{M}_{n\times 2}(k)$. Indeed, $\mathfrak{l}(2n+3)$ can be written as a unified product between $\mathfrak{l}(2n+1)$ and the vector space V with basis $\{E_{n+1}, F_{n+1}\}$ corresponding to the extending structure with only one non-trivial map, namely $\rightharpoonup : V \times \mathfrak{l}(2n+1) \to V$ given by $E_{n+1} \rightharpoonup G = E_{n+1}$ and $F_{n+1} \rightharpoonup G = -F_{n+1}$. Now if $(\sigma, r) \in \mathrm{GL}_k(V) \times \mathrm{Hom}_k(V, \mathfrak{l}(2n+1))$ a careful analysis of the compatibility conditions (G1)–(G4) yields:

$$\sigma(E_{n+1}) = aE_{n+1}, \qquad \sigma(F_{n+1}) = bF_{n+1}, \; a,\, b \in k, \; ab \neq 0$$

$$r(E_{n+1}) = \sum_{i=1}^{n} \alpha_i\, E_i, \qquad r(F_{n+1}) = \sum_{i=1}^{n} \beta_i\, F_i, \; \alpha_i, \beta_i \in k, \; i = 1, 2, \cdots n.$$

Thus, the pairs $(\sigma, r) \in \mathrm{GL}_k(V) \times \mathrm{Hom}_k(V, \mathfrak{l}(2n+1))$ are parameterized by $(k^* \times k^*) \times \mathcal{M}_{n\times 2}(k)$ and the conclusion follows by Theorem 3.6.5.

In the sequel we consider two general examples. The first one computes the Galois group of the extension $\mathfrak{g}' \subset \mathfrak{g}$ for a special class of Lie algebras $\mathfrak{g}$, namely the non-perfect ones with $C_{\mathfrak{g}}(\mathfrak{g}') = \{0\}$, where $\mathfrak{g}' = [\mathfrak{g}, \mathfrak{g}]$ is the derived algebra of $\mathfrak{g}$ and $C_{\mathfrak{g}}(\mathfrak{g}')$ denotes the centralizer of $\mathfrak{g}'$ in $\mathfrak{g}$. A generic example of such a Lie algebra is for instance $\mathfrak{g} := \mathfrak{gl}(n, k) \rtimes k^n$, the semidirect product of Lie algebras corresponding to the canonical action of $\mathfrak{gl}(n, k)$ on k^n.

Example 3.6.10 *Let $\mathfrak{g}$ be a non-perfect Lie algebra such that $C_{\mathfrak{g}}(\mathfrak{g}') = \{0\}$. Then there exists an isomorphism of groups*

$$\mathrm{Gal}\,(\mathfrak{g}/\mathfrak{g}') \cong \mathrm{GL}_k(V)$$

where V is a complement as vector spaces of $\mathfrak{g}'$ in $\mathfrak{g}$. First we write $\mathfrak{g}$ as the unified product between $\mathfrak{g}'$ and V associated to the Lie extending system whose non-trivial maps are given as follows: $x \rightharpoonup g = [x, g]$ and $\theta(x, y) = [x, y]$, for all $g \in \mathfrak{g}'$, x, $y \in V$. Now let $(\sigma, r) \in \mathrm{GL}_k(V) \times \mathrm{Hom}_k(V, \mathfrak{g})$ satisfying the compatibility conditions (G1)–(G4). One can easily see that (G1) and (G3) are trivially fulfilled while (G2) and (G4) come down to the following compatibilities:

$$[r(x) + \sigma(x) - x, g] = 0, \qquad [r(x) + \sigma(x), r(y) + \sigma(y)] = [x, y]$$

for all $g \in \mathfrak{g}'$, x, $y \in V$. We obtain that $r(x) + \sigma(x) - x \in C_{\mathfrak{g}}(\mathfrak{g}') = \{0\}$, for all $x \in V$. Thus $r = \mathrm{Id}_V - \sigma$ and hence the second equation is now *trivially fulfilled. Therefore, the pairs* $(\sigma, r) \in \mathrm{GL}_k(V) \times \mathrm{Hom}_k(V, \mathfrak{g})$ *satisfying the compatibility conditions (G1)–(G4) are of the form* $(\sigma, \mathrm{Id}_V - \sigma)$ *with* $\sigma \in \mathrm{GL}_k(V)$. *In this case the multiplication given by equation (0.2) becomes* $(\sigma, \mathrm{Id}_V - \sigma) \cdot (\sigma', \mathrm{Id}_V - \sigma') = (\sigma \circ \sigma', \mathrm{Id}_V - \sigma \circ \sigma')$ *and thus* $\mathbb{G}^V_{\mathfrak{g}'}(\rightharpoonup, \theta)$ *is isomorphic to* $\mathrm{GL}_k(V)$. *The conclusion now follows from Theorem 3.6.5.*

The next example deals with the holomorph of a Lie algebra.

Example 3.6.11 *Let* $\mathfrak{g}$ *be a complete Lie algebra (i.e.,* $\mathfrak{g}$ *has trivial center and only inner derivations; see [141] for further details) and* $\mathfrak{H}(\mathfrak{g})$ *its holomorph. Then there exists an isomorphism of groups:*

$$\mathrm{Gal}\left(\mathfrak{H}(\mathfrak{g})/\mathfrak{g}\right) \cong \mathrm{Aut}_{\mathrm{Lie}}(\mathfrak{g})$$

To start with, we point out that since $\mathfrak{g}$ *is complete all derivations are inner, i.e.,* $\mathrm{Der}(\mathfrak{g}) = \{\mathrm{ad}_x \mid x \in \mathfrak{g}\}$. *It can be easily seen that* $\mathfrak{h}(\mathfrak{g})$ *is a unified product between* $\mathfrak{g}$ *and* $\mathrm{Der}(\mathfrak{g})$ *corresponding to the extending system whose non-trivial maps are given as follows:*

$$\mathrm{ad}_x \rightharpoonup g = [x, g], \qquad \{\mathrm{ad}_x, \mathrm{ad}_y\} = \mathrm{ad}_{[x, y]}$$

for all $g, x, y \in \mathfrak{g}$. *Consider now* $(\sigma, r) \in \mathrm{GL}_k(\mathrm{Der}(\mathfrak{g})) \times \mathrm{Hom}_k(\mathrm{Der}(\mathfrak{g}), \mathfrak{g})$ *satisfying the compatibility conditions (G1)–(G4). As for any* $x \in \mathfrak{g}$ *we have* $\sigma(\mathrm{ad}_x) \in \mathrm{Der}(\mathfrak{g})$ *it follows that* $\sigma(\mathrm{ad}_x) = \mathrm{ad}_{\tau(x)}$ *for some bijective linear map* $\tau : \mathfrak{g} \to \mathfrak{g}$. *One can easily see that (G1) is trivially fulfilled, (G3) comes down to* τ *being a Lie algebra map while(G2) yields:*

$$[r(ad_x) + \tau(x) - x, g] = 0, \text{ for all } x, g \in \mathfrak{g}$$

Hence $r(ad_x) + \tau(x) - x \in Z(\mathfrak{g})$ *for all* $x \in \mathfrak{g}$, *where* $Z(\mathfrak{g})$ *denotes the center of the Lie algebra* $\mathfrak{g}$; *as* $\mathfrak{g}$ *is complete we obtain* $r(\mathrm{ad}_x) = x - \tau(x)$ *for all* $x \in \mathfrak{g}$. *Under these assumptions (G4) is also trivially fulfilled. To summarize, we proved that any pair* $(\sigma, r) \in \mathrm{GL}_k(\mathrm{Der}(\mathfrak{g})) \times \mathrm{Hom}_k(\mathrm{Der}(\mathfrak{g}), \mathfrak{g})$ *satisfying the compatibility conditions (G1)–(G4) is implemented by a Lie algebra automorphism* $\tau : \mathfrak{g} \to \mathfrak{g}$ *as follows:*

$$\sigma(\mathrm{ad}_x) = \mathrm{ad}_{\tau(x)}, \quad r(\mathrm{ad}_x) = x - \tau(x)$$

for all $x \in \mathfrak{g}$. *An easy computation shows that the map which sends each pair* $(\sigma, r) \in \mathrm{GL}_k(\mathrm{Der}(\mathfrak{g})) \times \mathrm{Hom}_k(\mathrm{Der}(\mathfrak{g}), \mathfrak{g})$ *to the corresponding Lie algebra automorphism* $\tau : \mathfrak{g} \to \mathfrak{g}$ *is a group automorphism between* $\mathbb{G}^{\mathrm{Der}(\mathfrak{g})}_{\mathfrak{g}}(\rightharpoonup, \theta)$ *and* $\mathrm{Aut}_{\mathrm{Lie}}(\mathfrak{g})$. *Now Theorem 3.6.5 is the last step in obtaining the desired conclusion.*

We now specialize the discussion to extensions of the form $\mathfrak{h}^G \subseteq \mathfrak{h}$, where G is a group acting on a Lie algebra $\mathfrak{h}$. Our approach has Artin's theorem [158, Theorem 1.8] as the source of inspiration: if $G \leq \mathrm{Aut}(K)$ is a finite group of automorphisms of a field K, then K is isomorphic to a crossed product algebra $k\#_{\sigma}\, k[G]^*$ between the field of invariants $k = K^G$ and the dual algebra of the group algebra $k[G]$, associated to some cocycle $\sigma : k[G]^* \otimes k[G]^* \to k$. In what follows we will prove the Lie algebra counterpart of this very important result. Let G be a finite group whose order $|G|$ is invertible in k and suppose G is acting on the Lie algebra $\mathfrak{h}$ via the group morphism $\varphi : G \to \mathrm{Aut}_{\mathrm{Lie}}(\mathfrak{h})$, $\varphi(g)(x) = g \triangleright x$, for all $g \in G$ and $x \in \mathfrak{h}$. Our goal is to describe the Galois group $\mathrm{Gal}\,(\mathfrak{h}/\mathfrak{h}^G)$ and to rebuild $\mathfrak{h}$ from the subalgebra of invariants $\mathfrak{h}^G$ and an extra set of data. We mention that if $G \leq \mathrm{Aut}_{\mathrm{Lie}}(\mathfrak{h})$ is a subgroup of the Lie algebra automorphism of $\mathfrak{h}$ acting on $\mathfrak{h}$ via the canonical action $\sigma \triangleright x := \sigma(x)$, for all $\sigma \in G \leq \mathrm{Aut}_{\mathrm{Lie}}(\mathfrak{h})$, then we have $G \subseteq \mathrm{Gal}\,(\mathfrak{h}/\mathfrak{h}^G)$—as opposed to the classical Artin's theorem, we will see that for Lie algebras we are far from having equality in the inclusion $G \subseteq \mathrm{Gal}\,(\mathfrak{h}/\mathfrak{h}^G)$. Since $|G|$ is invertible in k, we can choose the trace map $t :\, \mathfrak{h} \to \mathfrak{h}^G$ defined by $t(x) := |G|^{-1} \sum_{\gamma\in G} \gamma \triangleright x$, for all $x \in \mathfrak{h}$ as a linear retraction of the inclusion $\mathfrak{h}^G \hookrightarrow \mathfrak{h}$. We shall compute the canonical extending system of $\mathfrak{h}^G$ through $V := \mathrm{Ker}(t)$ associated to the trace map t using the formulas (3.6)–(3.7). For any $x \in V$ and $g \in \mathfrak{h}^G$ we have:

$$\begin{aligned} x \rightharpoonup g &= t([x,\, g]) = |G|^{-1} \sum_{\gamma\in G} \gamma \triangleright [x,\, g] = |G|^{-1} \sum_{\gamma\in G} [\gamma \triangleright x,\, \gamma \triangleright g] \\ &= |G|^{-1} \sum_{\gamma\in G} [\gamma \triangleright x,\, g] = [t(x),\, g] = 0 \end{aligned}$$

where the equalities in the last line follow from $g \in \mathfrak{h}^G$ and $x \in V = \mathrm{Ker}(t)$, respectively. Moreover, we can easily see that the action $\leftharpoonup$, the cocycle θ and the quasi-bracket $\{-,\, -\}$ on V take the form:

$$x \leftharpoonup g = [x,\, g], \quad \theta(x,\, y) = |G|^{-1} \sum_{\gamma\in G} [\gamma \triangleright x,\, \gamma \triangleright y] \tag{3.70}$$

$$\{x,\, y\} = [x,\, y] - |G|^{-1} \sum_{\gamma\in G} [\gamma \triangleright x,\, \gamma \triangleright y] \tag{3.71}$$

for all $x,\, y \in V$ and $g \in \mathfrak{h}^G$. The left action $\rightharpoonup$ being the trivial map has an important consequence: it follows from Section 3.3 that the unified product $\mathfrak{h}^G \natural\, V$ associated to this canonical extending system of $\mathfrak{h}^G$ by V is in fact a skew crossed product $\mathfrak{h}^G \#^{\bullet}\, V$ and the map defined for any $g \in \mathfrak{h}^G$ and $x \in V$ by:

$$\varphi : \mathfrak{h}^G \#^{\bullet}\, V \to \mathfrak{h}, \qquad \varphi(g, x) := g + x \tag{3.72}$$

is an isomorphism of Lie algebras. The Lie bracket on $\mathfrak{h}^G \#^{\bullet}\, V$ given by equation (3.21) takes the following form:

$$\begin{aligned} [(g,\, x),\, (g',\, x')] := \Big(& [g,\, g'] + |G|^{-1} \sum_{\gamma\in G} [\gamma \triangleright x,\, \gamma \triangleright x'], \\ & [x,\, x'] - |G|^{-1} \sum_{\gamma\in G} [\gamma \triangleright x,\, \gamma \triangleright x'] + [x,\, g'] - [x',\, g] \Big) \end{aligned} \tag{3.73}$$

for all $g, g' \in \mathfrak{h}^G$ and $x, x' \in V$. Given a group G acting on a Lie algebra $\mathfrak{h}$, the isomorphism given in equation (3.72) provides the reconstruction of $\mathfrak{h}$ from the subalgebra of invariants $\mathfrak{h}^G$. We continue our investigation in order to describe the Galois group $\mathrm{Gal}\,(\mathfrak{h}/\mathfrak{h}^G)$. Since the components of the canonical extending system given by equations (3.70)–(3.71) are implemented only by the action φ of G on $\mathfrak{h}$, we shall denote the group $\mathbb{G}_{\mathfrak{h}^G}^V\,(\leftharpoonup, \rightharpoonup, \theta, \{-, -\})$ constructed in Theorem 3.6.5 by $\mathbb{G}_{\mathfrak{h}^G}^V\,(\varphi)$. Thus $\mathbb{G}_{\mathfrak{h}^G}^V\,(\varphi)$ consists of the set of all pairs $(\sigma, r) \in \mathrm{GL}_k(V) \times \mathrm{Hom}_k(V, \mathfrak{h}^G)$ satisfying the following compatibility conditions for all $g \in \mathfrak{h}^G$ and $x, y \in V$:

$$\sigma([x, g]) = [\sigma(x), g], \qquad r([x, g]) = [r(x), g]$$
$$\sigma([x, y]) - [\sigma(x), \sigma(y)] = [\sigma(x), r(y)] + [r(x), \sigma(y)] + $$
$$+|G|^{-1} \sum_{\gamma\in G} \sigma([\gamma \triangleright x, \gamma \triangleright y]) - |G|^{-1} \sum_{\gamma\in G} [\gamma \triangleright \sigma(x), \gamma \triangleright \sigma(y)]$$
$$r([x, y]) - [r(x), r(y)] = |G|^{-1} \sum_{\gamma\in G} r([\gamma \triangleright x, \gamma \triangleright y]) + $$
$$+|G|^{-1} \sum_{\gamma\in G} [\gamma \triangleright \sigma(x), \gamma \triangleright \sigma(y)] - |G|^{-1} \sum_{\gamma\in G} [\gamma \triangleright x, \gamma \triangleright y]$$

which is exactly what is left from axioms (G1)–(G4) after using equations (3.70)–(3.71) and the fact that $\rightharpoonup$ is the trivial action. We note that the first two compatibilities above show that σ and r are morphisms of right Lie $\mathfrak{h}^G$-modules while the last two compatibilities measures how far they are from being Lie algebra maps. $\mathbb{G}_{\mathfrak{h}^G}^V\,(\varphi)$ is a group with multiplication given by equation (0.2). We record all these facts in the following:

Theorem 3.6.12 ***(Artin's Theorem for Lie algebras)*** *Let G be a finite group of invertible order in k acting on a Lie algebra $\mathfrak{h}$ via $\varphi : G \to \mathrm{Aut}_{\mathrm{Lie}}(\mathfrak{h})$. Let $\mathfrak{h}^G \subseteq \mathfrak{h}$ be the subalgebra of invariants and $V = \mathrm{Ker}(t)$, where $t : \mathfrak{h} \to \mathfrak{h}^G$ is the trace map. Then:*

(1) The map defined for any $g \in \mathfrak{h}^G$ and $x \in V$ by:

$$\varphi : \mathfrak{h}^G \#^{\bullet} V \to \mathfrak{h}, \qquad \varphi(g, x) := g + x \tag{3.74}$$

is an isomorphism of Lie algebras, where $\mathfrak{h}^G \#^{\bullet} V$ is the skew crossed product of Lie algebras having the bracket given by (3.73).

(2) The map defined for any $(\sigma, r) \in \mathbb{G}_{\mathfrak{h}^G}^V\,(\varphi)$, $g \in \mathfrak{h}^G$ and $x \in V$ by:

$$\Omega : \mathbb{G}_{\mathfrak{h}^G}^V\,(\varphi) \to \mathrm{Gal}\,(\mathfrak{h}/\mathfrak{h}^G), \quad \Omega(\sigma, r)(g + x) := g + r(x) + \sigma(x) \tag{3.75}$$

is an isomorphism of groups.

Even if G is a finite subgroup of $\mathrm{Aut}_{\mathrm{Lie}}(\mathfrak{h})$, Theorem 3.6.12 (2) shows that we are far away from having $\mathrm{Gal}\,(\mathfrak{h}/\mathfrak{h}^G) \cong G$ as in the case of fields. We present a relevant example below:

Example 3.6.13 *Let k be a field of characteristic $\neq 2$ and $\varphi : k^* \to \mathrm{Aut}_{\mathrm{Lie}}(\mathfrak{sl}(2,k))$ the action of k^* on $\mathfrak{sl}(2,k)$ given by equation (3.67). The subalgebra of invariants $\mathfrak{sl}(2,k)^{k^*}$ of this action is just ke_3 and Example 3.6.7 shows that the Galois group $\mathrm{Gal}\,(\mathfrak{sl}(2,k)/\mathfrak{sl}(2,k)^{k^*})$ is isomorphic to k^*. This situation occurs rarely. Indeed, if we consider $G := U_2(k) = \{\pm 1\}$ the cyclic subgroup of k^* of roots of unity of order two and the same action of $U_2(k) \leq k^*$ on $\mathfrak{sl}(2,k)$ as above we obtain the same subalgebra of invariants, namely $\mathfrak{sl}(2,k)^{U_2(k)} = ke_3$. Hence we have that $\mathrm{Gal}\,(\mathfrak{sl}(2,k)/\mathfrak{sl}(2,k)^{U_2(k)}) \cong k^* \neq U_2(k)$.*

A crucial step in applying Theorem 3.6.12 is the description of the kernel of the trace map $t : \mathfrak{h} \to \mathfrak{h}^G$, which heavily depends of the group G and on the action φ. In the case of cyclic groups acting on fields this kernel is described by Hilbert's theorem [158, Theorem 6.3]. As a nice surprise, its counterpart for Lie algebras is also true but the proof uses completely different techniques.

Theorem 3.6.14 ***(Hilbert's Theorem 90 for Lie algebras)*** *Let G be a finite cyclic group generated by an element γ whose order n is invertible in k. Let $\varphi : G \to \mathrm{Aut}_{\mathrm{Lie}}(\mathfrak{h})$ be a morphism of groups and $t : \mathfrak{h} \to \mathfrak{h}^G$ the trace map. Then $\mathrm{Ker}(t) = \{y - \gamma \triangleright y \,|\, y \in \mathfrak{h}\}$.*

Proof: It is straightforward to see that $t(y - g \triangleright y) = 0$, for all $y \in \mathfrak{h}$. Conversely, let $a \in \mathrm{Ker}(t)$. We define recursively the sequence of elements $(d_i)_{i\geq 0}$ of $\mathrm{Hom}_k(\mathfrak{h}, \mathfrak{h})$ by the formulas:

$$d_0(y) := a + y, \qquad d_{i+1}(y) := a + \gamma \triangleright d_i(y),$$

for all $i \geq 0$ and $y \in \mathfrak{h}$. Thus, we have $d_1(y) = a + \gamma \triangleright a + \gamma \triangleright y, \cdots$, $d_{n-2}(y) = a + \gamma \triangleright a + \cdots + \gamma^{n-2} \triangleright a + \gamma^{n-2} \triangleright y$ and using $t(a) = 0$ we obtain $d_{n-1}(y) = \gamma^{n-1} \triangleright y$ and hence $d_n(y) = a + \gamma \triangleright (\gamma^{n-1} \triangleright y) = a + y = d_0(y)$. Therefore, $d_n = d_0$ i.e., the sequence $(d_i)_{i\geq 0}$ is periodic. Now, in the abelian group $\mathrm{Hom}_k(\mathfrak{h}, \mathfrak{h})$ we add all the equalities listed below:

$$d_1 = a + \gamma \triangleright d_0, \quad d_2 = a + \gamma \triangleright d_1, \quad \cdots, d_{n-1} = a + \gamma \triangleright d_{n-2}, \quad d_n = a + \gamma \triangleright d_{n-1}$$

and using $d_n = d_0$, we obtain $\sum_{i=0}^{n-1} d_i = n\,a + \gamma \triangleright \left(\sum_{i=0}^{n-1} d_i\right)$. If $d_0 + d_1 + \cdots d_{n-1} = 0$ in the abelian group $\mathrm{Hom}_k(\mathfrak{h}, \mathfrak{h})$, we obtain using the invertibility of n in k, that $a = 0 = 0 - \gamma \triangleright 0$ and we are done. On the other hand, if $d_0 + d_1 + \cdots d_{n-1} \neq 0$ we can pick some $z \in \mathfrak{h}$ such that $y := \sum_{i=0}^{n-1} d_i(z) \neq 0$. Then, using $d_n(z) = d_0(z)$, we have:

$$n\,a + \gamma \triangleright y = n\,a + \sum_{i=0}^{n-1} \gamma \triangleright d_i(z) = \sum_{i=0}^{n-1} \big(a + \gamma \triangleright d_i(z)\big) = \sum_{i=0}^{n-1} d_{i+1}(z) = y$$

This shows that $a = n^{-1}\left(y - \gamma \triangleright y\right)$ and the proof is finished. □

Now we introduce the following:

Definition 3.6.15 *Let $\mathfrak{h}$ be a Lie algebra, G a group, $\varphi : G \to \mathrm{Aut}_{\mathrm{Lie}}(\mathfrak{h})$ a morphism of groups, $\gamma \in G$ and $\mathfrak{h}_\gamma := \{y - \gamma \triangleright y \,|\, y \in \mathfrak{h}\}$. The action φ is called γ-abelian if:*

$$[g \triangleright z,\, g' \triangleright z'] = 0 \tag{3.76}$$

for all $g \neq g' \in G$ and $z,\, z' \in \mathfrak{h}_\gamma$.

The structure theorem for cyclic Galois extensions of fields [158, Theorem 6.2] can be rephrased as follows: if $G \leq \mathrm{Aut}(K)$ is a cyclic subgroup of order n of the group of automorphisms of a field K of characteristic zero and $k := K^G$, then K is isomorphic to the splitting field over k of a polynomial of the form $X^n - a \in k[X]$. The Lie algebra counterpart of this result now follows by replacing the splitting field with the semidirect product of Lie algebras:

Corollary 3.6.16 *Let $\mathfrak{h}$ be a Lie algebra, G a finite cyclic group generated by an element γ whose order n is invertible in k and $\varphi : G \to \mathrm{Aut}_{\mathrm{Lie}}(\mathfrak{h})$ a γ-abelian morphism of groups. Then, the map defined for any $g \in \mathfrak{h}^G$ and $x \in \mathfrak{h}_\gamma$ by:*

$$\varphi : \mathfrak{h}^G \rtimes \mathfrak{h}_\gamma \to \mathfrak{h}, \qquad \varphi(g, x) := g + x \tag{3.77}$$

is an isomorphism of Lie algebras, where $\mathfrak{h}^G \rtimes \mathfrak{h}_\gamma$ is the semidirect product of Lie algebras associated to the right action $\leftharpoonup : \mathfrak{h}_\gamma \times \mathfrak{h}^G \to \mathfrak{h}_\gamma$, given by $x \leftharpoonup g := [x,\, g]$.

Proof: Using Theorem 3.6.14 together with Theorem 3.6.12 we only need to prove that the cocycle $\theta : \mathfrak{h}_\gamma \times \mathfrak{h}_\gamma \to \mathfrak{h}^G$ given by (3.70) is the trivial map. Moreover, in this case it also follows that the bracket $\{-,\, -\}$ on $\mathfrak{h}_\gamma$ depicted in (3.71) coincides with the Lie bracket on $\mathfrak{h}$, i.e., $\{x,\, y\} = [x,\, y]$, for all x, $y \in \mathfrak{h}_\gamma$ and $\mathfrak{h}_\gamma$ is an ideal of $\mathfrak{h}$. Indeed, let $y - \gamma \triangleright y$ and $y' - \gamma \triangleright y'$ be two elements of $\mathfrak{h}_\gamma$, for some y, $y \in \mathfrak{h}$. Then we have:

$$\begin{aligned}
\theta(y - \gamma \triangleright y,\, y' - \gamma \triangleright y') &= n^{-1} \sum_{\delta \in G} [\delta \triangleright (y - \gamma \triangleright y),\, \delta \triangleright (y' - \gamma \triangleright y')] \\
&= n^{-1} \sum_{i=0}^{n-1} [\gamma^i \triangleright (y - \gamma \triangleright y),\, \gamma^i \triangleright (y' - \gamma \triangleright y')] \\
&= n^{-1} [\sum_{i=0}^{n-1} \gamma^i \triangleright (y - \gamma \triangleright y),\, \sum_{i=0}^{n-1} \gamma^i \triangleright (y - \gamma \triangleright y)] = 0
\end{aligned}$$

where in the third equality we used the fact that φ is a γ-abelian action while the final equality holds due to the following trivial identity: $\sum_{i=0}^{n-1} \gamma^i \triangleright (y - \gamma \triangleright y) = 0$. □

Remark 3.6.17 *Under the assumptions of Corollary 3.6.16 we can provide an simpler description of the Galois group $\mathrm{Gal}\,(\mathfrak{h}/\mathfrak{h}^G) \cong \mathbb{G}_{\mathfrak{h}^G}^{\mathfrak{h}_\gamma}\,(\varphi)$. Indeed, in this case the group $\mathbb{G}_{\mathfrak{h}^G}^{\mathfrak{h}_\gamma}\,(\varphi)$ as defined in Theorem 3.6.12 consists of the set* *of all pairs* $(\sigma, r) \in \mathrm{GL}_k(\mathfrak{h}_\gamma) \times \mathrm{Hom}_k(\mathfrak{h}_\gamma, \mathfrak{h}^G)$ *satisfying the following compatibility conditions for any* $g \in \mathfrak{h}^G$ *and* $x, y \in \mathfrak{h}_\gamma$*:*

$$\sigma([x, g]) = [\sigma(x), g], \qquad r([x, g]) = [r(x), g], \qquad r([x, y]) = [r(x), r(y)]$$
$$\sigma([x, y]) - [\sigma(x), \sigma(y)] = [\sigma(x), r(y)] + [r(x), \sigma(y)]$$

which is a subgroup in the semidirect product of groups $\mathrm{GL}_k(\mathfrak{h}_\gamma) \rtimes \mathrm{Hom}_k(\mathfrak{h}_\gamma, \mathfrak{h}^G)$.

Applications and examples

We end this chapter with some applications and explicit examples of Galois groups. The simplest case is that of extensions $\mathfrak{g} \subseteq \mathfrak{h}$ for which the codimension of $\mathfrak{g}$ in $\mathfrak{h}$ is equal to 1. In this case we will show that the Galois group $\mathrm{Gal}\,(\mathfrak{h}/\mathfrak{g})$ is metabelian. To this end, consider $\mathfrak{g} \subseteq \mathfrak{h}$ to be an extension of Lie algebras such that $\mathfrak{g}$ has codimension 1 in $\mathfrak{h}$. Thus, we can write $\mathfrak{h} = \mathfrak{g} + V$, where $V := kx$, for a fixed element $x \in \mathfrak{h} \backslash \mathfrak{g}$. We choose the map p defined by $p(x) := 0$ and $p(g) = g$, for all $g \in \mathfrak{g}$, as a retraction of the inclusion map $\mathfrak{g} \hookrightarrow \mathfrak{h}$. Now recall that by Proposition 3.2.4 the space of all Lie extending systems of $\mathfrak{g}$ through $V = kx$ is parameterized by the set $\mathrm{TwDer}(\mathfrak{g})$ of all twisted derivations of $\mathfrak{g}$ (Definition 3.2.2). Let $(\lambda, \Delta) \in \mathrm{TwDer}(\mathfrak{g})$ be the twisted derivation associated to the canonical Lie extending system of $\mathfrak{g}$ through V arising from p via (3.11) and denote by $\mathfrak{g}_{(\lambda,\Delta)} := \mathfrak{g} \natural kx$ the corresponding unified product. Furthermore, recall that the Lie algebra $\mathfrak{g}_{(\lambda,\Delta)}$ has $\mathfrak{g} \times k$ as the underlying vector space and the bracket is given as follows for all $x, y \in \mathfrak{g}$ and $a, b \in k$:

$$\{(x, a), (y, b)\} := \Big([x, y] + b\,\Delta(x) - a\,\Delta(y),\; b\,\lambda(x) - a\,\lambda(y)\Big) \tag{3.78}$$

Of course $\mathfrak{g}_{(\lambda,\Delta)}$ contains $\mathfrak{g} \cong \mathfrak{g} \times \{0\}$ as a subalgebra of codimension 1.

Continuing our investigation it follows that the Galois group $\mathrm{Gal}\,(\mathfrak{h}/\mathfrak{g}) \cong \mathrm{Gal}\,(\mathfrak{g}_{(\lambda,\Delta)}/\mathfrak{g})$, which is a special case of the isomorphism given by (3.68). We denote by $\mathbb{G}_{\mathfrak{g}}\,(\lambda, \Delta)$ the set of all pairs $(u, g_0) \in k^* \times \mathfrak{g}$ satisfying the following compatibility condition for all $g \in \mathfrak{g}$:

$$\lambda(g)\,g_0 = [g_0, g] + (u - 1)\,\Delta(g) \tag{3.79}$$

Then $\mathbb{G}_{\mathfrak{g}}\,(\lambda, \Delta)$ is a subgroup in the metabelian group $\mathbb{G}_{\mathfrak{g}} := k^* \rtimes \mathfrak{g}$ whose multiplication is given by equation (0.3), that is $(u, g) \cdot (u', g') := (uu', u'g + g')$, for all $u, u' \in k^*$ and $g, g' \in \mathfrak{g}$. We can now prove the following:

Corollary 3.6.18 *Let* $\mathfrak{g} \subseteq \mathfrak{h}$ *be a Lie subalgebra of codimension* 1 *in* $\mathfrak{h}$ *and* $(\lambda, \Delta) \in \mathrm{TwDer}(\mathfrak{g})$ *the twisted derivation defined by equation (3.11) for a* *fixed* $x \in \mathfrak{h} \setminus \mathfrak{g}$. *Then there exists an isomorphism of groups given for any* $(u, g_0) \in \mathbb{G}_{\mathfrak{g}}(\lambda, \Delta)$, $g \in \mathfrak{g}$ *and* $\alpha \in k$ *by:*

$$\Omega : \mathbb{G}_{\mathfrak{g}}(\lambda, \Delta) \to \mathrm{Gal}(\mathfrak{h}/\mathfrak{g}), \quad \Omega(u, g_0)(g + \alpha x) := g + \alpha g_0 + u\alpha x \quad (3.80)$$

In particular, the Galois group $\mathrm{Gal}(\mathfrak{h}/\mathfrak{g})$ *is metabelian and hence solvable.*

Proof: We apply Theorem 3.6.5: since $V = kx$, any linear automorphism $\sigma : V \to V$ is uniquely determined by an invertible element $u \in k^*$ via $\sigma(x) := ux$ while a linear map $r : V \to \mathfrak{g}$ is implemented by an element $g_0 \in \mathfrak{g}$ via $r(x) := g_0$. Now, the axioms (G1), (G3) and (G4) which define the group $\mathbb{G}_{\mathfrak{g}}^{V}(\leftharpoonup, \rightharpoonup, \theta, \{-, -\})$ from Theorem 3.6.5 are trivially fulfilled, while axiom (G2) comes down to the compatibility condition (3.79). Finally, the group $\mathrm{Gal}(\mathfrak{h}/\mathfrak{g})$ is metabelian due to its embedding in the metabelian group $k^* \rtimes \mathfrak{g}$. □

Example 3.6.19 *Let* $n \in \mathbb{N}^*$ *be a positive integer and consider the extension of Lie algebras* $\mathfrak{h}^{2n+1} \subseteq \mathfrak{t}^{2n+2}$, *where* $\mathfrak{h}^{2n+1}$ *is the* $(2n+1)$*-dimensional Heisenberg Lie algebra from Example 3.6.8 and* $\mathfrak{t}^{2n+2}$ *is the Lie algebra with basis* $\{x_1, \cdots, x_n, y_1, \cdots, y_n, w, u\}$ *and bracket given for all* $i = 1, \cdots, n$ *by:* $[x_i, y_i] = w$, $[u, x_i] = w + u$, $[u, y_i] = w + u$. *Then there exists an isomorphism of groups:*

$$\mathrm{Gal}(\mathfrak{t}^{2n+2}/\mathfrak{h}^{2n+1}) \cong (k^*, \cdot).$$

First observe that the Lie algebra $\mathfrak{t}^{2n+2}$ *is isomorphic to* $\mathfrak{h}^{2n+1}_{(\lambda,\Delta)}$, *where the twisted derivation* (λ, Δ) *of the Heisenberg Lie algebra* $\mathfrak{h}^{2n+1}$ *is given by:* $\lambda(w) := 0$, $\lambda(x_i) = \lambda(y_i) := 1$, $\Delta(w) := 0$, $\Delta(x_i) = \Delta(y_i) := w$, *for all* $i = 1, 2, \cdots, n$. *Now a straightforward computation shows that* $\mathbb{G}_{\mathfrak{h}^{2n+1}}(\lambda, \Delta) = \{(\alpha, (\alpha - 1)w) \mid \alpha \in k^*\}$ *and the map* $\varphi : (\mathbb{G}_{\mathfrak{h}^{2n+1}}(\lambda, \Delta), \cdot) \to k^*$ *given by* $\varphi(\alpha, (\alpha - 1)w) = \alpha$ *is a group isomorphism where* $\cdot$ *is the multiplication defined by (0.3). The conclusion follows by Corollary 3.6.18.*

Given that supersolvable Lie algebras provide examples of flag extensions, we propose the following definition as the counterpart for Lie algebras of normal radical extensions of fields:

Definition 3.6.20 *An extension* $\mathfrak{g} \subseteq \mathfrak{h}$ *of Lie algebras is called a radical extension if there exists a chain of subalgebras as in (3.8) such that each* $\mathfrak{h}_{i-1}$ *is invariant with respect to any element* $\tau \in \mathrm{Gal}(\mathfrak{h}_i/\mathfrak{g})$, *i.e.,* $\tau(\mathfrak{h}_{i-1}) \subseteq \mathfrak{h}_{i-1}$, *for all* $\tau \in \mathrm{Gal}(\mathfrak{h}_i/\mathfrak{g})$ *and* $i = 1, \cdots, m$.

If $\mathfrak{g}$ has codimension 1 in $\mathfrak{h}$, then $\mathfrak{h}/\mathfrak{g}$ is a radical extension. Based on Theorem 3.6.5 and Corollary 3.6.18, exactly as in the classical case of radical extensions of fields, we can prove the following:

Theorem 3.6.21 *Let* $\mathfrak{g} \subseteq \mathfrak{h}$ *be a radical extension of finite-dimensional Lie algebras. Then the Galois group* $\mathrm{Gal}(\mathfrak{h}/\mathfrak{g})$ *is solvable.*

Proof: Consider a finite chain of subalgebras as in (3.8). We will proceed by induction on m. If $m = 1$ the conclusion follows by Corollary 3.6.18. Now let $m > 1$ and assume the statement to be true for $m - 1$, that is the group $\mathrm{Gal}\,(\mathfrak{h}_{m-1}/\mathfrak{g})$ is solvable. Then, the map:

$$\Gamma : \mathrm{Gal}\,(\mathfrak{h}/\mathfrak{g}) \to \mathrm{Gal}\,(\mathfrak{h}_{m-1}/\mathfrak{g}), \qquad \Gamma(\tau) := \tau_{|_{\mathfrak{h}_{m-1}}}$$

where $\tau_{|_{\mathfrak{h}_{m-1}}}$ is the restriction of τ to $\mathfrak{h}_{m-1}$ is well defined since the extension is radical, the Lie algebras are finite-dimensional and Γ is a morphism of groups. Now $\mathrm{Ker}(\Gamma) = \mathrm{Gal}(\mathfrak{h}/\mathfrak{h}_{m-1})$ which is a metabelian (in particular solvable) group again by Corollary 3.6.18. Thus, we obtain an isomorphism of groups $\mathrm{Gal}\,(\mathfrak{h}/\mathfrak{g})/\mathrm{Gal}\,(\mathfrak{h}/\mathfrak{h}_{m-1}) \cong \mathrm{Im}(\Gamma)$, and $\mathrm{Im}(\Gamma)$ is a solvable group as a subgroup in such a group. To conclude, we have obtained that $\mathrm{Gal}\,(\mathfrak{h}/\mathfrak{g})$ is an extension of a solvable group $\mathrm{Im}(\Gamma)$ by the solvable group $\mathrm{Gal}\,(\mathfrak{h}/\mathfrak{h}_{m-1})$, hence $\mathrm{Gal}\,(\mathfrak{h}/\mathfrak{g})$ is solvable too. □

The compatibility condition (3.79) which describes the elements of the group $\mathrm{Gal}\,(\mathfrak{g}_{(\lambda,\Delta)}/\mathfrak{g})$ is crucial and deserves a thorough analysis. First, observe that $(1,\, 0) \in \mathbb{G}_{\mathfrak{g}}\,(\lambda,\, \Delta)$. On the other hand, if $(u,\, g_0) \in \mathbb{G}_{\mathfrak{g}}\,(\lambda,\, \Delta)$, for some $u \neq 1$, then (3.79) implies that Δ is given by the formula $\Delta(g) = (u - 1)^{-1}\,\big(\lambda(g)\, g_0 - [g_0,\, g]\big)$, for all $g \in \mathfrak{g}$. A straightforward computation shows that the compatibility condition (3.10) is trivially fulfilled, being equivalent to the Jacobi identity. The center of a Lie algebra $\mathfrak{g}$ will be denoted by $Z(\mathfrak{g}) := \{g \in \mathfrak{g} \,|\, [g,\, -] = 0\}$. Then $Z(\mathfrak{g})$ is an abelian subgroup of $(\mathfrak{g}, +)$ and it can be realized as a Galois group of the following type of Lie algebra extensions:

Corollary 3.6.22 *Let $\mathfrak{g}$ be a Lie algebra and $\Delta \in \mathrm{Der}(\mathfrak{g})$ a derivation that is not inner. Then there exists an isomorphism of groups* $\mathrm{Gal}\,(\mathfrak{g}_{(\Delta)}/\mathfrak{g}) \cong Z(\mathfrak{g})$.

Proof: Using (3.79) for $\lambda := 0$, we obtain that $(u,\, g_0) \in \mathbb{G}_{\mathfrak{g}}\,(\Delta) := \mathbb{G}_{\mathfrak{g}}\,(0,\, \Delta)$ if and only if $(u - 1)\,\Delta(g) = [g,\, g_0]$, for all $g \in \mathfrak{g}$. Hence, $(1,\, g_0) \in \mathbb{G}_{\mathfrak{g}}\,(\Delta)$ if and only if $g_0 \in Z(\mathfrak{g})$. On the other hand, since Δ is not inner, it follows that $\mathbb{G}_{\mathfrak{g}}\,(\Delta)$ does not contain elements of the form $(u,\, g_0)$, with $u \neq 1$. Now, we apply Corollary 3.6.18. □

Example 3.6.23 *Let $n \in \mathbb{N}^*$ be a positive integer and consider $\mathfrak{h}^{2n+1}$ to be the $(2n+1)$-dimensional Heisenberg Lie algebra from Example 3.6.8. Then $\Delta : \mathfrak{h}^{2n+1} \to \mathfrak{h}^{2n+1}$ given by $\Delta(x_i) := y_i$, $\Delta(y_i) = \Delta(w) := 0$, for all $i = 1, 2, \cdots, n$ is a derivation of $\mathfrak{h}^{2n+1}$ that is not inner. Furthermore, we denote by $\mathfrak{b}^{2n+2}$ the Lie algebra $\mathfrak{h}^{2n+1}_{(\Delta)}$: it has the k-basis $\{x_1, \cdots, x_n,\, y_1, \cdots, y_n,\, w,\, z\}$ and bracket given for any $i = 1, \cdots, n$ by $[x_i,\, y_i] = w$, $[z,\, x_i] = y_i$. By applying Corollary 3.6.22 and taking into account that $Z(\mathfrak{h}^{2n+1}) = kw \cong (k, +)$ we obtain that there exists an isomorphism of groups* $\mathrm{Gal}\,(\mathfrak{b}^{2n+2}/\mathfrak{h}^{2n+1}) \cong (k, +)$.

Corollary 3.6.24 *Let $\mathfrak{g}$ be a sympathetic Lie subalgebra of codimension* 1 *in a Lie algebra $\mathfrak{h}$. Then there exists an isomorphism of groups* $\mathrm{Gal}\,(\mathfrak{h}/\mathfrak{g}) \cong k^*$. *In particular, if k is a field of characteristic zero then* $\mathrm{Gal}\,(\mathfrak{gl}(m,k)/\mathfrak{sl}(m,k)) \cong k^*$.

Proof: Indeed, since $\mathfrak{g}$ is perfect we obtain that $\mathfrak{h} \cong \mathfrak{g}_{(\Delta)}$, for some derivation Δ of $\mathfrak{g}$ [20, Proposition 2.1]. Let $\delta \in \mathfrak{g}$ such that $\Delta = [\delta, -]$. By applying the compatibility condition (3.79) for $\lambda := 0$ and $\Delta := [\delta, -]$ we obtain that $(u, g_0) \in \mathbb{G}_{\mathfrak{g}}\,(\Delta)$ if and only if $[g, g_0 + (u-1)\delta] = 0$, for all $g \in \mathfrak{g}$. Since $\mathrm{Z}(\mathfrak{g}) = \{0\}$, this is equivalent to the fact that $g_0 = (1-u)\delta$. Hence $\mathbb{G}_{\mathfrak{g}}\,(\Delta)$ consists of all elements of the form $(u, (1-u)\delta)$, for any $u \in k^*$ and there exists an isomorphism of groups $\mathbb{G}_{\mathfrak{g}}\,(\Delta) \cong k^*$. Now we apply Corollary 3.6.18. □

All examples of Lie algebra extensions $\mathfrak{h}/\mathfrak{g}$ presented so far have a non-trivial Galois group. We end the paper with a Lie algebra extension whose Galois group is trivial:

Example 3.6.25 *Let k be a field of characteristic $\neq 2$ and consider $\mathfrak{g}$ to be the perfect* 5*-dimensional Lie algebra with the basis* $\{e_1, e_2, e_3, e_4, e_5\}$ *and bracket given by:*

$$[e_1, e_2] = e_3, \quad [e_1, e_3] = -2e_1, \quad [e_1, e_5] = [e_3, e_4] = e_4$$
$$[e_2, e_3] = 2e_2, \quad [e_2, e_4] = e_5, \quad [e_3, e_5] = -e_5$$

It was proven in [20, Example 3.7] that the derivation given in matrix form by $\Delta := e_{11} - e_{41} - e_{22} + e_{53} - e_{44} - 2\,e_{55}$ *is not inner, where* $e_{ij} \in \mathcal{M}_5(k)$ *is the matrix having* 1 *in the* $(i,j)^{th}$ *position and zeros elsewhere. On the other hand, a straightforward computation shows that* $Z(\mathfrak{g}) = \{0\}$. *By applying Corollary 3.6.22 it follows that the extension* $\mathfrak{g} \subseteq \mathfrak{g}_{(\Delta)}$ *has trivial Galois group* $\{\mathrm{Id}_{\mathfrak{g}_{(\Delta)}}\}$.

Bibliographical Notes

The material presented in this chapter is part of the author's papers [19], [26] and respectively [20].

Chapter 4

Associative algebras

At the level of unitary associative algebras, Hölder's *extension problem* was first considered by Everett [100] and Hochschild [129] and is still an open and notoriously difficult problem. In its full generality it can be rephrased as follows:

Let A be an algebra, E a vector space and $\pi : E \to A$ a k-linear epimorphism of vector spaces. Describe and classify all algebra structures that can be defined on E such that $\pi : E \to A$ becomes a morphism of algebras.

The partial answer to the extension problem was given in [129, Theorem 6.2]: all algebra structures $\cdot$ on E such that $W := \mathrm{Ker}(\pi)$ is a two-sided ideal of null square are classified by the second Hochschild cohomological group $\mathrm{H}^2(A, W)$. The elementary manner used to state the extension problem allows us to consider its categorical dual, called the extending structures problem, which is the subject of the present chapter:

Let A be a unitary associative algebra and E a vector space containing A as a subspace. Describe and classify the set of all unitary associative algebra structures "$\cdot$" that can be defined on E such that A becomes a subalgebra of $(E, \cdot)$.

Compared to the Lie (resp. Leibniz) algebra situation considered in the previous chapter, we will see that the extending structures (ES) problem in the associative algebra setting turns out to be more difficult due to the algebra structure's lack of anti-symmetry which simplified the construction performed for Lie algebras. If we fix V a complement of A in E, then the ES problem can be rephrased equivalently as follows: describe and classify all unitary associative algebras containing A as a subalgebra of codimension equal to $\dim(V)$. The ES problem is the dual of the extension problem and it also generalizes the following so-called *radical embedding problem* introduced by Hall [127]: given a finite-dimensional nilpotent algebra A, describe the family of all unitary algebras with radical isomorphic to A; for more details see [111]. The ES problem is a very difficult question: if $A = k$ then the ES problem asks in fact for the classification of all algebra structures on a given vector space E. For this reason, from now on we will assume that $A \neq k$. Regardless of the difficulty of the ES problem, we can still provide detailed answers to it in certain special cases which depend mainly on the choice of the algebra A and on the codimension of A in E. Moreover, a new class of algebras, which we call *supersolvable* algebras, appear on the route as the associative algebra counterpart of supersolvable Lie algebras [46]: these are finite-dimensional algebras E for which there exists a finite chain of subalgebras $E_0 := k \subset E_1 \subset \cdots \subset E_m := E$ such that E_i has codimension 1 in E_{i+1}, for all $i = 0, \cdots, m-1$. All supersolvable algebras of a given dimension can be classified using a recursive type algorithm in which the key step is the one performed for codimension 1: the crucial role will be played by the characters of the algebra A and by some twisted derivations of A satisfying certain compatibility conditions.

The chapter is organized as follows. In Section 4.1 we will perform the abstract construction of the unified product $A \ltimes V$: it is associated to an algebra A, a vector space V and a system of data $\Omega(A, V) = \big(\triangleleft, \triangleright, \leftharpoonup, \rightharpoonup, f, \cdot\big)$ called an extending datum of A through V. Theorem 4.1.2 establishes the set of axioms that need to be satisfied by $\Omega(A, V)$ such that $A \ltimes V$ with a certain multiplication becomes an associative unitary algebra, i.e., a unified product. In this case, $\Omega(A, V) = \big(\triangleleft, \triangleright, \leftharpoonup, \rightharpoonup, f, \cdot\big)$ will be called an *algebra extending structure* of A through V. Compared to the Lie algebra situation, the construction of the unified product $A \ltimes V$ for algebras requires two more actions $\leftharpoonup : A \times V \to A$ and $\rightharpoonup : A \times V \to V$ which connects A and V. These two actions are missing in the case of Lie algebras as a consequence of the anti-symmetry on the bracket. Now let A be an algebra, E a vector space containing A as a subspace and V a given complement of A in E. Theorem 4.1.4 provides the answer to the description part of the ES problem: there exists an algebra structure $\cdot$ on E such that A is a subalgebra of $(E, \cdot)$ if and only if there exists an isomorphism of algebras $(E, \cdot) \cong A \ltimes V$, for some algebra extending structure $\Omega(\mathfrak{g}, V) = \big(\triangleleft, \triangleright, \leftharpoonup, \rightharpoonup, f, \cdot\big)$ of A through V. Moreover, the algebra isomorphism $(E, \cdot) \cong A \ltimes V$ can be chosen in such a way that it stabilizes A and co-stabilizes V. Based on this result we are able to give the theoretical answer to the ES problem in Theorem 4.1.8: all algebra structures on E containing A as a subalgebra are classified by two non-abelian cohomological type objects which are explicitly constructed. The first one is denoted by $\mathcal{AH}^2_A(V, A)$ and will classify all such structures up to an isomorphism that stabilizes A. We also indicate the bijection between the elements of $\mathcal{AH}^2_A(V, A)$ and the isomorphism classes of all extending structures of A to E. Having in mind that we want to *extend* the algebra structure on A to a bigger vector space E this is in fact the object responsible for the classification of the ES problem. If $V = k^n$, the object $\mathcal{AH}^2_A(k^n, A)$ classifies up to an isomorphism all algebras which contain and stabilize A as a subalgebra of codimension n. Hence, by computing $\mathcal{AH}^2_A(k^n, A)$, for a given algebra A, we obtain important information regarding the classification of finite-dimensional algebras. On the other hand, in order to comply with the traditional way of approaching the extension problem [129], we also construct a second classifying object, denoted by $\mathcal{AH}^2(V, A)$, which provides the classification of the ES problem up to an isomorphism of algebras that stabilizes A and co-stabilize V. Thus, the object $\mathcal{AH}^2(V, A)$, whose construction is simpler than the one of $\mathcal{AH}^2_A(V, A)$, appears as a sort of dual of the classical Hochschild cohomological group. There exists a canonical projection $\mathcal{AH}^2(V, A) \twoheadrightarrow \mathcal{AH}^2_A(V, A)$ between these two classifying objects. Computing the classifying objects $\mathcal{AH}^2_A(V, A)$ and

$\mathcal{AH}^2(V, A)$ is a highly nontrivial problem. In Section 4.2 we shall identify a way of computing both objects for what we have called *flag extending structures* of A to E as defined in Definition 4.2.1. All flag extending structures of A to E can be completely described by a recursive reasoning where the key step is the case when A has codimension 1 as a subspace of E. This case is considered in Theorem 4.2.5, where $\mathcal{AH}^2_A(k, A)$ and $\mathcal{AH}^2(k, A)$ are described. The key players in this context are the so-called flag datums of an algebra A introduced in Definition 4.2.2: in the definition of a flag datum two characters of the algebra A are involved as well as two twisted derivations satisfying certain compatibility conditions. As an application, in Corollary 4.2.6 we shall compute the Galois group for any algebra extension $A \subseteq B$ for which A has codimension 1 in B. Theorem 4.2.5 proves itself to be efficient for the classification of all supersolvable algebras. Corollary 4.2.7 classifies and counts the number of types of isomorphisms of algebras of dimension 2 over an arbitrary field k. By iterating the algorithm, we can increase the dimension by 1 at each step, obtaining in this way the classification of all supersolvable algebras in dimension 3, 4, etc. (see Theorem 4.2.12 for dimension 3). We mention that all classification results presented in this paper are over an arbitrary field k, including the case of characteristic 2 whose difficulty is illustrated.

Section 4.3 contains several special cases of the unified product and we emphasize the problem for which each of these products is responsible. Let $i : A \hookrightarrow E$ be an inclusion of algebras. Corollary 4.3.1, Corollary 4.3.3 and Corollary 4.3.5 give necessary and sufficient conditions for i to have a retraction which is a left/right A-linear map, an A-bimodule map and respectively an algebra map. In the latter case, the associated unified product has a simple form which we call the *semidirect product* by analogy with the group and Lie algebra cases since it describes the split monomorphisms in the category of algebras. We also show that the classical crossed products [196] and their generalizations [58, 69] as well as the Ore extensions are all special cases of the unified product. Definition 4.3.6 introduces a new concept, namely the *matched pair* of two algebras A and V. As a special case of the unified product, we define the *bicrossed product* $A \bowtie V$ associated to a matched pair of algebras. Even if the definition is a lot more complicated than the one given in the case of groups [224] or Lie algebras [172], Corollary 4.3.7 shows that it plays the same role, namely it provides the tool to answer the factorization problem for algebras. Finally, we end the section by considering the ES problem for commutative algebras: in this case the unified product and its axioms simplify considerably.

Although the results presented in Section 4.1 are of a rather technical nature, their efficiency and applicability will be shown in several examples. As a first surprising application of the classifying object $\mathcal{AH}^2_A(V, A)$, we will describe the Galois group $\mathrm{Gal}(B/A)$ of an arbitrary extension $A \subseteq B$ of associative algebras as a subgroup of a semidirect product of groups $\mathrm{GL}_k(V) \rtimes \mathrm{Hom}_k(V, A)$, where the vector space V is isomorphic to the quotient vector space B/A (Corollary 4.4.1). In particular, for any extension of finite-dimensional algebras $A \subsetneq B$, the Galois group $\mathrm{Gal}(B/A)$ embeds in $\mathrm{GL}(m, k) \rtimes \mathrm{M}_{n\times m}(k)$, for some positive integers m and n, where $\mathrm{M}_{n\times m}(k)$ is the additive group of $(n \times m)$-matrices over k. This shows that, for a potential generalization of the classical Galois theory at the level of associative algebras the role of the symmetric groups S_n has to be taken by these canonical semidirect products of groups $\mathrm{GL}(m, k) \rtimes \mathrm{M}_{n\times m}(k)$.

The last section of this chapter deals with the classifying complements problem (CCP) in the context of associative algebras. For an extension of algebras $A \subset E$, if X is a given A-complement of E then Theorem 4.5.5 provides the description of all complements of A in E: any A-complement of E is isomorphic to an r-deformation of X, as defined by (4.42). In other words, given X an A-complement of E all the other A-complements of E are deformations of the algebra X by certain maps $r : X \to A$ associated with the canonical matched pair which arises from the factorization $E = A + X$. The theoretical answer to the (CCP) is given in Theorem 4.5.8 where we explicitly construct a cohomological type object $\mathcal{HA}^2(X, A \,|\, (\triangleright, \triangleleft, \leftharpoonup, \rightharpoonup))$ which parameterizes all A-complements of E. We introduce the factorization index $[E : A]^f$ of a given extension $A \subset E$ as the cardinal of the (possibly empty) isomorphism classes of all A-complements. Moreover, we prove that the factorization index is computed by the formula: $[E : A]^f = |\, \mathcal{HA}^2(X, A \,|\, (\triangleright, \triangleleft, \leftharpoonup, \rightharpoonup)) \,|$. Several explicit examples are provided. More precisely, we indicate associative algebra extensions whose factorization index is 1, 2 or 3. We end the paper with an extension of index at least 4.

4.1 Unified products for algebras

In this section we introduce the unified product for algebras which provides the answer to the ES problem. First we need the following:

Definition 4.1.1 *Let A be an algebra and V a vector space. An extending datum of A through V is a system $\Omega(A, V) = \big(\triangleleft, \triangleright, \leftharpoonup, \rightharpoonup, f, \cdot\big)$ consisting of six bilinear maps*

$$\triangleleft : V \times A \to V, \qquad \triangleright : V \times A \to A, \quad \leftharpoonup : A \times V \to A, \quad \rightharpoonup : A \times V \to V$$
$$f : V \times V \to A, \quad \cdot : V \times V \to V$$

The extension datum $\Omega(A, V)$ is called normalized *if for any $x \in V$ we have:*

$$x \triangleright 1_A = 0, \quad x \triangleleft 1_A = x, \quad 1_A \leftharpoonup x = 0, \quad 1_A \rightharpoonup x = x \tag{4.1}$$

Let $\Omega(A, V) = \big(\triangleleft, \triangleright, \leftharpoonup, \rightharpoonup, f, \cdot\big)$ be an extending datum. We denote by $A \ltimes_{\Omega(A,V)} V = A \ltimes V$ the vector space $A \times V$ together with the bilinear map $\bullet$ defined by:

$$(a, x) \bullet (b, y) := \big(ab + a \leftharpoonup y + x \triangleright b + f(x, y),\ a \rightharpoonup y + x \triangleleft b + x \cdot y\big) \tag{4.2}$$

for all $a, b \in A$ *and* $x, y \in V$*. The object* $A \ltimes V$ *is called the unified product or the* dual Hochschild product *of* A *and* $\Omega(A, V)$ *if it is an associative algebra with the multiplication given by (4.2) and the unit* $(1_A, 0_V)$*. In this case the extending datum* $\Omega(A, V) = \big(\triangleleft, \triangleright, \leftharpoonup, \rightharpoonup, f, \cdot\big)$ *is called an algebra extending structure of* A *through* V*. The maps* $\triangleleft$, $\triangleright$, $\leftharpoonup$ *and* $\rightharpoonup$ *are called the actions of* $\Omega(A, V)$ *and* f *is called the cocycle of* $\Omega(A, V)$*.*

The multiplication given by (4.2) has a rather complicated formula; however, for some specific elements we obtain easier forms which will be useful for future computations:

$$(a, 0) \bullet (b, y) = (ab + a \leftharpoonup y,\ a \rightharpoonup y) \tag{4.3}$$
$$(0, x) \bullet (b, y) = (x \triangleright b + f(x, y),\ x \triangleleft b + x \cdot y) \tag{4.4}$$
$$(a, x) \bullet (0, y) = (a \leftharpoonup y + f(x, y),\ a \rightharpoonup y + x \cdot y) \tag{4.5}$$
$$(a, x) \bullet (b, 0) = (ab + x \triangleright b,\ x \triangleleft b) \tag{4.6}$$

for all $a, b \in A$ and $x, y \in V$. In particular, for any $a, b \in A$ and $x, y \in V$ we have:

$$(a, 0) \bullet (b, 0) = (ab,\ 0), \qquad (0, x) \bullet (0, y) = (f(x, y),\ x \cdot y) \tag{4.7}$$
$$(a, 0) \bullet (0, x) = (a \leftharpoonup x,\ a \rightharpoonup x), \quad (0, x) \bullet (b, 0) = (x \triangleright b,\ x \triangleleft b) \tag{4.8}$$

The next theorem provides the necessary and sufficient conditions that need to be fulfilled by an extending datum $\Omega(A, V)$ such that $A \ltimes V$ is a unified product.

Theorem 4.1.2 *Let* A *be an algebra,* V *a vector space and* $\Omega(A, V) = \big(\triangleleft, \triangleright, \leftharpoonup, \rightharpoonup, f, \cdot\big)$ *an extending datum of* A *by* V*. The following statements are equivalent:*

(1) $A \ltimes V$ *is a unified product;*

(2) *The following compatibilities hold for any* $a, b \in A$, $x, y, z \in V$*:*

(A1) $\Omega(A, V)$ *is a normalized extending datum and* $(V, \rightharpoonup, \triangleleft) \in {}_A\mathcal{M}_A$ *is an* A*-bimodule;*

(A2) $x \cdot (y \cdot z) - (x \cdot y) \cdot z = f(x, y) \rightharpoonup z - x \triangleleft f(y, z)$*;*

(A3) $f(x, y \cdot z) - f(x \cdot y, z) = f(x, y) \leftharpoonup z - x \triangleright f(y, z)$*;*

(A4) $a \rightharpoonup (x \cdot y) = (a \rightharpoonup x) \cdot y + (a \leftharpoonup x) \rightharpoonup y$*;*

(A5) $(a \leftharpoonup x) \leftharpoonup y = a \leftharpoonup (x \cdot y) + a f(x, y) - f(a \rightharpoonup x, y)$*;*

(A6) $(ab) \leftharpoonup x = a(b \leftharpoonup x) + a \leftharpoonup (b \rightharpoonup x)$*;*

(A7) $x \triangleright (ab) = (x \triangleright a)b + (x \triangleleft a) \triangleright b$*;*

(A8) $x \triangleright (y \triangleright a) = (x \cdot y) \triangleright a + f(x, y)a - f(x, y \triangleleft a)$*;*

(A9) $(x \cdot y) \triangleleft a = x \triangleleft (y \triangleright a) + x \cdot (y \triangleleft a)$;

(A10) $a(x \triangleright b) + a \leftharpoonup (x \triangleleft b) = (a \leftharpoonup x)b + (a \rightharpoonup x) \triangleright b$;

(A11) $x \triangleright (a \leftharpoonup y) + f(x,\, a \rightharpoonup y) = (x \triangleright a) \leftharpoonup y + f(x \triangleleft a,\, y)$;

(A12) $x \triangleleft (a \leftharpoonup y) + x \cdot (a \rightharpoonup y) = (x \triangleright a) \rightharpoonup y + (x \triangleleft a) \cdot y$.

Before going into the proof of the theorem, we have a few observations on the compatibilities in Theorem 4.1.2. Although they look rather complicated at first sight, they are in fact quite natural and can be interpreted as follows: (A2) measures how far $(V, \cdot)$ is from being an associative algebra and is called the *twisted associativity condition*. The compatibility condition (A3) is a 2-cocycle type condition. (A5) and (A8) are deformations of the usual module conditions and they can be called *twisted module conditions* for the actions $\leftharpoonup$ and $\triangleright$. (A4), (A6), (A7), (A9), (A10) and (A12) are compatibility conditions between the actions $(\triangleleft, \triangleright, \leftharpoonup, \rightharpoonup)$ of $\Omega(A, V)$. They will be used in the next section for defining the notion of *matched pair* of algebras.

Proof: We can easily check that $(1_A, 0_V)$ is a unit for the multiplication given by (4.2) if and only if the extending datum $\Omega(A, V)$ is normalized. The rest of the proof relies on a detailed analysis of the associativity condition for the multiplication given by (4.2):

$$\big[(a,\, x) \bullet (b,\, y)\big] \bullet (c,\, z) = (a,\, x) \bullet \big[(b,\, y) \bullet (c,\, z)\big] \tag{4.9}$$

where $a,\, b,\, c \in A$ and $x,\, y,\, z \in V$. Furthermore, since in $A \ltimes V$ we have $(a, x) = (a, 0) + (0, x)$ it follows that (4.9) holds if and only if it holds for all generators of $A \ltimes V$, i.e., for the set $\{(a,\, 0) \mid a \in A\} \cup \{(0,\, x) \mid x \in V\}$. However, since the computations are rather long but straightforward, we will only indicate the main steps of the proof. We will start by proving that (A2) and (A3) hold if and only if (4.9) holds for the triple $(0, x)$, $(0, y)$, $(0, z)$ with $x,\, y,\, z \in V$. Indeed, we have:

$$\big[(0,\, x) \bullet (0,\, y)\big] \bullet (0,\, z) = \big(f(x,\, y) \leftharpoonup z + f(x \cdot y,\, z),\, f(x,\, y) \rightharpoonup z + (x \cdot y) \cdot z\big)$$
$$(0,\, x) \bullet \big[(0,\, y) \bullet (0,\, z)\big] = \big(x \triangleright f(y,\, z) + f(x,\, y \cdot z),\, x \triangleleft f(y,\, z) + x \cdot (y \cdot z)\big)$$

Therefore, (A2) and (A3) hold if and only if (4.9) holds for the triple $(0, x)$, $(0, y)$, $(0, z)$ with $x,\, y,\, z \in V$. In the same manner we can prove the following: (A4) and (A5) hold if and only if (4.9) holds for the triple $(a, 0)$, $(0, x)$, $(0, y)$ with $a \in A$, $x,\, y \in V$. Furthermore, (4.9) holds for the triple $(a, 0)$, $(b, 0)$, $(0, x)$ if and only if (A6) holds and $\rightharpoonup$ is a left A-module structure on V. (A7) holds, together with the fact that $\triangleleft$ is a right A-module structure on V, if and only if (4.9) holds for the triple $(0, x)$, $(a, 0)$, $(b, 0)$. (A8) and (A9) hold if and only if (4.9) holds for the triple $(0, x)$, $(0, y)$, $(a, 0)$. (4.9) holds for the triple $(a, 0)$, $(0, x)$, $(b, 0)$ if and only if (A10) holds as well as the compatibility condition which makes V an A-bimodule with respect to the actions $\rightharpoonup$ and $\triangleleft$. Finally, (A11) and (A12) hold if and only if (4.9) holds for the triple $(0, x)$, $(a, 0)$, $(0, y)$. □

From now on, an algebra extending structure of A through a vector space V will be viewed as a system $\Omega(A, V) = \big(\triangleleft, \triangleright, \leftharpoonup, \rightharpoonup, f, \cdot\big)$ satisfying the compatibility conditions (A1)-(A12). We denote by $\mathcal{AE}(A, V)$ the set of all algebra extending structures of A through V.

Example 4.1.3 *We provide the first example of an algebra extending structure and the corresponding unified product. More examples will be given in Section 4.3 and Section 4.2.*

Let $\Omega(A, V) = \big(\triangleleft, \triangleright, \leftharpoonup, \rightharpoonup, f, \cdot\big)$ be an extending datum of A through V such that $\triangleright$, $\leftharpoonup$ and $\cdot$ are the trivial maps. Then, $\Omega(A, V)$ is an algebra extending structure of A through V if and only if $(V, \rightharpoonup, \triangleleft) \in {}_A\mathcal{M}_A$ and the following compatibilities are fulfilled:

$$\begin{aligned} f(x,\, y) \rightharpoonup z &= x \triangleleft f(y,\, z), \qquad a f(x,\, y) = f(a \rightharpoonup x,\, y) \\ f(x,\, y \triangleleft a) &= f(x,\, y)a, \quad\; f(x,\, a \rightharpoonup y) = f(x \triangleleft a,\, y) \end{aligned}$$

for all $a \in A$, x, y and $z \in V$. In this case, the associated unified product $A \ltimes V$ has the multiplication defined for any a, $b \in A$ and x, $y \in V$ by:

$$(a,\, x) \bullet (b,\, y) := \big(ab + f(x, y),\; a \rightharpoonup y + x \triangleleft b\big)$$

that is, $A \ltimes V$ is a cocycle deformation of the usual trivial extension of A by V, dual to the one considered in [129].

Let $\Omega(A, V) = \big(\triangleleft, \triangleright, \leftharpoonup, \rightharpoonup, f, \cdot\big) \in \mathcal{AE}(A, V)$ be an algebra extending structure and $A \ltimes V$ the associated unified product. Then the canonical inclusion

$$i_A : A \to A \ltimes V, \qquad i_A(a) = (a,\, 0)$$

is an injective algebra map. Therefore, we can see A as a subalgebra of $A \ltimes V$ through the identification $A \cong i_A(A) = A \times \{0\}$.

Before proving the description part of the ES problem we need to introduce some new concepts. Let A be an algebra, E a vector space such that A is a subspace of E and V a complement of A in E, i.e., V is a subspace of E such that $E = A + V$ and $A \cap V = \{0\}$. For a linear map $\varphi : E \to E$ we consider the diagram:

$$\begin{array}{ccccc} A & \xrightarrow{\;i\;} & E & \xrightarrow{\;\pi\;} & V \\ {\scriptstyle Id}\downarrow & & \downarrow{\scriptstyle \varphi} & & \downarrow{\scriptstyle Id} \\ A & \xrightarrow{\;i\;} & E & \xrightarrow{\;\pi\;} & V \end{array} \tag{4.10}$$

where $\pi : E \to V$ is the canonical projection of $E = A + V$ on V and $i : A \to E$ is the inclusion map. We say that $\varphi : E \to E$ *stabilizes* A (resp. *co-stabilizes* V) if the left square (resp. the right square) of diagram (4.10) is commutative.

Two algebra structures $\cdot$ and $\cdot'$ on E containing A as a subalgebra are called *equivalent* and we denote this by $(E, \cdot) \equiv (E, \cdot')$, if there exists an algebra isomorphism $\varphi : (E, \cdot) \to (E, \cdot')$ which stabilizes A. The algebra structures $\cdot$ and $\cdot'$ on E are called *cohomologous* and we denote this by $(E, \cdot) \approx (E, \cdot')$, if there exists an algebra isomorphism $\varphi : (E, \cdot) \to (E, \cdot')$ which stabilizes A and co-stabilizes V, i.e., diagram (4.10) is commutative. Then $\equiv$ and $\approx$ are both equivalence relations on the set of all algebra structures on E containing A as a subalgebra and we denote by $\mathrm{Extd}\,(E,\, A)$ (resp. $\mathrm{Extd}'\,(E,\, A)$) the set of all equivalence classes via $\equiv$ (resp. $\approx$). $\mathrm{Extd}\,(E,\, A)$ is the classifying object for the ES problem: by explicitly computing $\mathrm{Extd}\,(E,\, A)$ we obtain a parameterization of the set of all isomorphism classes of algebra structures on E which contain and stabilize A as a subalgebra. $\mathrm{Extd}'\,(E,\, A)$ gives a more restrictive classification of the ES problem, similar to the approach used in the case of the extension problem. Any two cohomologous algebra structures on E are of course equivalent, hence there exists a canonical projection $\mathrm{Extd}'\,(E,\, A) \twoheadrightarrow \mathrm{Extd}\,(E,\, A)$.

Theorem 4.1.4 *Let A be an algebra, E a vector space containing A as a subspace and $*$ an algebra structure on E such that A is a subalgebra in $(E, *)$. Then there exists an algebra extending structure $\Omega(A, V) = \big(\triangleleft, \triangleright, \leftharpoonup, \rightharpoonup, f, \cdot\big)$ of A through a subspace V of E and an isomorphism of algebras $(E, *) \cong A \ltimes V$ that stabilizes A and co-stabilizes V.*

Proof: Since k is a field, there exists a linear map $p : E \to A$ such that $p(a) = a$, for all $a \in A$. Then $V := \mathrm{Ker(p)}$ is a complement of A in E. We define the extending datum $\Omega(A, V) = \big(\triangleleft = \triangleleft_p,\, \triangleright = \triangleright_p,\, \leftharpoonup = \leftharpoonup_p,\, \rightharpoonup = \rightharpoonup_p,\, f = f_p,\, \cdot = \cdot_p\big)$ of A through V by the following formulas:

$$\triangleright : V \times A \to A, \qquad x \triangleright a := p(x * a), \qquad \triangleleft : V \times A \to V,$$
$$x \triangleleft a := x * a - p(x * a)$$
$$\leftharpoonup : A \times V \to A, \qquad a \leftharpoonup x := p(a * x), \qquad \rightharpoonup : A \times V \to V,$$
$$a \rightharpoonup x := a * x - p(a * x)$$
$$f : V \times V \to A, \qquad f(x, y) := p(x * y), \qquad \cdot : V \times V \to V,$$
$$x \cdot y := x * y - p(x * y)$$

for any $a \in A$ and $x,\, y \in V$. We shall prove that $\Omega(A, V) = \big(\triangleleft_p, \triangleright_p, \leftharpoonup_p, \rightharpoonup_p, f_p, \cdot_p\big)$ is an algebra extending structure of A through V and

$$\varphi : A \ltimes V \to E, \qquad \varphi(a, x) := a + x$$

is an isomorphism of algebras that stabilizes A and co-stabilizes V. Instead of proving the compatibility conditions (A1)-(A12), which require a very long and laborious computation, we use the following trick combined with Theorem 4.1.2: $\varphi : A \times V \to E$, $\varphi(a,\, x) = a + x$ is a linear isomorphism between the algebra $(E, *)$ and the direct product of vector spaces $A \times V$ with the inverse given by $\varphi^{-1}(y) := \big(p(y),\, y - p(y)\big)$, for all $y \in E$. Thus, there exists a unique algebra structure on $A \times V$ such that φ is an isomorphism of algebras and this unique multiplication $\circ$ on $A \times V$ is given by $(a, x) \circ (b, y) := \varphi^{-1}\big(\varphi(a, x) * \varphi(b, y)\big)$, for all a, $b \in A$ and x, $y \in V$. The proof is finished if we prove that this multiplication is the one defined by (4.2) associated to the system $\big(\triangleleft_p, \triangleright_p, \leftharpoonup_p, \rightharpoonup_p, f_p, \cdot_p\big)$. Indeed, for any a, $b \in A$ and x, $y \in V$ we have:

$$\begin{aligned}
(a, x) \circ (b, y) &= \varphi^{-1}\big(\varphi(a, x) * \varphi(b, y)\big) = \varphi^{-1}\big((a + x) * (b + y)\big)\\
&= \varphi^{-1}(ab + a * y + x * b + x * y)\\
&= \big(ab + p(a * y) + p(x * b) + p(x * y),\\
&\qquad a * y - p(a * y) + x * b - p(x * b) + x * y - p(x * y)\big)\\
&= \big(ab + a \leftharpoonup y + x \triangleright b + f(x, y),\ a \rightharpoonup y + x \triangleleft b + x \cdot y\big)\\
&= (a, x) \bullet (b, y)
\end{aligned}$$

as needed. Moreover, the following diagram is commutative

$$\begin{array}{ccccc}
A & \xrightarrow{i} & A \ltimes V & \xrightarrow{q} & V\\
{\scriptstyle Id}\downarrow & & \downarrow{\scriptstyle \varphi} & & \downarrow{\scriptstyle Id}\\
A & \xrightarrow{i} & E & \xrightarrow{\pi} & V
\end{array}$$

where $\pi : E \to V$ is the projection of $E = A + V$ on the vector space V and $q : A \ltimes V \to V$, $q(a, x) := x$ is the canonical projection. The proof is now finished. □

Based on Theorem 4.1.4, the classification of all algebra structures on E that contain A as a subalgebra reduces to the classification of all unified products $A \ltimes V$, associated to all algebra extending structures $\Omega(A, V) = \big(\triangleleft, \triangleright, \leftharpoonup, \rightharpoonup, f, \cdot\big)$, for a fixed complement V of A in E. Next we will construct explicitly the non-abelian cohomological type objects $\mathcal{AH}^2_A\,(V,\,A)$ and $\mathcal{AH}^2\,(V,\,A)$ which will parameterize the classifying sets $\mathrm{Extd}\,(E, A)$ and respectively $\mathrm{Extd}'\,(E, A)$. First we need the following:

Lemma 4.1.5 *Let* $\Omega(A, V) = \big(\triangleleft, \triangleright, \leftharpoonup, \rightharpoonup, f, \cdot\big)$ *and* $\Omega(A, V) = \big(\triangleleft', \triangleright', \leftharpoonup', \rightharpoonup', f', \cdot'\big)$ *be two algebra extending structures of* A *through* V *and* $A \ltimes V$, *respectively* $A \ltimes' V$ *the associated unified products. Then there exists a bijection between the set of all morphisms of algebras* $\psi : A \ltimes V \to A \ltimes' V$ *which stabilize* A *and the set of pairs* $(r,\, v)$, *where* $r : V \to A$, $v : V \to V$ *are linear maps satisfying the following compatibility conditions for any* $a \in A$, x, $y \in V$*:*

(M1) $r(x \cdot y) = r(x)r(y) + f'(v(x),\, v(y)) - f(x,\, y) + r(x) \leftharpoonup' v(y) + v(x) \triangleright' r(y)$*;*

(M2) $v(x \cdot y) = r(x) \rightharpoonup' v(y) + v(x) \triangleleft' r(y) + v(x) \cdot' v(y)$*;*

(M3) $r(x \triangleleft a) = r(x)a - x \triangleright a + v(x) \triangleright' a$*;*

(M4) $v(x \triangleleft a) = v(x) \triangleleft' a;$

(M5) $r(a \rightharpoonup x) = ar(x) - a \leftharpoonup x + a \leftharpoonup' v(x);$

(M6) $v(a \rightharpoonup x) = a \rightharpoonup' v(x).$

Under the above bijection the morphism of algebras $\psi = \psi_{(r,v)} : A \ltimes V \to A \ltimes' V$ corresponding to (r, v) is given by $\psi(a, x) = (a + r(x), v(x))$, for all $a \in A$ and $x \in V$. Moreover, $\psi = \psi_{(r,v)}$ is an isomorphism if and only if $v : V \to V$ is an isomorphism and $\psi = \psi_{(r,v)}$ co-stabilizes V if and only if $v = \mathrm{Id}_V$.

Proof: For a linear map $\psi : A \ltimes V \to A \ltimes' V$ which stabilizes A we have $\psi(a, 0) = (a, 0)$ for all $a \in A$. Therefore, ψ is uniquely determined by two linear maps $r : V \to A$, $v : V \to V$ such that $\psi(0, x) = (r(x), v(x))$ for all $x \in V$. Then, for all $a \in A$ and $x \in V$ we have $\psi(a, x) = (a + r(x), v(x))$. Let $\psi = \psi_{(r,v)}$ be such a linear map. We will prove that ψ is an algebra map if and only if the compatibility conditions (M1)-(M6) hold. It is enough to prove that the following compatibility holds for all generators of $A \ltimes V$:

$$\psi\big((d, w) \cdot (e, t)\big) = \psi\big((d, w)\big) \cdot' \psi\big((e, t)\big) \tag{4.11}$$

By a straightforward computation it follows that (4.11) holds for the pair $(a, 0)$, $(b, 0)$ if and only if (M1) and (M2) are fulfilled while (4.11) holds for the pair $(0, x)$, $(a, 0)$ if and only if (M3) and (M4) are satisfied. Finally, (4.11) holds for the pair $(a, 0)$, $(0, x)$ if and only if (M5) and (M6) hold.

Assume now that v is bijective. Then $\psi_{(r,v)}$ is an isomorphism of algebras with the inverse given by $\psi^{-1}_{(r,v)}(a, x) = \big(a - r\big(v^{-1}(x)\big), v^{-1}(x)\big)$ for all $a \in A$ and $x \in V$. Conversely, assume that $\psi_{(r,v)}$ is bijective. Then v is obviously surjective. Consider now $x \in V$ such that $v(x) = 0$. We obtain $\psi_{(r,v)}(0, 0) = (0, 0) = (0, v(x)) = \psi_{(r,v)}(-r(x), x)$. As $\psi_{(r,v)}$ is bijective we get $x = 0$. Therefore v is also injective and hence bijective. Finally, it is straightforward to see that ψ co-stabilizes V if and only if $v = Id$ and the proof is now finished. □

In order to construct the object that parameterizes Extd (E, A) we need the following:

Definition 4.1.6 *Let A be an algebra and V a vector space. Two algebra extending structures of A by V, $\Omega(A, V) = \big(\triangleleft, \triangleright, \leftharpoonup, \rightharpoonup, f, \cdot\big)$ and $\Omega(A, V) = \big(\triangleleft', \triangleright', \leftharpoonup',$
$\rightharpoonup', f', \cdot'\big)$ are called* equivalent, *and we denote this by $\Omega(A, V) \equiv \Omega'(A, V)$, if there exists a pair (r, v) of linear maps, where $r : V \to A$ and $v \in \mathrm{Aut}_k(V)$ such that $\big(\triangleleft', \triangleright', \leftharpoonup', \rightharpoonup', f', \cdot'\big)$ is implemented from $\big(\triangleleft, \triangleright, \leftharpoonup, \rightharpoonup, f, \cdot\big)$ using (r, v) via:*

$$\begin{aligned}
a \rightharpoonup' x &= v\big(a \rightharpoonup v^{-1}(x)\big)\\
a \leftharpoonup' x &= r\big(a \rightharpoonup v^{-1}(x)\big) - a r\big(v^{-1}(x)\big) + a \leftharpoonup v^{-1}(x)\\
x \triangleleft' a &= v\big(v^{-1}(x) \triangleleft a\big)\\
x \triangleright' a &= r\big(v^{-1}(x) \triangleleft a\big) - r\big(v^{-1}(x)\big)a + v^{-1}(x) \triangleright a\\
x \cdot' y &= v\big(v^{-1}(x) \cdot v^{-1}(y)\big) - v\Big(r\big(v^{-1}(x)\big) \rightharpoonup v^{-1}(y)\Big)\\
&\quad - v\Big(v^{-1}(x) \triangleleft r\big(v^{-1}(y)\big)\Big)\\
f'(x,y) &= r\big(v^{-1}(x) \cdot v^{-1}(y)\big) + f\big(v^{-1}(x),\, v^{-1}(y)\big)\\
&\quad - r\Big(r\big(v^{-1}(x)\big) \rightharpoonup v^{-1}(y)\Big) - - r\big(v^{-1}(x)\big) \leftharpoonup v^{-1}(y)\\
&\quad - r\Big(v^{-1}(x) \triangleleft r\big(v^{-1}(y)\big)\Big) + r\big(v^{-1}(x)\big) r\big(v^{-1}(y)\big) -\\
&\quad - v^{-1}(x) \triangleright r\big(v^{-1}(y)\big)
\end{aligned}$$

for all $a \in A$, $x,\ y \in V$.

On the other hand, in order to parameterize $\mathrm{Extd}'\,(E, A)$ we need the following:

Definition 4.1.7 *Let* A *be an algebra and* V *a vector space. Two algebra extending structures* $\Omega(A, V) = (\triangleleft,\, \triangleright,\, \leftharpoonup,\, \rightharpoonup,\, f,\, \cdot)$ *and* $\Omega(A, V) = (\triangleleft',\, \triangleright',\, \leftharpoonup',\, \rightharpoonup',\, f',\, \cdot')$ *are called* cohomologous, *and we denote this by* $\Omega(A, V) \approx \Omega'(A, V)$ *if and only if* $\triangleleft' = \triangleleft$, $\rightharpoonup' = \rightharpoonup$ *and there exists a linear map* $r : V \to A$ *such that*

$$\begin{aligned}
a \leftharpoonup' x &= r(a \rightharpoonup x) - a r(x) + a \leftharpoonup x\\
x \triangleright' a &= r(x \triangleleft a) - r(x)a + x \triangleright a\\
x \cdot' y &= x \cdot y - r(x) \rightharpoonup y - x \triangleleft r(y)\\
f'(x,y) &= r(x \cdot y) + f(x,\, y) - r\Big(r(x) \rightharpoonup y\Big) - r(x) \leftharpoonup y - r\big(x \triangleleft r(y)\big) +\\
&\quad + r(x)r(y) - x \triangleright r(y)
\end{aligned}$$

for all $a \in A$, $x,\ y \in V$.

As a conclusion of this section, the theoretical answer to the ES problem now follows:

Theorem 4.1.8 *Let* A *be an algebra,* E *a vector space which contains* A *as a subspace and* V *a complement of* A *in* E*. Then:*

(1) $\equiv$ *is an equivalence relation on the set* $\mathcal{AE}(A, V)$ *of all algebra extending structures of* A *through* V*. If we denote by* $\mathcal{AH}^2_A\,(V,\, A) := \mathcal{AE}(A, V)/\equiv$, *then the map*

$$\mathcal{AH}^2_A\,(V,\, A) \to \mathrm{Extd}\,(E, A), \qquad \overline{(\triangleleft,\, \triangleright,\, \leftharpoonup,\, \rightharpoonup,\, f,\, \cdot)} \longmapsto (A \ltimes V,\, \cdot)$$

is a bijection between $\mathcal{AH}^2_A(V, A)$ and the isomorphism classes of all algebra structures on E that contain and stabilize A as a subalgebra.

(2) *$\approx$ is an equivalence relation on the set $\mathcal{AE}(A, V)$. If we denote by $\mathcal{AH}^2(V, A) := \mathcal{AE}(A, V)/\approx$, then the map*

$$\mathcal{AH}^2(V, A) \to \mathrm{Extd}'(E, A), \qquad \overline{\overline{(\triangleleft, \triangleright, \leftharpoonup, \rightharpoonup, f, \cdot)}} \longmapsto (A \ltimes V, \cdot)$$

is a bijection between $\mathcal{AH}^2(V, A)$ and the isomorphism classes of all algebra structures on E which stabilize A and co-stabilize V.[1]

Proof: The proof follows from Theorem 4.1.2, Theorem 4.1.4 and Lemma 4.1.5 once we observe that $\Omega(\mathfrak{g}, V) \equiv \Omega'(\mathfrak{g}, V)$ in the sense of Definition 4.1.6 if and only if there exists an isomorphism of algebras $\psi : A \ltimes V \to A \ltimes' V$ which stabilizes A. Therefore, $\equiv$ is an equivalence relation on the set $\mathcal{AE}(A, V)$ and the first part follows. In the same way $\Omega(\mathfrak{g}, V) \approx \Omega'(\mathfrak{g}, V)$ as defined in Definition 4.1.7 if and only if there exists an isomorphism of algebras $\psi : A \ltimes V \to A \ltimes' V$ which stabilizes A and co-stabilizes V and this proves the second part of the theorem. □

4.2 Flag and supersolvable algebras: Examples

The challenge that remains after providing the theoretical answer to the ES problem in Theorem 4.1.8 is a purely computational one: for a given algebra A that is a subspace in a vector space E with a given complement V, we have to compute the classifying object $\mathcal{AH}^2_A(V, A)$ and then to list all algebra structures on E which extend the one of A. For the sake of completeness, we can also compute the space $\mathcal{AH}^2(V, A)$. In what follows we provide a way of answering this problem for a large class of such structures.

Definition 4.2.1 *Let A be an algebra and E a vector space containing A as a subspace. An algebra structure on E is called a* flag extending structure *of A to E if there exists a finite chain of subalgebras of E*

$$E_0 := A \subset E_1 \subset \cdots \subset E_m := E \tag{4.12}$$

such that E_i has codimension 1 in E_{i+1}, for all $i = 0, \cdots, m-1$. An algebra E that is a flag extending structure of k will be called a supersolvable *algebra.*

The notion introduced in the above definition is just the associative algebra counterpart of the well-known concept of supersolvable Lie algebra ([46]). All

[1] $\overline{(\triangleleft, \triangleright, \leftharpoonup, \rightharpoonup, f, \cdot)}$ (resp. $\overline{\overline{(\triangleleft, \triangleright, \leftharpoonup, \rightharpoonup, f, \cdot)}}$ denotes the equivalence class of $(\triangleleft, \triangleright, \leftharpoonup, \rightharpoonup, f, \cdot)$ via $\equiv$ (resp. $\approx$).

flag extending structures of A to E can be completely described by a recursive reasoning where the key step is $m = 1$. This step describes and classifies all unified products $A \ltimes V_1$, for a 1-dimensional vector space V_1. Then, by replacing the initial algebra A with such a unified product $A \ltimes V_1$, which will be explicitly described in terms of A only, we can iterate the process: after m steps, we obtain the description of all flag extending structures of A to E. A special case of interest for the classification of finite-dimensional algebras is the case when $A = k$, i.e., to classify all m-dimensional supersolvable algebras. In this context we recall that the classification of solvable Lie algebras, over arbitrary fields, was achieved only up to dimension four [88]. First we need to introduce the following concept which plays the key role in the classification of flag extending structures:

Definition 4.2.2 *Let A be an algebra. A* flag datum *of A is a 6-tuple $(\Lambda, \lambda, D, d, a_0, u)$, where Λ, $\lambda : A \to k$ are morphisms of algebras, D, $d : A \to A$ are linear maps, $a_0 \in A$, $u \in k$ satisfying the following compatibilities:*

$$\Lambda(a_0) = \lambda(a_0), \quad D(a_0) = d(a_0), \quad \lambda \circ d = 0, \quad \Lambda \circ D = 0 \tag{4.13}$$
$$d(ab) = a\, d(b) + d(a)\, \lambda(b), \quad D(ab) = \Lambda(a)\, D(b) + D(a)\, b \tag{4.14}$$
$$d^2(a) = u\, d(a) + a\, a_0 - \lambda(a)\, a_0, \quad D^2(a) = u\, D(a) + a_0\, a - \Lambda(a)\, a_0 \tag{4.15}$$
$$D\big(d(a)\big) - d\big(D(a)\big) = \big(\Lambda(a) - \lambda(a)\big)\, a_0 \tag{4.16}$$
$$\Lambda\big(d(a)\big) - \lambda\big(D(a)\big) = \big(\Lambda(a) - \lambda(a)\big)\, u \tag{4.17}$$
$$a\, D(b) + \Lambda(b)\, d(a) = d(a)\, b + \lambda(a)\, D(b) \tag{4.18}$$

for all a, $b \in A$. We denote by $\mathcal{F}(A) \subseteq \mathrm{Alg}(A,k)^2 \times \mathrm{Hom}_k(A,A)^2 \times A \times k$ the set of all flag datums of A.

$\mathcal{F}(A)$ can be the empty set: for instance, if the algebra A has no characters, like in the case of the matrix algebra $A = M_n(k)$, for $n \geq 2$. The compatibilities (4.14) show that d and D are twisted derivations of the algebra A. Applying these compatibilities for $a = b = 1_A$ we obtain that $D(1_A) = d(1_A) = 0$, for any $(\Lambda, \lambda, D, d, a_0, u) \in \mathcal{F}(A)$.

Proposition 4.2.3 *Let A be an algebra and V a vector space of dimension 1 with basis $\{x\}$. Then there exists a bijection between the set $\mathcal{AE}(A,V)$ of all algebra extending structures of A through V and the set $\mathcal{F}(A)$ of all flag datums of A.*

Through the above bijection, the unified product corresponding to $(\Lambda, \lambda, D, d, a_0, u) \in \mathcal{F}(A)$ will be denoted by $A \ltimes_{(\Lambda, \lambda, D, d, a_0, u)} x$ and has the multiplication given for any a, $b \in A$ by:

$$(a,0) \bullet (b,0) = (ab,\, 0), \qquad (0,x) \bullet (0,x) = (a_0,\, u\, x) \tag{4.19}$$
$$(a,0) \bullet (0,x) = (d(a),\, \lambda(a)\, x), \quad (0,x) \bullet (a,0) = (D(a),\, \Lambda(a) x) \tag{4.20}$$

i.e., $A \ltimes_{(\Lambda,\, \lambda,\, D,\, d,\, a_0,\, u)} x$ *is the algebra generated by the algebra* A *and* x *subject to the relations:*

$$x^2 = a_0 + u\,x, \quad ax = d(a) + \lambda(a)\,x, \quad xa = D(a) + \Lambda(a)\,x \tag{4.21}$$

for all $a \in A$.

Proof: We have to compute the set of all bilinear maps $\triangleleft : V \times A \to V$, $\triangleright : V \times A \to A$, $\leftharpoonup : A \times V \to A$, $\rightharpoonup : A \times V \to V$, $f : V \times V \to A$ and $\cdot : V \times V \to V$ satisfying the compatibility conditions (A1)-(A12) of Theorem 4.1.2. Since V has dimension 1, there exists a bijection between the set of all extending datums of A through V and the set of all 6-tuples $(\Lambda,\, \lambda,\, D,\, d,\, a_0,\, u)$ consisting of four linear maps $\Lambda,\, \lambda : A \to k$, $D,\, d : A \to A$ and two elements $a_0 \in A$ and $u \in k$. The bijection is given such that the extending datum $\Omega(A, V) = \big(\triangleleft,\, \triangleright,\, \leftharpoonup,\, \rightharpoonup,\, f,\, \cdot\big)$ corresponding to $(\Lambda,\, \lambda,\, D,\, d,\, a_0,\, u)$ is given by:

$$\begin{aligned} x \triangleleft a &:= \Lambda(a)x, \quad x \triangleright a := D(a), \quad a \leftharpoonup x := d(a), \quad a \rightharpoonup x := \lambda(a)x \\ f(x, x) &:= a_0, \qquad x \cdot x := u\,x \end{aligned}$$

for all $a \in A$. Now, by a straightforward computation one can see that the axioms (A1)-(A12) of Theorem 4.1.2 are equivalent to the fact that $\Lambda,\, \lambda : A \to k$ are algebra maps and the compatibility conditions (4.13)-(4.18) hold. For instance, the fact that $(V, \rightharpoonup, \triangleleft)$ is an A-bimodule is equivalent to the fact that λ and $\Lambda : A \to k$ are algebra maps. The axiom (A2) holds if and only if $\Lambda(a_0) = \lambda(a_0)$, while the axiom (A4) is equivalent to $\lambda(d(a)) = 0$, for all $a \in A$. The remaining details are left to the reader. □

Example 4.2.4 *Let* A *be an algebra. Then there is a bijection between the set of all algebra extending systems of* A *through* k *and the set of all 4-tuples* $(\Lambda,\, \Delta,\, f_0,\, u) \in A^* \times \mathrm{End}_k(A) \times A \times k$ *satisfying the following compatibilities for any* $a,\, b \in A$*:*

(FA1) $\Lambda : A \to k$ *is an algebra map and* $\Lambda \circ \Delta = 0$

(FA2) $\Delta(ab) = a\,\Delta(b) + \Lambda(b)\,\Delta(a)$

(FA3) $\Delta^2(a) = u\,\Delta(a) + f_0\,a - \Lambda(a)\,f_0$

The bijection is given such that the algebra extending system $\Omega(A, k) = \big(\triangleleft,\, \triangleright,\, f,\, \cdot\big)$ *associated to* $(\Lambda,\, \Delta,\, f_0,\, u)$ *is defined for any* $x,\, y \in k$ *and* $a \in A$ *by:*

$$x \triangleleft a := x\,\Lambda(a), \quad x \triangleright a := x\,\Delta(a), \quad f(x,\, y) := xy\,f_0, \quad x \cdot y := xyu \tag{4.22}$$

A 4-tuple $(\Lambda,\, \Delta,\, f_0,\, u)$ *satisfying (FA1)-(FA3) is called a* flag datum *of* A *and we denote by* $\mathcal{F}(A)$ *the set of all flag datums of* A*. The unified product* $A \ltimes_{\Omega(A,k)} k$ *associated to a flag datum* $(\Lambda,\, \Delta,\, f_0,\, u)$ *will be denoted by*

$A_{(\Lambda,\,\Delta,\,f_0,\,u)}$ *and coincides with the vector space* $A\times k$ *having the multiplication given for any* $a,\ b\in A,\ x,\ y\in k$ *by:*

$$(a,\,x)\bullet(b,\,y) := \big(ab + x\Delta(b) + y\Delta(a) + xy\,f_0,\ x\Lambda(b) + y\Lambda(a) + xyu\big)$$

An algebra B *contains* A *as a subalgebra of codimension* 1 *if and only if* $B\cong A_{(\Lambda,\,\Delta,\,f_0,\,u)}$*, for some flag datum* $(\Lambda,\ \Delta,\ f_0,\ u)\in\mathcal{F}(A)$.

Proposition 4.2.3 provides an explicit description of all algebras which contain A as a subalgebra of codimension 1: they are isomorphic to an algebra defined by (4.21), for some flag datum $(\Lambda,\ \lambda,\ D,\ d,\ a_0,\ u)$ of A. The existence of this type of algebras depends essentially on the algebra A. Next we will classify this type of algebras by providing the first explicit classification result of the ES problem:

Theorem 4.2.5 *Let* A *be an algebra of codimension* 1 *in the vector space* E. *Then:*

(1) $\mathrm{Extd}\,(E,A)\ \cong\ \mathcal{AH}^2_A\,(k,A)\ \cong\ \mathcal{F}\,(A)/\equiv$, *where* $\equiv$ *is the equivalence relation on the set* $\mathcal{F}\,(A)$ *defined as follows:* $(\Lambda,\ \lambda,\ D,\ d,\ a_0,\ u)\equiv(\Lambda',\ \lambda',\ D',\ d',\ a_0',\ u')$ *if and only if* $\Lambda=\Lambda'$, $\lambda=\lambda'$ *and there exists a pair* $(q,\alpha)\in k^*\times A$ *such that:*

$$D(a) = q\,D'(a) + \alpha\,a - \Lambda(a)\,\alpha \tag{4.23}$$
$$d(a) = q\,d'(a) + a\,\alpha - \lambda(a)\,\alpha \tag{4.24}$$
$$a_0 = q^2\,a_0' + \alpha^2 - u\,\alpha + q\,d'(\alpha) + qD'(\alpha) \tag{4.25}$$
$$u = q\,u' + \lambda'(\alpha) + \Lambda'(\alpha) \tag{4.26}$$

for all $a\in A$. *The bijection between* $\mathcal{F}\,A)/\equiv$ *and* $\mathrm{Extd}\,(E,A)$ *is given by:*

$$\overline{(\Lambda,\ \lambda,\ D,\ d,\ a_0,\ u)}\mapsto A\ltimes_{(\Lambda,\,\lambda,\,D,\,d,\,a_0,\,u)}\,x$$

where $\overline{(\Lambda,\ \lambda,\ D,\ d,\ a_0,\ u)}$ *is the equivalence class of* $(\Lambda,\ \lambda,\ D,\ d,\ a_0,\ u)$ *via the relation* $\equiv$ *and* $A\ltimes_{(\Lambda,\,\lambda,\,D,\,d,\,a_0,\,u)}\,x$ *is the algebra defined by (4.21).*

(2) $\mathrm{Extd}'\,(E,A)\cong\mathcal{AH}^2(k,A)\cong\mathcal{F}\,(A)/\approx$, *where* $\approx$ *is the following relation on the set* $\mathcal{F}\,(A)$: $(\Lambda,\ \lambda,\ D,\ d,\ a_0,\ u)\approx(\Lambda',\ \lambda',\ D',\ d',\ a_0',\ u')$ *if and only if* $\Lambda=\Lambda'$, $\lambda=\lambda'$ *and there exists* $\alpha\in A$ *such that (4.23)-(4.26) are fulfilled for* $q=1$. *The bijection between* $\mathcal{F}\,(A)/\approx$ *and* $\mathrm{Extd}'\,(E,A)$ *is given by:*

$$\overline{\overline{(\Lambda,\ \lambda,\ D,\ d,\ a_0,\ u)}}\mapsto A\ltimes_{(\Lambda,\,\lambda,\,D,\,d,\,a_0,\,u)}\,x$$

where $\overline{\overline{(\Lambda,\ \lambda,\ D,\ d,\ a_0,\ u)}}$ *is the equivalence class of* $(\Lambda,\ \lambda,\ D,\ d,\ a_0,\ u)$ *via* $\approx$.

Proof: Let $(\Lambda,\ \lambda,\ D,\ d,\ a_0,\ u)$, $(\Lambda',\ \lambda',\ D',\ d',\ a_0',\ u')\in\mathcal{F}\,(A)$ and $\Omega(A,V)$, respectively $\Omega'(A,V)$ the corresponding algebra extending structures constructed in the proof of Proposition 4.2.3. The proof relies on Proposition 4.2.3 and Theorem 4.1.8. Since $\dim_k(V)=1$, any linear map $r:V\to A$ is uniquely determined by an element $\alpha \in A$ such that $r(x) = \alpha$, where $\{x\}$ is a basis in V. On the other hand, any automorphism v of V is uniquely determined by a non-zero scalar $q \in k^*$ such that $v(x) = qx$. Based on these facts, a little computation shows that the compatibility conditions from Definition 4.1.6 and respectively Definition 4.1.7 take precisely the form given in (1) and (2) above and hence the proof is finished. □

Using the first statement of Theorem 4.2.5 we are now able to describe the Galois group $\mathrm{Gal}\,(B/A)$ of an algebra extension $A \subseteq B$ for which A has codimension 1 in B.

Corollary 4.2.6 *Let A be an algebra and $(\Lambda,\, \lambda,\, D,\, d,\, a_0,\, u) \in \mathcal{F}\,(A)$ a flag datum of A. Then there exists an isomorphism of groups*

$$\mathrm{Gal}\,\big(A \ltimes_{(\Lambda,\, \lambda,\, D,\, d,\, a_0,\, u)} x/A\big) \cong \mathbb{G}_A\,\big(\Lambda,\, \lambda,\, D,\, d,\, a_0,\, u\big)$$

where $\mathbb{G}_A\,\big(\Lambda,\, \lambda,\, D,\, d,\, a_0,\, u\big)$ is the set of all pairs $(\alpha,\, q) \in A \times k^*$ such that for any $a \in A$:

$$(1 - q)\, D(a) = \alpha\, a - \Lambda(a)\, \alpha, \qquad (1 - q)\, d(a) = a\, \alpha - \lambda(a)\, \alpha$$
$$(1 - q^2)\, a_0 = \alpha^2 - u\, \alpha + q\, d(\alpha) + qD(\alpha), \quad (1 - q)\, u = \lambda(\alpha) + \Lambda(\alpha)$$

with the multiplication given by $(\alpha,\, q)\cdot(\alpha',\, q') := (\alpha' + q'\alpha,\, qq')$, for all $(\alpha,\, q)$, $(\alpha',\, q') \in \mathbb{G}_A\,\big(\Lambda,\, \lambda,\, D,\, d,\, a_0,\, u\big)$.

Proof: It follows from Theorem 4.2.5 taking into account Proposition 4.2.3 and Corollary 4.4.1. The multiplication on $\mathbb{G}_A\,\big(\Lambda,\, \lambda,\, D,\, d,\, a_0,\, u\big)$ is the one given by (4.38) which in our context comes down to the desired formula. □

Next we will highlight the efficiency of Theorem 4.2.5 in classifying supersolvable algebras. We start with $A = k$: by computing $\mathcal{AH}^2_k\,(k, k)$ we will classify in fact all 2-dimensional algebras over an arbitrary field k since any algebra map between two 2-dimensional algebras automatically stabilizes k. Thus the next corollary originates in [200, 215], where all 2-dimensional algebras over the field of complex numbers $\mathbb{C}$ were classified. By replacing $\mathbb{C}$ with an arbitrary field k the situation changes: the number of isomorphism types of 2-dimensional algebras depends heavily on the characteristic of k as well as on the set $k \setminus k^2$, where $k^2 = \{q^2 \,|\, q \in k\}$. First we set the notations which will play the key role in the classification of flag algebras:

If $k^2 \neq k$, we shall fix $S \subseteq k \setminus k^2$ a system of representatives for the following relation on $k \setminus k^2$: $d \equiv d'$ if and only if there exists $q \in k^*$ such that $d = q^2 d'$. Hence, $|S| = [k^* : (k^2)^*] - 1$, where $[k^* : (k^2)^*]$ is the index of $(k^2)^*$ in the multiplicative group $(k^*, \cdot)$.

If $\mathrm{char}(k) = 2$ and $k^2 \neq k$ we denote by $R \subseteq k \setminus k^2$ a system of representatives for the following new equivalence relation on $k \setminus k^2$: $\delta \equiv_1 \delta'$ if and only if there exists $q \in k^*$ such that $\delta - q^2\delta' \in k^2$. Then, $|R| \leq [k^* : (k^2)^*] - 1$.

If $\mathrm{char}(k) = 2$ we also consider the following equivalence relation on k: $c \equiv_2 c'$ if and only if there exists $\alpha \in k$ such that $c - c' = \alpha^2 - \alpha$. We shall fix $T \subseteq k$, a system of representatives for this relation such that $0 \in T$.

Based on Theorem 4.2.5 we can prove the following results that classify all 2-dimensional algebras over an arbitrary field. Of course, (2) and (3) below are well-known results. For (4) and (5) we are not able to indicate a reference in the literature.

Corollary 4.2.7 *Let k be an arbitrary field. Then:*

(1) There exists a bijection $\mathcal{AH}_k^2(k, k) \cong k \times k/\equiv$, where $\equiv$ is the equivalence relation on $k \times k$ defined as follows: $(a, b) \equiv (a', b')$ if and only if there exists a pair $(q, \alpha) \in k^* \times k$ such that:

$$a = q^2 a' + \alpha^2 - b\alpha, \qquad b = q b' + 2\alpha \tag{4.27}$$

The bijection between $\overline{k \times k/\equiv}$ and the isomorphism classes of all 2-dimensional algebras is given by $\overline{(a, b)} \mapsto k_{(a,b)}$, where $k_{(a,b)}$ is the algebra having $\{1, x\}$ as a basis and the multiplication given by $x^2 = a + bx$.

(2) If $\mathrm{char}(k) \neq 2$ and $k = k^2$, then the factor set $k \times k/\equiv$ is equal to $\{\overline{(0,0)}, \overline{(0,1)}\}$. Thus, there exist only two types of 2-dimensional algebras, namely $k_{(0,0)}$ and $k_{(0,1)} \cong k \times k$.

(3) If $\mathrm{char}(k) \neq 2$ and $k \neq k^2$ then the factor set $k \times k/\equiv$ is equal to $\{\overline{(0,0)}, \overline{(0,1)}\} \cup \{\overline{(d,0)} \mid d \in S\}$. Thus, in this case there exist $1 + [k^* : (k^2)^*]$ types of isomorphisms of 2-dimensional algebras namely $k_{(0,0)}$, $k_{(0,1)}$ and $k_{(d,0)}$, for some $d \in S$.

(4) If $\mathrm{char}(k) = 2$ and $k = k^2$ then the factor set $k \times k/\equiv$ is equal to $\{\overline{(0,0)}\} \cup \{\overline{(c,1)} \mid c \in T\}$. Thus, in this case there exist $1 + |T|$ types of isomorphisms of 2-dimensional algebras namely $k_{(0,0)}$ and $k_{(c,1)}$, for some $c \in T$.

(5) If $\mathrm{char}(k) = 2$ and $k \neq k^2$ then the factor set $k \times k/\equiv$ is equal to $\{\overline{(0,0)}\} \cup \{\overline{(c,1)} \mid c \in T\} \cup \{\overline{(\delta,0)} \mid \delta \in R\}$. Thus, in this case there exist $1 + |T| + |R|$ types of isomorphisms of 2-dimensional algebras namely $k_{(0,0)}$, $k_{(c,1)}$, $k_{(\delta,0)}$, for some $c \in T$ and $\delta \in R$.

The algebra $k_{(d,0)}$, for some $d \in S$ is denoted by $k(\sqrt{d})$ and is a quadratic field extension of k.

Proof: (1) Since there exists only one algebra map $k \to k$, namely the identity map, and any linear map $D : k \to k$ with $D(1) = 0$ is the trivial map it follows that $\mathcal{F}(k) \cong k \times k$. Through this identification it is straightforward to see that the equivalence relation in Theorem 4.2.5 takes the form given by (4.27).

(2) The fact that $k \times k/\equiv$ has only two elements, namely $\{\overline{(0,0)}, \overline{(0,1)}\}$ follows trivially. Let $(a, b) \in k \times k$. If $a + 4^{-1}b^2 = 0$, then we can we denote $b = 2\alpha$, with $\alpha \in k$. Then it follows that $(a, b) = (-\alpha^2, 2\alpha)$, for some $\alpha \in k$ and thus $(a, b) \equiv (0, 0)$. On the other hand, if $a + 4^{-1}b^2 \neq 0$, then we can pick $T \in k^*$ such that $a + 4^{-1}b^2 = T^2$, since $k = k^2$. Again, we denote $b = 2\alpha$, with $\alpha \in k$ and we obtain $(a, b) = (T^2 - \alpha^2, 2\alpha)$, and hence $(a, b) \equiv (0, 1)$.

(3) If $k \neq k^2$ we will prove that $k \times k/\equiv$ coincides with $\{\overline{(0,0)}, \overline{(0,1)}\} \cup \{\overline{(d,0)} \mid d \in S\}$. Indeed, consider $(a, b) \in k \times k$. Besides from the two possibilities already studied in (2) we can also have $a + 4^{-1}b^2 = d$, with $d \in k \setminus k^2$. As before, we denote $b = 2\alpha$, with $\alpha \in k$. It follows that $(a, b) = (d - \alpha^2, 2\alpha)$ and therefore $(a, b) \equiv (d, 0)$.

(4) and (5) The proof is based on the following observations. $(a, b) \equiv (0, 0)$ if and only if $b = 0$ and $a \in k^2$. Thus, if $k = k^2$, then we have that $(a, 0) \equiv (0, 0)$, for any $a \in k$. If $k \neq k^2$, then $(a, 0)$ is either equivalent to $(0, 0)$ if $a \in k^2$ or to $(\delta, 0)$, for some $\delta \in R$ in the case that $a \in k \setminus k^2$. We take into account that $(\delta, 0) \equiv (\delta', 0)$ if and only if $\delta \equiv_1 \delta'$.

Let now $(a, b) \in k$, with $b \neq 0$. Then $(a, b) \equiv (ab^{-2}, 1)$. If $a = 0$ the latter is equivalent to $(0, 1)$ and, if $a \neq 0$, $(ab^{-2}, 1)$ is equivalent to $(c, 1)$, for some $c \in T$ since $(c, 1) \equiv (c', 1)$ if and only if $c \equiv_2 c'$. This finishes the proof. □

Theorem 4.2.5 provides the necessary tool for describing and classifying supersolvable algebras in a purely computational and algorithmic way. Having described all 2-dimensional algebras over arbitrary fields in Corollary 4.2.7 we can now take a step further and describe all supersolvable algebras of dimension 3, i.e., all algebras E for which there exists a chain of subalgebras

$$k = E_0 \subset E_1 \subset E_2 = E \tag{4.28}$$

such that $\dim(E_1) = 2$. The algebras described in this way will be classified up to an isomorphism that stabilizes E_1 by using Theorem 4.2.5. First we should notice that since E_1 has dimension 2, it should coincide with one the following algebras: $k_{(0,0)}$, $k_{(0,1)}$, $k_{(d,0)} = k(\sqrt{d})$, or $k_{(c,1)}$. Now, the algebra $k(\sqrt{d})$ has no characters (i.e., there exist no algebra maps $k(\sqrt{d}) \to k$). This means that $\mathcal{F}(k(\sqrt{d}))$ is empty, i.e., there exist no 3-dimensional algebras containing $k(\sqrt{d})$ as a subalgebra. Thus, any 3-dimensional supersolvable algebra E has the middle term E_1 isomorphic to $k_{(0,0)}$, $k_{(0,1)}$ if $\mathrm{char}(k) \neq 2$ or to $k_{(0,0)}$, $k_{(c,1)}$, for some $c \in T$ in the case when $\mathrm{char}(k) = 2$. In what follows we describe all 3-dimensional supersolvable algebras E that contain and stabilize $k_{(0,0)}$ as a subalgebra. First, we need to describe the set of all flag datums of the algebra $k_{(0,0)}$:

Lemma 4.2.8 *Let k be a field. Then $\mathcal{F}(k_{(0,0)})$ is the coproduct of the following sets:*

$$\mathcal{F}(k_{(0,0)}) := \mathcal{F}_1(k_{(0,0)}) \sqcup \mathcal{F}_2(k_{(0,0)}) \sqcup \mathcal{F}_3(k_{(0,0)}), \qquad \text{where :}$$

$\mathcal{F}_1(k_{(0,0)}) \cong k \times k^* \times k$, *with the bijection given such that the flag datum* $(\Lambda, \lambda, D, d, a_0, u)$ *corresponding to* $(D_1, a_{01}, u) \in k \times k^* \times k$ *is defined by:*

$$\lambda(x) = \Lambda(x) := 0, \quad D(x) = d(x) := D_1 x, \quad a_0 := D_1^2 - u\,D_1 + a_{01}\,x, \quad u := u$$

$\mathcal{F}_2(k_{(0,0)}) \cong k^2$, *with the bijection given such that the flag datum* $(\Lambda, \lambda, D, d, a_0, u)$ *corresponding to* $(D_1, u) \in k^2$ *is defined by:*

$$\lambda(x) = \Lambda(x) := 0, \quad D(x) = d(x) := D_1 x, \quad a_0 := D_1^2 - u\,D_1, \quad u := u$$

$\mathcal{F}_3(k_{(0,0)}) \cong \{(D_1, d_1) \in k \times k \mid D_1 \neq d_1\}$ with the bijection given such that the flag datum $(\Lambda, \lambda, D, d, a_0, u)$ corresponding to $(D_1, d_1) \in k \times k$ (with $D_1 \neq d_1$) is defined by:

$$\lambda(x) = \Lambda(x) := 0, \quad D(x) := D_1 x, \quad d(x) := d_1 x, \quad a_0 := -D_1 d_1, \quad u := D_1 + d_1$$

Proof: The algebra $k_{(0,0)}$ has only one character, namely the map $\Lambda : k_{(0,0)} \to k$, defined by $\Lambda(1) = 1$ and $\Lambda(x) = 0$. A straightforward computation shows that the set $\mathcal{F}(k_{(0,0)})$ of all flag datums of $k_{(0,0)}$ identifies with the set of all quadruples $(D_1, d_1, a_{01}, u) \in k^4$ satisfying the following two compatibilities:

$$a_{01}\, D_1 = a_{01}\, d_1, \qquad D_1^2 - u\, D_1 = d_1^2 - u\, d_1$$

Under this bijection the flag datum $(\Lambda, \lambda, D, d, a_0, u)$ associated to the quadruple $(D_1, d_1, a_{01}, u) \in k^4$ is given by:

$$\lambda(x) = \Lambda(x) := 1, \quad D(x) := D_1 x, \quad d(x) := d_1 x, \quad a_0 := D_1^2 - u\, D_1 + a_{01}\, x$$

A detailed discussion on these coefficients (if $a_{01} = 0$ or $a_{01} \neq 0$) allows us to write $\mathcal{F}(k_{(0,0)})$ as the disjoint union of the sets mentioned in the statement. □

The next result classifies all 3-dimensional supersolvable algebras that contain and stabilize $k_{(0,0)}$ as a subalgebra. The classification does not depend on the characteristic of the field k.

Corollary 4.2.9 *Let k be an arbitrary field.*

(1) If $k = k^2$, then there exist five isomorphism classes of 3-dimensional supersolvable algebras that contain and stabilize $k_{(0,0)}$ as a subalgebra, each of them having the k-basis $\{1, x, y\}$ and relations given as follows:

$$\begin{array}{llll} A_1^0: & x^2 = 0, & y^2 = y, & xy = x, \quad yx = 0; \\ A_2^0: & x^2 = 0, & y^2 = y, & xy = yx = 0; \\ A_3^0: & x^2 = 0, & y^2 = 0, & xy = yx = 0; \\ A_4^0: & x^2 = 0, & y^2 = x, & xy = yx = 0; \\ A_5^0: & x^2 = 0, & y^2 = x + y, & xy = yx = 0 \end{array}$$

(2) If $k \neq k^2$, then there exist $4 + [k^* : (k^2)^*]$ isomorphism classes of 3-dimensional supersolvable algebras that contain and stabilize $k_{(0,0)}$ as a subalgebra, namely those from (1) and an additional family defined for any $d \in S$ by the relations:[2]

$$A^0(d): \qquad x^2 = 0, \quad y^2 = d\, x, \quad xy = yx = 0$$

[2] We recall that $|S| = [k^* : (k^2)^*] - 1$.

Proof: The proof is similar to the one of Corollary 4.2.7. Indeed, in Lemma 4.2.8 we have described the set of all flag datums $\mathcal{F}\,(k_{(0,0)})$. The equivalence relation from Theorem 4.2.5 takes the following form for each of the sets $\mathcal{F}_i\,(k_{(0,0)})$, $i = 1, 2, 3$:

• For $\mathcal{F}_3\,(k_{(0,0)})$: $(D_1, d_1) \equiv (D_1', d_1')$ if and only if there exists $(q, \alpha) \in k^* \times k$ such that

$$D_1 = qD_1' + \alpha, \qquad d_1 = qd_1' + \alpha$$

In this case the factor set $\mathcal{F}_3\,(k_{(0,0)})/\equiv$ is a singleton having $\overline{(0,1)}$ as the only element. The algebra associated to $\overline{(0,1)}$, i.e., the unified product $k_{(0,0)} \ltimes_{(\Lambda,\, \lambda,\, D,\, d,\, a_0,\, u)} y$ from Theorem 4.2.5, is precisely the noncommutative algebra A_1^0.

• For $\mathcal{F}_2\,(k_{(0,0)})$: $(D_1, u) \equiv (D_1', u')$ if and only if there exists $(q, \alpha) \in k^* \times k$ such that

$$D_1 = qD_1' + \alpha, \qquad u = qu' + 2\alpha$$

We can easily show that, regardless of the characteristic of k, the factor set $\mathcal{F}_2\,(k_{(0,0)})/\equiv$ has two elements namely $\{\overline{(0,0)}, \overline{(0,1)}\}$. The associated unified products $k_{(0,0)} \ltimes_{(\Lambda,\, \lambda,\, D,\, d,\, a_0,\, u)} y$ are the algebras A_2^0 and A_3^0.

• For $\mathcal{F}_1\,(k_{(0,0)})$: $(D_1, a_{01}, u) \equiv (D_1', a_{01}', u') \in k \times k^* \times k$ if and only if there exists $(q, \alpha_0, \alpha_1) \in k^* \times k \times k$ such that

$$D_1 = qD_1' + \alpha_0, \quad u = qu' + 2\alpha_0, \quad a_{01} = q^2\, a_{01}' - q\alpha_1 u' + 2q\alpha_1 D_1' \qquad (4.29)$$

Suppose first that $\text{char}(k) \neq 2$. In order to compute the factor set $\mathcal{F}_1\,(k_{(0,0)})/\equiv$ we distinguish two cases. If $k = k^2$, we can easily show that $\mathcal{F}_1\,(k_{(0,0)})/\equiv$ has two elements namely $\{\overline{(0,1,0)}, \overline{(0,1,1)}\}$; the unified products $k_{(0,0)} \ltimes_{(\Lambda,\, \lambda,\, D,\, d,\, a_0,\, u)} y$ associated to $\overline{(0,1,0)}$ and $\overline{(0,1,1)}$ are precisely the algebras A_4^0 and A_5^0.

On the other hand, if $k \neq k^2$, it is straightforward to see that $\mathcal{F}_1\,(k_{(0,0)})/\equiv$ is precisely the set $\{\overline{(0,1,0)}, \overline{(0,1,1)}\} \cup \{\overline{(0,d,0)} \mid d \in S\}$. Moreover, the unified product associated to $\overline{(0,d,0)}$, for some $d \in S$, is the algebra $A^0(d)$.

Assume now that $\text{char}(k) = 2$. Then the equivalence relation given by (4.29) takes the form:

$$D_1 = qD_1' + \alpha_0, \quad u = qu', \quad a_{01} = q^2\, a_{01}' - q\alpha_1 u'$$

The factor set $\mathcal{F}_1\,(k_{(0,0)})/\equiv$ is the same as in the case $\text{char}(k) \neq 2$. This can be easily seen from the following observations: for any $u \neq 0$, we have that $(D_1, a_{01}, u) \equiv (0,1,1)$ and $(D_1, a_{01}, 0) \equiv (0,1,0)$ if and only if $a_{01} \in (k^*)^2$. Finally, for $d \in k \setminus k^2$ we have that $(D_1, d, 0) \equiv (0, d, 0)$ and the proof is finished since $(0, d, 0) \equiv (0, d', 0)$ if and only if there exists $q \in k^*$ such that $d = q^2 d'$. □

Next we will classify all supersolvable algebras of dimension 3 that contain and stabilize $k_{(0,1)}$ as a subalgebra: remark that even in characteristic 2 this algebra still exists as an intermediary algebra corresponding to $k_{(c,1)}$, for

$c := 0 \in T$, since we have chosen a system of representatives T which contains 0. First we need the following:

Lemma 4.2.10 *Let k be a field. Then $\mathcal{F}(k_{(0,1)})$ is the coproduct of the following sets:*

$$\mathcal{F}(k_{(0,1)}) := \mathcal{F}_1(k_{(0,1)}) \sqcup \mathcal{F}_2(k_{(0,1)}) \sqcup \mathcal{F}_3(k_{(0,1)}) \sqcup \mathcal{F}_4(k_{(0,1)}), \qquad \text{where :}$$

$\mathcal{F}_1(k_{(0,1)}) \cong k^3$ *with the bijection given such that the flag datum* $(\Lambda,\, \lambda,\, D,\, d,\, a_0,\, u)$ *corresponding to* $(D_1, a_{01}, u) \in k^3$ *is defined by*

$$\Lambda(x) = \lambda(x) := 0, \quad D(x) = d(x) := D_1\, x, \quad a_0 := D_1^2 - u\, D_1 - a_{01} + a_{01}\, x,$$

$$u := u$$

$\mathcal{F}_2(k_{(0,1)}) \cong k^3$ *with the bijection given such that the flag datum* $(\Lambda,\, \lambda,\, D,\, d,\, a_0,\, u)$ *corresponding to* $(D_1, a_{01}, u) \in k^3$ *is defined by*

$$\Lambda(x) = \lambda(x) := 1,\; D(x) = d(x) := D_1(1-x),\; a_0 := D_1^2 + uD_1 + a_{01}\, x,\; u := u$$

$\mathcal{F}_3(k_{(0,1)}) \cong k^2$ *with the bijection given such that the flag datum* $(\Lambda,\, \lambda,\, D,\, d,\, a_0,\, u)$ *corresponding to* $(D_1, d_1) \in k^2$ *is defined by*

$$\Lambda(x) := 0,\; \lambda(x) := 1,\; D(x) := D_1\, x,\; d(x) := d_1(1-x),\; a_0 := D_1\, d_1,$$

$$u := D_1 - d_1$$

$\mathcal{F}_4(k_{(0,1)}) \cong k^2$ *with the bijection given such that the flag datum* $(\Lambda,\, \lambda,\, D,\, d,\, a_0,\, u)$ *corresponding to* $(D_1, d_1) \in k^2$ *is defined by*

$$\Lambda(x) := 1,\; \lambda(x) := 0,\; D(x) := D_1(1-x),\; d(x) := d_1\, x,\; a_0 := D_1\, d_1,$$

$$u := d_1 - D_1$$

Proof: Since $x^2 = x$ it follows that the algebra $k_{(0,1)}$ has only two characters, namely the maps that send $x \mapsto 0$ and respectively $x \mapsto 1$. Thus, in order to compute $\mathcal{F}(k_{(0,1)})$ we distinguish four cases depending on the characters (Λ, λ). More precisely, $\mathcal{F}_1(k_{(0,1)})$ will parameterize all flag datums for which $\Lambda(x) = \lambda(x) = 0$ while $\mathcal{F}_2(k_{(0,1)})$ those for which $\Lambda(x) = \lambda(x) = 1$. We are left with two more cases: $\mathcal{F}_3(k_{(0,1)})$ (resp. $\mathcal{F}_4(k_{(0,1)})$) parameterizes all flag datums for which $\Lambda(x) := 0$ and $\lambda(x) = 1$ (resp. $\Lambda(x) = 1$ and $\lambda(x) = 0$). The conclusion follows in a straightforward manner by applying Definition 4.2.2. □

Now we shall classify all 3-dimensional supersolvable algebras that contain and stabilize $k_{(0,1)}$ as a subalgebra. This time, the classification depends heavily on the base field k.

Corollary 4.2.11 *The isomorphism classes of all 3-dimensional supersolvable algebras that contain and stabilize $k_{(0,1)}$ as a subalgebra are given as follows:*

(1) *If* $\operatorname{char}(k) \neq 2$ *and* $k = k^2$, *then there are five such isomorphism classes, namely the algebras with* k*-basis* $\{1, x, y\}$ *and relations given as follows:*

$$\begin{array}{llll} A_1^1: & x^2 = x, & y^2 = 0, & xy = yx = 0 \\ A_2^1: & x^2 = x, & y^2 = x - 1, & xy = yx = 0 \\ A_3^1: & x^2 = x, & y^2 = 0, & xy = yx = y \\ A_4^1: & x^2 = x, & y^2 = x, & xy = yx = y \\ A_5^1: & x^2 = x, & y^2 = 0, & xy = y, \quad yx = 0 \end{array}$$

(2) *If* $\operatorname{char}(k) \neq 2$ *and* $k \neq k^2$, *then there exist* $3 + 2\,[k^* : (k^2)^*]$ *such isomorphism classes. These are the five types from (1) and two additional families defined by the following relations for any* $d \in S$:

$$\begin{array}{llll} B_1(d): & x^2 = x, & y^2 = d\,(x-1), & xy = yx = 0 \\ B_2(d): & x^2 = x, & y^2 = d\,x, & xy = yx = y \end{array}$$

(3) *If* $\operatorname{char}(k) = 2$ *and* $k = k^2$, *then there exist* $3 + 2\,|T|$ *such isomorphism classes, namely the algebras having* $\{1, x, y\}$ *as a* k*-basis and subject to the following relations for any* $c \in T$:

$$\begin{array}{rllll} C_1^1 & : & x^2 = x, & y^2 = 0, & xy = yx = 0 \\ C_2^1(c) & : & x^2 = x, & y^2 = c + (c+1)\,x, & xy = yx = 0 \\ C_3^1 & : & x^2 = x, & y^2 = 0, & xy = yx = y \\ C_4^1(c) & : & x^2 = x, & y^2 = y + c\,x, & xy = yx = y \\ C_5^1 & : & x^2 = x, & y^2 = 0, & xy = y, \quad yx = 0 \end{array}$$

(4) *If* $\operatorname{char}(k) = 2$ *and* $k \neq k^2$, *then there exist* $3 + 2\,|T| + 2|R|$ *such isomorphism classes. These are the types from (3) and two additional families defined by the following relations for any* $\delta \in R$:

$$\begin{array}{llll} D_1(\delta): & x^2 = x, & y^2 = \delta\,(x+1), & xy = yx = 0 \\ D_2(\delta): & x^2 = x, & y^2 = \delta\,x, & xy = yx = y \end{array}$$

Proof: We shall use the description of $\mathcal{F}\,(k_{(0,1)})$ given in Lemma 4.2.10. First of all we remark that, due to symmetry, the unified products associated to the flag datums of $\mathcal{F}_4\,(k_{(0,1)})$ are isomorphic to the ones associated to $\mathcal{F}_3\,(k_{(0,1)})$. Thus we only have to analyze the cases $\mathcal{F}_i\,(k_{(0,1)})$, with $i = 1, 2, 3$. The equivalence relation from Theorem 4.2.5, applied to each of the sets k^3 and k^2, takes the following form:

• For $\mathcal{F}_1\,(k_{(0,1)}) \cong k^3$: $(D_1, a_{01}, u) \equiv (D_1', a_{01}', u')$ if and only if there exists $(q, \alpha_0, \alpha_1) \in k^* \times k \times k$ such that

$$\begin{aligned} D_1 &= q\,D_1' + \alpha_0 + \alpha_1 && (4.30)\\ a_{01} &= q^2\,a_{01}' + \alpha_1^2 - q\,u'\alpha_1 + 2q\,\alpha_1 D_1' && (4.31)\\ u &= q\,u' + 2\,\alpha_0 && (4.32) \end{aligned}$$

Suppose first that $\mathrm{char}(k) \neq 2$. Then we have: $(D_1, a_{01}, u) \equiv (0,0,0)$ if and only if $a_{01} = (D_1 - 2^{-1}\,u)^2$ and $(D_1, a_{01}, u) \equiv (0,1,0)$ if and only if $a_{01} - (D_1 - 2^{-1}\,u)^2 \in (k^*)^2$. These two observations show that if $k = k^2$ then the factor set $k^3/\equiv$ has two elements, namely $\{\overline{(0,0,0)}, \overline{(0,1,0)}\}$. The unified products $k_{(0,1)} \ltimes_{(\Lambda,\,\lambda,\,D,\,d,\,a_0,\,u)} y$ associated to $\overline{(0,0,0)}$ and respectively $\overline{(0,1,0)}$ are the algebras A_1^1 and A_2^1. On the other hand, if $k \neq k^2$, then the factor set $k^3/\equiv$ is equal to $\{\overline{(0,0,0)}, \overline{(0,1,0)}\} \cup \{\overline{(0,d,0)}\,|\, d \in S\}$. This can be easily seen from the following observations: $(D_1, a_{01}, u) \equiv (0,d,0)$ if and only if $a_{01} - (D_1 - 2^{-1}\,u)^2 = q^2 d$, for some $q \in k^*$ while $(0,d,0) \equiv (0,d',0)$ if and only if $d = q^2 d'$, for some $q \in k^*$. The unified product associated to $\overline{(0,d,0)}$ is the algebra $B_1(d)$.

Assume now that $\mathrm{char}(k) = 2$. Then the equivalence relation on k^3 given by (4.30)-(4.32) takes the form: $(D_1, a_{01}, u) \equiv (D_1', a_{01}', u')$ if and only if there exists $(q, \alpha_0, \alpha_1) \in k^* \times k \times k$ such that

$$D_1 = q\,D_1' + \alpha_0 + \alpha_1, \qquad a_{01} = q^2\,a_{01}' + \alpha_1^2 - q\,u'\alpha_1, \qquad u = q\,u'$$

We will prove the following: if $k = k^2$, then the factor set $k^3/\equiv$ is equal to $\{\overline{(0,0,0)}, \overline{(0,c,1)}\,|\, c \in T\}$ while if $k \neq k^2$, then the factor set $k^3/\equiv$ turns out to be $\{\overline{(0,0,0)},\ \overline{(0,c,1)}\,|\, c \in T\} \cup \{\overline{(0,\delta,0)}\,|\, \delta \in R\}$. Indeed, the above equalities are consequences of the following observations: $(D_1, a_{01}, u) \equiv (0,0,0)$ if and only if $u = 0$ and $a_{01} \in k^2$; $(D_1, a_{01}, u) \equiv (0,\delta,0)$ if and only if $u = 0$ and $a_{01} \equiv_1 d$. Now, for any $u \neq 0$ we have that $(D_1, a_{01}, u) \equiv (u^{-1}\,D_1, u^{-2}\,a_{01}, 1)$, and moreover $(D_1, a_{01}, 1) \equiv (0, c, 1)$, for some $c \in T$. Finally, $\overline{(0,c,1)} \equiv \overline{(0,c',1)}$ if and only if $c \equiv_2 c'$. The algebras associated to $\overline{(0,0,0)}$, $\overline{(0,c,1)}$ and $\overline{(0,\delta,0)}$ are C_1^1, $C_2^1(c)$ and respectively $D_1(\delta)$.

• For $\mathcal{F}_2\,(k_{(0,1)}) \cong k^3$: $(D_1, a_{01}, u) \equiv (D_1', a_{01}', u')$ if and only if there exists $(q, \alpha_0, \alpha_1) \in k^* \times k \times k$ such that

$$\begin{aligned} D_1 &= q\,D_1' - \alpha_0\\ a_{01} &= q^2\,a_{01}' - \alpha_1^2 - q\,\alpha_1 u' - 2q\alpha_1 D_1'\\ u &= q\,u' + 2(\alpha_0 + \alpha_1) \end{aligned}$$

Suppose first that $\mathrm{char}(k) \neq 2$. Then we have: $(D_1, a_{01}, u) \equiv (0,0,0)$ if and only if $a_{01} = -(D_1 + 2^{-1}\,u)^2$ and $(D_1, a_{01}, u) \equiv (0,1,0)$ if and only if $a_{01} = -(D_1 + 2^{-1}\,u)^2 + q^2$, for some $q \in k^*$. These two observations show that if $k = k^2$ then the factor set $k \times k \times k/\equiv$ has two elements, namely $\overline{(0,0,0)}$ and $\overline{(0,1,0)}$. The unified products $k_{(0,1)} \ltimes_{(\Lambda,\,\lambda,\,D,\,d,\,a_0,\,u)} y$ associated to $\overline{(0,0,0)}$ and $\overline{(0,1,0)}$ are the algebras A_3^1 and respectively A_4^1. On the other hand, if $k \neq k^2$

then the factor set $k \times k \times k/\equiv$ is equal to $\{\overline{(0,0,0)}, \overline{(0,1,0)}\} \cup \{\overline{(0,d,0)} \,|\, d \in S\}$. Indeed, let $(D_1, a_{01}, u) \in k^3$ such that $a_{01} = -(D_1 + 2^{-1}u)^2 + d$, for some $d \neq 0$. Then $(D_1, a_{01}, u) \equiv (0, d, 0)$ and moreover $(0, d, 0) \equiv (0, d', 0)$ if and only if there exists $q \in k^*$ such that $d = q^2 d'$. The unified product $k_{(0,1)} \ltimes_{(\Lambda, \lambda, D, d, a_0, u)} y$ associated to $\overline{(0,d,0)}$, for some $d \in S$, is the algebra $B_2(d)$.

Consider now the case $\mathrm{char}(k) = 2$. We have: $(D_1, a_{01}, u) \equiv (0,0,0)$ if and only if $u = 0$ and $a_{01} \in k^2$; $(D_1, a_{01}, u) \equiv (0, \delta, 0)$ if and only if $u = 0$ and $a_{01} \equiv_1 \delta$. Let $(D_1, a_{01}, u) \in k^3$ with $u \neq 0$, then $(D_1, a_{01}, u) \equiv (u^{-1}D_1, u^{-2}a_{01}, 1)$. Finally, $(D_1, a_{01}, 1) \equiv (D_1', a_{01}', 1)$ if and only if $a_{01} \equiv_2 a_{01}'$. These observations show that if $k = k^2$, then the factor set $k^3/\equiv$ is equal to $\{\overline{(0,0,0)}, \overline{(0,c,1)} \,|\, c \in T\}$ while if $k \neq k^2$, the factor set $k^3/\equiv$ coincides with $\{\overline{(0,0,0)}, \overline{(0,c,1)} \,|\, c \in T\} \cup \{\overline{(0,\delta,0)} \,|\, \delta \in R\}$. The algebras corresponding to $\overline{(0,0,0)}$, $\overline{(0,c,1)}$ and $\overline{(0,\delta,0)}$ are C_3^1, $C_4^1(c)$ and respectively $D_2(\delta)$.

• For $\mathcal{F}_3\left(k_{(0,1)}\right) \cong k^2$: $(D_1, d_1) \equiv (D_1', d_1')$ if and only if there exists a triple $(q, \alpha_0, \alpha_1) \in k^* \times k \times k$ such that

$$D_1 = q\,D_1' + \alpha_0 + \alpha_1, \quad d_1 = q\,d_1' - \alpha_0$$

Regardless of the characteristic of k, the factor set $k \times k/\equiv$ contains only one element, namely $\overline{(0,0)}$. The algebra associated to it is A_5^1, if $\mathrm{char}(k) \neq 2$ or C_5^1, if $\mathrm{char}(k) = 2$. The proof is now finished. □

In order to complete the description of all 3-dimensional supersolvable algebras we are left to study one more case, namely the one when $\mathrm{char}(k) = 2$ and the intermediary algebra is isomorphic to $k_{(c,1)}$, for some $c \in T$. The case $c = 0$ is treated in Corollary 4.2.11. Suppose now that $c \neq 0$. We will see that the set $\mathcal{F}\left(k_{(c,1)}\right)$ of all flag datums is empty (i.e., there exists no 3-dimensional algebras containing $k_{(c,1)}$ as a subalgebra) since the algebra $k_{(c,1)}$ has no characters. Indeed, the algebra $k_{(c,1)}$ is generated by an element x such that $x^2 = c + x$. Therefore, the characters of this algebra are in bijection with the solutions *in* k of the equation $y^2 + y + c = 0$. Now, since $c \in T$ and $c \neq 0$ this equation has no solutions in k: if α is such a solution then $c = \alpha + \alpha^2$, i.e., $c \equiv_2 0$ and we obtain $c = 0$ since $c \in T$ a system o representatives for $\equiv_2$ containing 0. As a conclusion of this section we arrive at the classification of all 3-dimensional supersolvable algebras:

Theorem 4.2.12 *Let k be an arbitrary field. Then any 3-dimensional supersolvable algebra is isomorphic to an algebra from the following list:*

(1) A_i^0 or A_i^1, for all $i = 1, \cdots, 5$, if $\mathrm{char}(k) \neq 2$ and $k = k^2$.

(2) A_i^0, A_i^1, $A^0(d)$, $B_1(d)$ or $B_2(d)$, for all $i = 1, \cdots, 5$ and $d \in S$, if $\mathrm{char}(k) \neq 2$ and $k \neq k^2$.

(3) A_i^0, C_1^1, $C_2^1(c)$, C_3^1, $C_4^1(c)$ or C_5^1, for all $i = 1, \cdots, 5$ and $c \in T$, if $\mathrm{char}(k) = 2$ and $k = k^2$.

(4) A_i^0, $A^0(d)$, C_1^1, $C_2^1(c)$, C_3^1, $C_4^1(c)$, C_5^1, $D_1(\delta)$ or $D_2(\delta)$, for all $i = 1, \cdots, 5$, $d \in S$, $c \in T$ and $\delta \in R$, if $\mathrm{char}(k) = 2$ and $k \neq k^2$.

4.3 Special cases of unified products for algebras

In this section we present certain special cases of unified products and we emphasize the problem for which each of these products is responsible. The following convention will be used: if one of the maps $\triangleright$, $\leftharpoonup$, f or $\cdot$ of an extending datum is trivial then we will omit it from the system $\Omega(A, V) = (\triangleleft, \triangleright, \leftharpoonup, \rightharpoonup, f, \cdot)$.

Relative split extensions and cocycle semidirect products of algebras

We will prove that several special cases of the unified product are responsible for the description of algebra extensions $A \subset E$ which split as morphisms of left/right A-modules, A-bimodule or as algebra maps. If $A \subset E$ is an inclusion of algebras, then E will be viewed as a left/right A-module via the restriction of scalars: $a \cdot x \cdot a' := axa'$, for all a, $a' \in A$ and $x \in E$. If $(V, \rightharpoonup, \triangleleft) \in {}_A\mathcal{M}_A$ is an A-bimodule, then $V \times V$ is viewed as an A-bimodule in the canonical way, i.e., the left (resp. right) action of A on $V \times V$ is implemented by $\rightharpoonup$ (resp. $\triangleleft$). A bilinear map $f : V \times V \to A$ is called A-balanced if $f(x, a \rightharpoonup y) = f(x \triangleleft a, y)$, for all $a \in A$, $x, y \in V$. Of course, A-bimodules and A-balanced maps $f : V \times V \to A$ are in bijection to the set of all A-bimodule maps $\tilde{f} : V \otimes_A V \to A$, where $\otimes_A$ is the tensor product over A.

First, we shall describe extensions of algebras $A \subset E$ that split in ${}_A\mathcal{M}$ (resp. $\mathcal{M}_A$), i.e., there exists a left (resp. right) A-module map $p : E \to A$ such that $p(a) = a$, for all $a \in A$.

Corollary 4.3.1 *An extension of algebras $A \subset E$ has a retraction that is a left (resp. right) A-module map if and only if there exists an isomorphism of algebras $E \cong A \ltimes V$, where $A \ltimes V$ is the unified product associated to an algebra extending structure $\Omega(A, V) \in \mathcal{AE}(A, V)$ having $\leftharpoonup : A \times V \to A$ (resp. $\triangleright : V \times A \to A$) the trivial map.*

Proof: Let $A \ltimes V$ be a unified product associated to $\Omega(A, V) \in \mathcal{AE}(A, V)$, for which $\leftharpoonup$ is the trivial map. Then the canonical projection $p_A : A \ltimes V \to A$ is a retraction of the inclusion $i_A : A \to A \ltimes V$ that is also left A-linear since, using (4.3), we have:

$$p_A(a \cdot (b, x)) = p_A\big((a, 0) \bullet (b, x)\big) = p_A(ab,\, a \rightharpoonup x) = ab = ap_A((b, x))$$

for all $a, b \in A$, $x \in V$. Conversely, let $A \subset E$ be an inclusion of algebras which has a retraction $p : E \to A$ that is also a left A-module map. It follows from the proof of Theorem 4.1.4 that the action $\leftharpoonup = \leftharpoonup_p$ associated to the retraction p is the trivial map since $a \leftharpoonup x = p(ax) = ap(x) = 0$, for all $a \in A$ and $x \in V = \mathrm{Ker}(p)$. Thus, there exists an isomorphism of algebras $E \cong A \ltimes V$, where $A \ltimes V$ is the unified product associated to $\Omega(A, V) \in \mathcal{AE}(A, V)$, for which $\leftharpoonup$ is the trivial map.

Analogously we can prove that the algebra extensions $A \subset E$ that split as right A-module maps are parameterized by the unified products $A \ltimes V$ associated to algebra extending structures $\Omega(A, V) \in \mathcal{AE}(A, V)$ for which the action $\triangleright : V \times A \to A$ is the trivial map. □

Examples 4.3.2 *1. The basic example of an algebra extension which splits as in Corollary 4.3.1 is a group algebras extension. Let $H \leq G$ be a subgroup of a group G. Then the group algebras extension $k[H] \subset k[G]$ has a retraction which is a left $k[H]$-module map. Hence, there exists an isomorphism of algebras $k[G] \cong k[H] \ltimes V$, for some algebra extending structure $\Omega(k[H], V) \in \mathcal{AE}(k[H], V)$ having $\leftharpoonup : k[H] \times V \to A$ the trivial map.*

2. The second class of split extensions in the sense of Corollary 4.3.1 are the classical crossed products of algebras [196]. Let A be an algebra, G be a group and $\alpha : G \to \mathrm{Aut}(A)$, $f : G \times G \to U(A)$ be two maps. We shall denote by $g \triangleright a := \alpha(g)(a)$, for all $g \in G$ and $a \in A$. Let $\overline{G}$ be a copy as a set of the group G and $A_\alpha^f[G]$ be the free left A-module having $\overline{G}$ as an A-basis with the multiplication given by:

$$(a\,\overline{g})(b\,\overline{h}) := a(g \triangleright b) f(g, h)\, \overline{gh} \tag{4.33}$$

for all a, $b \in A$ and g, $h \in G$. $A_\alpha^f[G]$ is called the crossed product *of A and G if it is an associative algebra with the unit $1_A\overline{1_G}$. This is equivalent ([196]) to the fact that $f(1_G, 1_G) = 1_A$ and the following compatibilities hold for any g, h, $l \in G$ and $a \in A$:*

$$g \triangleright (h \triangleright a) = f(g, h)\big((gh) \triangleright a\big) f(g, h)^{-1},$$
$$f(g, h) f(gh, l) = \big(g \triangleright f(h, l)\big) f(g, hl).$$

Any crossed product $A_\alpha^f[G]$ is an extension of A via the canonical map $i_A : A \to A_\alpha^f[G]$, $i_A(a) := a\overline{1_G}$. This extension splits in the category of left A-modules: the left A-linear map that splits i_A being the augmentation map $\pi_A : A_\alpha^f[G] \to A$, $\pi_A(a\overline{g}) := a$, for all $a \in A$ and $g \in G$. Thus, using Corollary 4.3.1 we obtain that any crossed product $A_\alpha^f[G]$ is isomorphic to a unified product $A \ltimes V$ associated to an algebra extending structure $\Omega(A, V) \in \mathcal{AE}(A, V)$ for which the action $\leftharpoonup : A \times V \to A$ is the trivial map.

3. Let A be an algebra and $(W, 1_W)$ a pointed vector space. All algebra structures $\cdot$ on the vector space $A \otimes W$ such that $(a \otimes 1_W) \cdot (b \otimes w) = ab \otimes w$, for all a, $b \in A$ and $w \in W$ and having $1_A \otimes 1_W$ as a unit are fully described in [58, Proposition 2.1]: they are parameterized by the set of all pairs (σ, R) consisting of two linear maps $\sigma : W \otimes W \to A \otimes W$, $R : W \otimes A \to A \otimes W$, satisfying a laborious set of axioms. Such an algebra structure, which is a very general construction, is denoted by $A \otimes_{R,\sigma} W$ and is called the Brzezinski's product*; they generalize the twisted tensor product algebras [69] and are classified, up to an isomorphism of algebras that stabilizes A, in [194]. Now, $i_A : A \to$*

$A \otimes_{R,\sigma} W$, $i_A(a) = a \otimes 1_W$ is an injective algebra map which has a retraction that is a left A-module map. Indeed, let $B = \{e_i \,|\, i \in I\}$ be a basis in W such that $1_W \in B$ and $\varepsilon : W \to k$, $\varepsilon(e_i) = 1$, for all $i \in I$. Then $\pi_A : A \otimes_{R,\sigma} W \to A$, $\pi_A := \mathrm{Id}_A \otimes \varepsilon$ is a left A-linear map and a retraction of i_A. Thus, $A \otimes_{R,\sigma} W \cong A \ltimes V$ for an algebra extending structure $\Omega(A, V) \in \mathcal{AE}(A, V)$ for which the action $\leftharpoonup: A \times V \to A$ is the trivial map.

4. The Ore extensions are also a special case of the unified product; in particular any Weyl algebra is a unified product. Indeed, let $\sigma : A \to A$ be an automorphism of the algebra A, $\delta : A \to A$ a σ-derivation and $A[X, \sigma, \delta]$ the Ore extension associated to (σ, δ), that is $A[X, \sigma, \delta]$ is the free left A-module having $\{X^n \,|\, n \geq 0\}$ as a basis and the multiplication given by $Xa = \sigma(a)X + \delta(a)$, for all $a \in A$. Then, the canonical embedding $i_A : A \to A[X, \sigma, \delta]$, $i_A(a) = a$, has a retraction $p_A : A[X, \sigma, \delta] \to A$ that is a left A-module map given by $p_A\big(\sum_{i=0}^{n} a_i X^i\big) := a_0$. Thus, there exists an isomorphism of algebras $A[X, \sigma, \delta] \cong A \ltimes V$, for an algebra extending structure $\Omega(A, V) \in \mathcal{AE}(A, V)$ having $\leftharpoonup: A \times V \to A$ the trivial map.

Using Corollary 4.3.1 we can describe the algebra extensions $A \subset E$ that admit a retraction $p : E \to A$ which is an A-bimodule map. In this case the axioms (A1)- (A12) which describe the corresponding unified products simplify considerably. Indeed, let $\Omega(A, V) = \big(\triangleleft, \triangleright, \leftharpoonup, \rightharpoonup, f, \cdot\big)$ be an extending datum such that $\leftharpoonup$ and $\triangleright$ are both the trivial maps. Then $\Omega(A, V) = \big(\triangleleft, \rightharpoonup, f, \cdot\big)$ is an algebra extending structure of A through V if and only if $(V, \rightharpoonup, \triangleleft) \in {}_A\mathcal{M}_A$ is an A-bimodule, $f : V \times V \to A$ is an A-balanced A-bimodule map and the following compatibilities hold for any a, $b \in A$, x, y, $z \in V$:

$$x \cdot (y \cdot z) - (x \cdot y) \cdot z = f(x,\, y) \rightharpoonup z \;-\; x \triangleleft f(y,\, z)$$
$$a \rightharpoonup (x \cdot y) = (a \rightharpoonup x) \cdot y, \qquad (x \cdot y) \triangleleft a = x \cdot (y \triangleleft a)$$
$$x \cdot (a \rightharpoonup y) = (x \triangleleft a) \cdot y, \qquad f(x,\, y \cdot z) = f(x \cdot y,\, z)$$

A system $\Omega(A, V) = \big(\triangleleft, \rightharpoonup, f, \cdot\big)$ satisfying these compatibilities will be called a *cocycle semidirect system* of algebras. The unified product associated to a cocycle semidirect system $\Omega(A, V) = \big(\triangleleft, \rightharpoonup, f, \cdot\big)$ will be denoted by $A\#^f V$ and will be called the *cocycle semidirect product of associative algebras*. Explicitly, $A\#^f V = A \times V$ (as vector spaces) with the multiplication given by:

$$(a,\, x) \bullet (b,\, y) := \big(ab + f(x, y),\; a \rightharpoonup y + x \triangleleft b + x \cdot y\big) \tag{4.34}$$

for all a, $b \in A$ and x, $y \in V$. Corollary 4.3.1 provides the following result:

Corollary 4.3.3 *An extension of algebras $A \subset E$ has a retraction that is an A-bimodule map if and only if there exists an isomorphism of algebras $E \cong A\#^f V$, where $A\#^f V$ is a cocycle semidirect product of algebras.*

Example 4.3.4 *Examples of algebras that split in the sense of Corollary 4.3.3 are the* twisted products *of algebras. A twisted product is a crossed product*

$A^f_\alpha[G]$ as defined in Example 4.3.2 for which the action α is the trivial action, that is $g \triangleright a = a$, for all $g \in G$ and $a \in A$. In this case the augmentation map π_A is also a right A-module map and thus any twisted product $A^f[G]$ is isomorphic to a cocycle semidirect product of algebras.

A cocycle semidirect system of algebras $\Omega(A, V) = (\triangleleft, \rightharpoonup, f, \cdot)$ for which f is the trivial map is called a *semidirect system* of algebras. Explicitly, $\Omega(A, V) = (\triangleleft, \rightharpoonup, \cdot)$ is a semidirect system of algebras if and only if $(V, \rightharpoonup, \triangleleft) \in {}_A\mathcal{M}_A$ is an A-bimodule, $(V, \cdot)$ is an associative (not-necessarily unitary) algebra and

$$a \rightharpoonup (x \cdot y) = (a \rightharpoonup x) \cdot y, \quad (x \cdot y) \triangleleft a = x \cdot (y \triangleleft a), \quad x \cdot (a \rightharpoonup y) = (x \triangleleft a) \cdot y$$

for all $a \in A$, x, $y \in V$. The cocycle semidirect product of algebras associated to a semidirect system $\Omega(A, V) = (\triangleleft, \rightharpoonup, \cdot)$ is called the *semidirect product* of algebras and will be denoted by $A\#V$. This is a classical construction: it appears in an equivalent form in [199, Lemma a, pg. 212]. We call it the semidirect product by analogy with the group and Lie algebra case where the semidirect product describes the split extensions. More precisely, we have:

Corollary 4.3.5 *An extension of algebras $A \subset E$ has a retraction that is an algebra map if and only if there exists an isomorphism of algebras $E \cong A\#V$, where $A\#V$ is a semidirect product of algebras.*

Proof: Indeed, the canonical projection $p_A : A\#V \to A$, $p_A(a, x) = a$ is a retraction of the inclusion $i_A : A \to A\#V$ and an algebra map. Conversely, from Theorem 4.1.4 it follows that if $p : E \to A$ is an algebra map then $\triangleright_p$, $\leftharpoonup_p$ and f_p constructed in the proof are all trivial maps, i.e., the corresponding unified product $A \ltimes V$ is a semidirect product $A\#V$. □

Matched pairs, bicrossed products and the factorization problem for algebras

The concept of a matched pair of groups was introduced in [224] while the one for Lie algebras in [172, Theorem 4.1] and independently in [170, Theorem 3.9]. As we have seen in the previous chapters, to any matched pair of Lie algebras a new Lie algebra, called the *bicrossed product*, is associated and it is responsible for the factorization problem. The corresponding concepts exist in the literature for several types of categories such as groupoids, Hopf algebras, local compact quantum groups, etc. In what follows we will introduce the corresponding notion for associative algebras. First, we set the terminology. If $(V, \cdot)$ is an associative (not-necessarily unitary) algebra, then the concept of left/right V-module or V-bimodule is defined as in the case of unitary algebras except of course for the unitary condition.

Definition 4.3.6 *A* matched pair *of algebras is a system* $(A,\, V,\, \triangleleft,\, \triangleright,\, \leftharpoonup,\, \rightharpoonup)$ *consisting of a unitary algebra* A*, a (not-necessarily unitary) associative algebra* $V = (V, \cdot)$ *and four bilinear maps*

$$\triangleleft : V \times A \to V, \quad \triangleright : V \times A \to A, \quad \leftharpoonup : A \times V \to A, \quad \rightharpoonup : A \times V \to V$$

such that $(V, \rightharpoonup, \triangleleft) \in {}_A\mathcal{M}_A$ *is an* A*-bimodule,* $(A, \triangleright, \leftharpoonup) \in {}_V\mathcal{M}_V$ *is a* V*-bimodule and the following compatibilities hold for any* $a,\, b \in A,\ x,\, y \in V$*:*

(MP1) $a \rightharpoonup (x \cdot y) = (a \rightharpoonup x) \cdot y + (a \leftharpoonup x) \rightharpoonup y$*;*

(MP2) $(ab) \leftharpoonup x = a(b \leftharpoonup x) + a \leftharpoonup (b \rightharpoonup x)$*;*

(MP3) $x \triangleright (ab) = (x \triangleright a)b + (x \triangleleft a) \triangleright b$*;*

(MP4) $(x \cdot y) \triangleleft a = x \triangleleft (y \triangleright a) + x \cdot (y \triangleleft a)$*;*

(MP5) $a(x \triangleright b) + a \leftharpoonup (x \triangleleft b) = (a \leftharpoonup x)b + (a \rightharpoonup x) \triangleright b$*;*

(MP6) $x \triangleleft (a \leftharpoonup y) + x \cdot (a \rightharpoonup y) = (x \triangleright a) \rightharpoonup y + (x \triangleleft a) \cdot y$*;*

We make a few comments on these compatibilities. If we apply (MP2) and (MP3) for $a = b = 1_A$, we obtain that $1_A \leftharpoonup x = 0$ and $x \triangleright 1_A = 0$, for all $x \in V$. The two compatibility conditions together with the unitary condition derived from the fact that $(V, \rightharpoonup, \triangleleft)$ is an A-bimodule show that the system $(\triangleleft,\, \triangleright,\, \leftharpoonup,\, \rightharpoonup)$ is normalized in the sense of Definition 4.1.1. Similar to the Lie (resp. Leibniz) algebra case, the above axioms can be derived from the ones of an algebra extending structure for which the cocycle f is the trivial map. More precisely, let $\Omega(A, V) = \big(\triangleleft,\, \triangleright,\, \leftharpoonup,\, \rightharpoonup,\, f,\, \cdot\big)$ be an extending datum such that f is the trivial map. Then $\Omega(A, V) = \big(\triangleleft,\, \triangleright,\, \leftharpoonup,\, \rightharpoonup,\, \cdot\big)$ is an algebra extending structure of A through V if and only if $(A,\, (V, \cdot),\, \triangleleft,\, \triangleright,\, \leftharpoonup,\, \rightharpoonup)$ is a matched pair of algebras. In this case, the associated unified product $A \ltimes V$ will be denoted, as in the case of groups, Lie algebras, Hopf algebras, etc. by $A \bowtie V$ and will be called the *bicrossed product* associated to the matched pair $(A,\, V,\, \triangleleft,\, \triangleright,\, \leftharpoonup,\, \rightharpoonup)$. Thus, $A \bowtie V = A \times V$, as a vector space, with a unitary associative algebra structure given by

$$(a,\, x) \bullet (b,\, y) := \big(ab + a \leftharpoonup y + x \triangleright b,\ a \rightharpoonup y + x \triangleleft b + x \cdot y\big) \tag{4.35}$$

for all $a,\, b \in A$ and $x,\, y \in V$. The bicrossed product of two algebras is the construction responsible for the so-called *factorization problem*, which in the case of associative algebras comes down to:

Let A *be a unitary algebra and* V *a (not-necessarily unitary) associative algebra. Describe and classify all unitary algebras* E *that factorize through* A *and* V*, i.e.,* E *contains* A *and* V *as subalgebras such that* $E = A + V$ *and* $A \cap V = \{0\}$.

Indeed, as a special case of Theorem 4.1.4 we have:

Corollary 4.3.7 *Let A be a unitary algebra and V a (not-necessarily unitary) associative algebra. Then, an algebra E factorizes through A and V if and only if there exists a matched pair of algebras $(A, V, \triangleleft, \triangleright, \leftharpoonup, \rightharpoonup)$ such that $E \cong A \bowtie V$.*

Proof: To start with, it is easy to see that any bicrossed product $A \bowtie V$ factorizes through $A \cong A \times \{0\}$ and $V \cong \{0\} \times V$, which are subalgebras in $A \bowtie V$. Conversely, assume that E factorizes through A and V. Let $p : E \to A$ be the k-linear projection of E on A, i.e., $p(a+x) := a$, for all $a \in A$ and $x \in V$. Now, we apply Theorem 4.1.4 for $V = \mathrm{Ker}(p)$. Since V is a subalgebra of E, the map $f = f_p$ constructed in the proof of Theorem 4.1.4 is the trivial map as $x \cdot y \in V = \mathrm{Ker}(p)$. Thus, the algebra extending structure $\Omega(A, V)$ constructed in the proof of Theorem 4.1.4 is precisely a matched pair of algebras and the unified product $A \ltimes V = A \bowtie V$ is the bicrossed product of the matched pair $(A, V, \triangleleft, \triangleright, \leftharpoonup, \rightharpoonup)$. □

More precisely, if E factorizes through A and X we can construct a matched pair of algebras as follows:

$$x \triangleright a + x \triangleleft a := x\, a, \qquad a \leftharpoonup x + a \rightharpoonup x := a\, x \tag{4.36}$$

for all $a \in A$, $x \in X$. Throughout, the above matched pair will be called *the canonical matched pair* associated with the factorization of E through A and X.

Corollary 4.3.7 shows that the factorization problem for algebras can be restated in a purely computational manner as follows: Let A and V be two given algebras. Describe the set of all matched pairs of algebras $(A, V, \triangleleft, \triangleright, \leftharpoonup, \rightharpoonup)$ and classify up to an isomorphism all bicrossed products $A \bowtie V$. Several examples of matched pairs of algebras together with the corresponding bicrossed product can be found in Section 4.5.

The commutative case

The case of commutative algebras will be treated distinctly due to its important implications for the next chapter. On the one hand, we obtain from (4.7) and (4.8) that a unified product $A \ltimes V$ is a commutative algebra if and only if A is commutative, $f : V \times V \to A$, $\cdot : V \times V \to V$ are symmetric bilinear maps, $a \leftharpoonup x = x \triangleright a$ and $a \rightharpoonup x = x \triangleleft a$, for all $a \in A$ and $x \in V$. On the other hand, if we look at the construction of the algebra extending structure from Theorem 4.1.4 in the case when $A \subseteq E$ is an extension of commutative algebras, we also obtain that $a \leftharpoonup_p x = x \triangleright_p a$ and $a \rightharpoonup_p x = x \triangleleft_p a$ for all $a \in A$ and $x \in V$. Thus, in the commutative case Definition 4.1.1 takes the following form:

Definition 4.3.8 *Let A be a commutative algebra and V a vector space. A commutative extending datum of A through V is a system $\Omega(A,V) = \big(\triangleleft, \triangleright, f, \cdot\big)$ consisting of four bilinear maps*

$$\triangleleft : V \times A \to V, \quad \triangleright : V \times A \to A, \quad f : V \times V \to A, \quad \cdot : V \times V \to V$$

such that f and $\cdot$ are symmetric. Let $\Omega(A,V) = \big(\triangleleft, \triangleright, f, \cdot\big)$ be a commutative extending datum. Then the multiplication on $A \ltimes V = A \times V$ given by (4.2) takes the form:

$$(a, x) \bullet (b, y) := \big(ab + y \triangleright a + x \triangleright b + f(x, y),\ y \triangleleft a + x \triangleleft b + x \cdot y\big) \tag{4.37}$$

for all a, $b \in A$ and x, $y \in V$. $A \ltimes V$ is a commutative unified product if it is a commutative associative algebra with the multiplication given by (4.37) and the unit $(1_A, 0_V)$. In this case the extending datum $\Omega(A,V) = \big(\triangleleft, \triangleright, f, \cdot\big)$ is called a commutative associative algebra extending structure of A through V.

In other words, a commutative algebra extending structure of a commutative algebra A through a vector space V is a commutative extending datum $\Omega(A,V) = \big(\triangleleft, \triangleright, f, \cdot\big)$ satisfying the axioms (A1)-(A12) of Theorem 4.1.2 in which we replace $a \leftharpoonup x := x \triangleright a$ and $a \rightharpoonup x := x \triangleleft a$ for all $a \in A$, $x \in V$. That is, the following compatibility conditions hold for any a, $b \in A$, x, y, $z \in V$:

(CA1) $(V, \triangleleft)$ is an A-module and $x \triangleright 1_A = 0$;

(CA2) $x \cdot (y \cdot z) - (x \cdot y) \cdot z = z \triangleleft f(x, y) \ - \ x \triangleleft f(y, z)$;

(CA3) $(x \cdot y) \triangleleft a = x \triangleleft (y \triangleright a) + x \cdot (y \triangleleft a)$;

(CA4) $x \triangleright (ab) = a(x \triangleright b) + (x \triangleleft b) \triangleright a$;

(CA5) $(x \cdot y) \triangleright a = x \triangleright (y \triangleright a) + f(x, y \triangleleft a) - f(x, y)a$;

(CA6) $f(x, y \cdot z) - f(x \cdot y, z) = z \triangleright f(x, y) - x \triangleright f(y, z)$;

The above axioms are derived from those of Theorem 4.1.2 by taking into account that A is commutative, f and $\cdot$ are symmetric bilinear maps and $a \leftharpoonup x = x \triangleright a$ and $a \rightharpoonup x = x \triangleleft a$ for all $a \in A$ and $x \in V$. Indeed, as $a \leftharpoonup x = x \triangleright a$ and $a \rightharpoonup x = x \triangleleft a$, the normalizing conditions come down to $x \triangleright 1_A = 0$ while (A1) reduces to $(V, \triangleleft)$ being an A-module. Moreover, (A4) follows from (A7) by using $(x \cdot y) \triangleright a = (y \cdot x) \triangleright a$. In the same manner we can derive (A8) out of (A3) by having in mind that $(x \cdot y) \triangleleft a = (y \cdot x) \triangleleft a$. (A6) and (A9) can be derived from (A5) by using the commutativity of A, more precisely $x \triangleright ab = x \triangleright ba$. Finally, (A10) comes out of (A4) by having in mind that $(x \cdot y) \triangleright a = (y \cdot x) \triangleright a$ while (A11) follows from (A3) by using $(x \cdot y) \triangleleft a = (y \cdot x) \triangleleft a$. Therefore, we are left with the independent set of 6 axioms listed above.

4.4 The Galois group of algebra extensions

As a first application of the theory developed in Section 4.1 we will describe the Galois group $\mathrm{Gal}\,(B/A)$ of an arbitrary extension $A \subseteq B$ of algebras as a certain subgroup of a semidirect product of groups $\mathrm{GL}_k(V) \rtimes \mathrm{Hom}_k(V,\, A)$, where the vector space V is a fixed complement of A in B, i.e., $V \cong B/A$ as vector spaces.

Let $A \subseteq B$ be an extension of algebras. We define the *Galois group* $\mathrm{Gal}\,(B/A)$ of this extension as a subgroup of $\mathrm{Aut}_{\mathrm{Alg}}(B)$ consisting of all algebra automorphisms of B that fix A, i.e.,

$$\mathrm{Gal}\,(B/A) := \{\sigma \in \mathrm{Aut}_{\mathrm{Alg}}(B) \,|\, \sigma(a) = a,\, \forall\, a \in A\}$$

For a given subgroup $G \leq \mathrm{Aut}_{\mathrm{Alg}}(B)$ we denote by B^G its subalgebra of invariants: $B^G := \{b \in B \,|\, \sigma(b) = b,\, \forall \sigma \in G\}$. Of course, we have that $A \subseteq B^{\mathrm{Gal}(B/A)}$. As in the classical Galois theory we say that $A \subseteq B$ is a *Galois extension* if $B^{\mathrm{Gal}(B/A)} = A$.

Let A and V be two vector spaces and denote $\mathbb{G}_A^V := \mathrm{Hom}_k(V,\, A) \times \mathrm{GL}_k(V)$, where $\mathrm{GL}_k(V) = \mathrm{Aut}_k(V)$ is the group of all linear automorphisms of V. Then, we can easily prove that $\mathbb{G}_A^V$ has a group structure with the multiplication given for any $r,\, r' \in \mathrm{Hom}_k(V,\, A)$ and $\sigma,\, \sigma' \in \mathrm{GL}_k(V)$ by:

$$(r,\, \sigma) \cdot (r',\, \sigma') := (r' + r \circ \sigma',\, \sigma \circ \sigma') \tag{4.38}$$

The unit of $\mathbb{G}_A^V$ is $1_{\mathbb{G}_A^V} = (0,\, \mathrm{Id}_V)$ and the inverse of $(r,\, \sigma)$ is given by $(r,\, \sigma)^{-1} := (-r \circ \sigma^{-1},\, \sigma^{-1})$. Moreover, one can easily see that $\mathrm{GL}_k(V) \cong \{0\} \times \mathrm{GL}_k(V)$ is a subgroup of $\mathbb{G}_A^V$ and the abelian group $\mathrm{Hom}_k(V,\, A) \cong \mathrm{Hom}_k(V,\, A) \times \{\mathrm{Id}_V\}$ is a normal subgroup of $\mathbb{G}_A^V$ since $(r,\, \sigma) \cdot (r',\, \mathrm{Id}_V) \cdot (r,\, \sigma)^{-1} = (r' \circ \sigma^{-1},\, \mathrm{Id}_V)$, for all $r,\, r' \in \mathrm{Hom}_k(V,\, A)$ and $\sigma \in \mathrm{GL}_k(V)$. On the other hand, the relation $(r,\, \sigma) = (0,\, \sigma) \cdot (r,\, \mathrm{Id}_V)$ gives an exact factorization $\mathbb{G}_A^V = \mathrm{GL}_k(V) \cdot \mathrm{Hom}_k(V,\, A)$ of the group $\mathbb{G}_A^V$ through the subgroup $\mathrm{GL}_k(V)$ and the abelian normal subgroup $\mathrm{Hom}_k(V,\, A)$. These observations show that $\mathbb{G}_A^V$ is a semidirect product $\mathrm{GL}_k(V) \rtimes \mathrm{Hom}_k(V,\, A)$ of groups, where the semidirect product is written in the right-hand side convention.

Now let $A \subseteq B$ be an extension of algebras. Assume that the codimension of A in B is $\mathfrak{c}$ and let V be a vector space of dimension $\mathfrak{c}$ that is a complement of A in B. It follows from Theorem 4.1.4 that $B \cong A \ltimes V$, for a canonical algebra extending structure $\Omega(A, V) = \big(\triangleleft,\, \triangleright,\, \leftharpoonup,\, \rightharpoonup,\, f,\, \cdot\big)$ of A by V associated to a given linear retraction $p : B \to A$ of the inclusion map $A \hookrightarrow B$. We fix such an algebra extending structure $\Omega(A, V) = \big(\triangleleft,\, \triangleright,\, \leftharpoonup,\, \rightharpoonup,\, f,\, \cdot\big)$ and we denote by $\mathbb{G}_A^V\,\big(\triangleleft,\, \triangleright,\, \leftharpoonup,\, \rightharpoonup,\, f,\, \cdot\big)$ the set of all pairs $(r,\, \sigma) \in \mathbb{G}_A^V = \mathrm{Hom}_k(V,\, A) \times \mathrm{GL}_k(V)$

satisfying the following compatibility conditions:

$$f(x,\, y) = f(\sigma(x),\, \sigma(y)) + r(x)r(y) - r(x \cdot y) + r(x) \leftharpoonup \sigma(y) + \sigma(x) \triangleright r(y)$$
$$\sigma(x \cdot y) = \sigma(x) \cdot \sigma(y) + r(x) \rightharpoonup \sigma(y) + \sigma(x) \triangleleft r(y)$$
$$r(x \triangleleft a) = r(x)a - x \triangleright a + \sigma(x) \triangleright a$$
$$r(a \rightharpoonup x) = a\, r(x) - a \leftharpoonup x + a \leftharpoonup \sigma(x)$$
$$\sigma(x \triangleleft a) = \sigma(x) \triangleleft a, \qquad \sigma(a \rightharpoonup x) = a \rightharpoonup \sigma(x)$$

for all $a \in A$, $x,\, y \in V$. Using Theorem 4.1.4 and Lemma 4.1.5 we obtain an explicit description for the Galois group of an algebra extension.

Corollary 4.4.1 *Let $A \subseteq B$ be an extension of algebras. We fix V a vector space and $\Omega(A, V) = \big(\triangleleft,\, \triangleright,\, \leftharpoonup,\, \rightharpoonup,\, f,\, \cdot\big)$ an algebra extending structure of A by V such that $B \cong A \ltimes V$. Then, $\mathbb{G}_A^V\big(\triangleleft,\, \triangleright,\, \leftharpoonup,\, \rightharpoonup,\, f,\, \cdot\big)$ is a subgroup of the semidirect product of groups $\mathbb{G}_A^V = \mathrm{GL}_k(V) \rtimes \mathrm{Hom}_k(V,\, A)$ and there exists an isomorphism of groups $\mathrm{Gal}(B/A) \cong \mathbb{G}_A^V\big(\triangleleft,\, \triangleright,\, \leftharpoonup,\, \rightharpoonup,\, f,\, \cdot\big)$.*

Proof: Indeed, it follows from Lemma 4.1.5 that there exists a bijective correspondence between elements of $\mathrm{Gal}\,(A \ltimes V/A)$ and the set $\mathbb{G}_A^V\big(\triangleleft,\, \triangleright,\, \leftharpoonup,\, \rightharpoonup,\, f,\, \cdot\big)$ defined such that $\psi_{(r,\sigma)} \in \mathrm{Gal}\,(A \ltimes V/A)$ corresponding to (r, σ) is given by $\psi_{(r,\,\sigma)}(a, x) = (a + r(x),\, \sigma(x))$, for all $a \in A$ and $x \in V$. The proof is finished once we observe that $\psi_{(r,\,\sigma)} \circ \psi_{(r',\,\sigma')} = \psi_{(r' + r\circ\sigma',\, \sigma\circ\sigma')}$, for all $(r,\, \sigma)$, $(r',\, \sigma') \in \mathbb{G}_A^V\big(\triangleleft,\, \triangleright,\, \leftharpoonup,\, \rightharpoonup,\, f,\, \cdot\big)$. □

The Galois group $\mathbb{G}_A^V\big(\triangleleft,\, \triangleright,\, \leftharpoonup,\, \rightharpoonup,\, f,\, \cdot\big)$ has an abelian normal subgroup corresponding to those automorphisms of $A \ltimes V$ which stabilize A and co-stabilize V. We denote this subgroup by $\mathbb{H}_A^V\big(\triangleleft,\, \triangleright,\, \leftharpoonup,\, \rightharpoonup,\, f,\, \cdot\big)$ – it can be identified with the set of all linear maps $r \in \mathrm{Hom}_k(V,\, A)$ such that for any $a \in A$, $x,\, y \in V$:

$$r(x \cdot y) = r(x)r(y) + r(x) \leftharpoonup y + x \triangleright r(y), \quad r(x) \rightharpoonup y = -x \triangleleft r(y)$$
$$r(x \triangleleft a) = r(x)a, \qquad r(a \rightharpoonup x) = a\, r(x)$$

In the finite-dimensional case we obtain:

Corollary 4.4.2 *Let $A \subseteq B$ be an extension of algebras such that $\dim_k(A) = n$ and $\dim_k(B) = n + m$, for two positive integers m and n. Then there exists a canonical embedding $\mathrm{Gal}\,(B/A) \hookrightarrow \mathrm{GL}(m,\, k) \rtimes \mathrm{M}_{n\times m}(k)$, where $\mathrm{M}_{n\times m}(k)$ is the additive group of $(n \times m)$-matrices over k.*

Thus, in light of Corollary 4.4.2, in order to compute the Galois groups for extensions of finite-dimensional algebras, the groups $\mathrm{GL}(m,\, k) \rtimes \mathrm{M}_{n\times m}(k)$ will play the role of the symmetric group S_n from the classical theory. A more detailed investigation of these concepts will be performed somewhere else, here we only indicate a few examples and in Corollary 4.2.6 we compute the Galois group for any algebra extension $A \subseteq B$ such that A has codimension 1 in B.

Examples 4.4.3 *1. Let* $A := k < x \,|\, x^2 = 0 >$ *and* B *be the* 3*-dimensional non-commutative algebra having* $\{1, x, y\}$ *as a basis and the multiplication defined by* $x^2 = 0$, $y^2 = y$, $xy = x$ *and* $yx = 0$*. Then* $\mathrm{Gal}\,(B/A) \cong (k, +)$ *and* $A \subseteq B$ *is a Galois extension of algebras.*

2. Consider B *to be the non-commutative algebra having* $\{1, x, y\}$ *as a basis and the multiplication defined by* $x^2 = x$, $y^2 = 0$, $xy = x$ *and* $yx = 0$ *. Then* $\mathrm{Gal}\,(B/k) \cong (k, +) \ltimes (k^*, \cdot)$*, where we denoted by* $(k, +) \ltimes (k^*, \cdot)$ *the semidirect product of groups associated to the group homomorphism* $\psi : (k^*, \cdot) \to \mathrm{Aut}((k, +))$ *given by* $\psi(h)(t) = ht$ *for all* $h \in k^*$, $t \in k$*. Furthermore, the extension of algebras* $k \subset B$ *is Galois.*

3. Let $A \subseteq B$ *be the extension of algebras where* B *is the* 4*-dimensional algebra having* $\{e_1, e_2, e_3, e_4\}$ *as a basis and the multiplication given by* $e_1e_1 = e_1$, $e_1e_3 = e_3e_1 = e_3$, $e_2e_2 = e_2$, $e_2e_4 = e_4$ *and* $e_4e_1 = e_4$ *(undefined operations are zero) and* A *is the subalgebra with* $\{e_1, e_2\}$ *as a basis. The unit is* $1_B = e_1 + e_2$*. Then, we can easily find that there exists an isomorphism of groups* $\mathrm{Gal}\,(B/A) \cong k^* \times k^*$ *given such that* $\sigma \in \mathrm{Gal}\,(B/A)$ *associated to* $(a, b) \in k^* \times k^*$ *is given by* $\sigma(e_1) = e_1$, $\sigma(e_2) = e_2$, $\sigma(e_3) = ae_3$ *and* $\sigma(e_4) = be_4$*. Moreover, the extension* $A \subseteq B$ *in Galois, since* $A = B^{\mathrm{Gal}(B/A)}$*.*

Moreover, if we consider the same algebra A *as above but this time viewed as a subalgebra of the* 4*-dimensional algebra* C*, having the basis* $\{e_1, e_2, e_3, e_4\}$ *and multiplication* $e_1e_1 = e_1$, $e_2e_2 = e_2$, $e_2e_3 = e_3$, $e_2e_4 = e_4$, $e_3e_1 = e_3$ *and* $e_4e_1 = e_4$ *then we will obtain that there exists an isomorphism of groups* $\mathrm{Gal}\,(C/A) \cong \mathrm{GL}(2, k)$ *given such that* $\sigma \in \mathrm{Gal}\,(B/A)$ *associated to the* 2×2 *invertible matrix* (a_{ij}) *is given by* $\sigma(e_3) = a_{11}e_3 + a_{21}e_4$ *and* $\sigma(e_4) = a_{12}e_3 + a_{22}e_4$*. The extension* $A \subseteq C$ *is also Galois, as shown by an elementary computation.*

4.5 Classifying complements for associative algebras

Throughout this section the field k has characteristic zero. Furthermore, the algebras considered will be associative not necessarily unitary. Hence the concept of left/right A-module or A-bimodule is defined as in the case of unitary algebras except of course for the unitary condition.

The main theme of this section is the classifying complements problem (CCP) in the context of associative algebras:

Classifying complements problem (CCP): *Let* $A \subset E$ *be a given subalgebra of* E*. If an* A*-complement of* E *exists, describe explicitly, classify all* A*-complements of* E *and compute the cardinal of the (possibly empty) isomorphism classes of all* A*-complements of* E *(which will be called the factorization index* $[E : A]^f$ *of* A *in* E*).*

A similar problem, called invariance under twisting, was studied in [194] for Brzezinski's crossed products.

We start by presenting several examples of matched pairs of algebras which appear quite naturally from minimal sets of data.

Examples 4.5.1 1) *Let A be an algebra and $(X, \rightharpoonup, \triangleleft) \in {}_A\mathcal{M}_A$ an A-bimodule. We see X as an algebra with the trivial multiplication, i.e., $xy = 0$ for all x, $y \in X$. It is straightforward to see that $(A, X, \triangleleft, \triangleright_0, \leftharpoonup_0, \rightharpoonup)$ is a matched pair of algebras, where $\triangleright_0$ and $\leftharpoonup_0$ are the trivial actions. The multiplication on the corresponding bicrossed product $A \bowtie X$ is given as follows:*

$$(a, x) \bullet (b, y) := \big(ab, \ a \rightharpoonup y + x \triangleleft b\big) \tag{4.39}$$

The above bicrossed product is precisely the trivial extension *of A by the A-bimodule X.*

2) *The previous example can be slightly generalized by considering A and X to be both algebras such that $(X, \rightharpoonup, \triangleleft) \in {}_A\mathcal{M}_A$ is an A-bimodule for which the following compatibilities hold*

$$a \rightharpoonup (x\,y) = (a \rightharpoonup x)\,y, \quad (x\,y) \triangleleft a = x\,(y \triangleleft a), \quad x\,(a \rightharpoonup y) = (x \triangleleft a)\,y \tag{4.40}$$

for all $a \in A$, x, $y \in X$. In [199, Definition, pg. 212] a bimodule X satisfying (4.40) is called a multiplicative A-bimodule. *Then, the bicrossed product associated with the matched pair $(A, X, \triangleleft, \triangleright_0, \leftharpoonup_0, \rightharpoonup)$, where $\triangleright_0$, $\leftharpoonup_0$ are the trivial actions will be called, following [25, pg. 20],* a semidirect product of A and X. *The multiplication on the corresponding bicrossed product $A \bowtie X$ is given as follows:*

$$(a, x) \bullet (b, y) := \big(ab, \ a \rightharpoonup y + x \triangleleft b + xy\big)$$

This construction originates in [199, Lemma a] where is presented in a different form.

Examples 4.5.2 1) *Let $n \in \mathbb{N}$, $n \geq 2$. It can be easily seen that $\mathcal{M}_n(k)$ factorizes through the subalgebra of strictly lower triangular matrices $A = \{(a_{ij})_{i,\, j=\overline{1,n}} \mid a_{ij} = 0 \text{ for } i \leq j\}$ and the subalgebra of upper triangular matrices $X = \{(x_{ij})_{i,\, j=\overline{1,n}} \mid x_{ij} = 0 \text{ for } i > j\}$. We denote by $\mathcal{B}_A := \{e_{ij} \mid i, j \in \overline{1,n},\, i > j\}$ and $\mathcal{B}_X := \{e_{ij} \mid i, j \in \overline{1,n},\, i \leq j\}$ the k-basis of A, respectively X. Then, the canonical matched pair associated with this factorization is given as follows:*

$$e_{ij} \leftharpoonup e_{lu} = \begin{cases} e_{iu}, & \text{if } i > u \geq j = l \\ 0, & \text{otherwise} \end{cases},$$

$$e_{ij} \rightharpoonup e_{lu} = \begin{cases} e_{iu}, & \text{if } l = j < i \leq u \\ 0, & \text{otherwise} \end{cases}$$

$$e_{rs} \triangleright e_{pt} = \begin{cases} e_{rt}, & \text{if } t < r \leq s = p \\ 0, & \text{otherwise} \end{cases},$$

$$e_{rs} \triangleleft e_{pt} = \begin{cases} e_{rt}, & \text{if } r \leq t < s = p \\ 0, & \text{otherwise} \end{cases}$$

2) *Consider* $n \in \mathbb{N}$, $n \geq 2$. *Then* $\mathcal{M}_n(k)$ *factorizes also through the subalgebras* $A = \{(a_{ij})_{i,\, j=\overline{1,n}} \mid a_{n\,u} = 0 \text{ for all } u = \overline{1,n}\}$ *and* $X = \{(x_{ij})_{i,\, j=\overline{1,n}} \mid x_{r\,l} = 0 \text{ for all } r = \overline{1,n-1} \text{ and } l = \overline{1,n}\}$. *We denote by* $\mathcal{B}_A := \{e_{ij} \mid i = \overline{1, n-1},\ j = \overline{1,n}\}$ *and* $\mathcal{B}_X := \{e_{n\,j} \mid j = \overline{1, n}\}$ *the* k-*basis of* A, *respectively* X. *The canonical matched pair associated with this factorization is given as follows:*

$$e_{n\,u} \triangleleft e_{v\,t} = \begin{cases} e_{n\,t}, & \text{if } u = v \\ 0, & \text{otherwise} \end{cases}, \qquad e_{v\,t} \leftharpoonup e_{n\,u} = \begin{cases} e_{v\,u}, & \text{if } t = n \\ 0, & \text{otherwise} \end{cases}$$

while the other two actions are both trivial.

3) *Let* R *and* S *be two algebras and* $M \in {}_R\mathcal{M}_S$. *Then the algebra* $E = \begin{pmatrix} R & M \\ 0 & S \end{pmatrix}$ *factorizes through the subalgebras* $A := \begin{pmatrix} R & 0 \\ 0 & S \end{pmatrix}$ *and* $X = \begin{pmatrix} 0 & M \\ 0 & 0 \end{pmatrix}$. *More precisely, the associated matched pair is given as follows for all* $r \in R$, $s \in S$, $m \in M$:

$$\begin{pmatrix} 0 & m \\ 0 & 0 \end{pmatrix} \triangleleft \begin{pmatrix} r & 0 \\ 0 & s \end{pmatrix} = \begin{pmatrix} 0 & ms \\ 0 & 0 \end{pmatrix}, \qquad \begin{pmatrix} r & 0 \\ 0 & s \end{pmatrix} \rightharpoonup \begin{pmatrix} 0 & m \\ 0 & 0 \end{pmatrix} = \begin{pmatrix} 0 & rm \\ 0 & 0 \end{pmatrix}$$

while the other two actions are both trivial.

In order to prove the main result of this section which answers the (CCP) for algebras we first need to introduce the following concept:

Definition 4.5.3 *Let* $(A, X, \triangleright, \triangleleft, \leftharpoonup, \rightharpoonup)$ *be a matched pair of algebras. A linear map* $r : X \to A$ *is called a* deformation map *of the matched pair* $(A, X, \triangleright, \triangleleft, \leftharpoonup, \rightharpoonup)$ *if the following compatibility holds for all* $x, y \in X$:

$$r(x)\, r(y) - r(x\,y) = r\big(r(x) \rightharpoonup y + x \triangleleft r(y)\big) - r(x) \leftharpoonup y - x \triangleright r(y) \qquad (4.41)$$

We denote by $\mathcal{DM}\,(A, X \,|\, (\triangleright, \triangleleft, \leftharpoonup, \rightharpoonup))$ the set of all deformation maps of the matched pair $(A, X, \triangleright, \triangleleft, \leftharpoonup, \rightharpoonup)$. The trivial map $r : X \to A$, $r(x) = 0$, for all $x \in X$ is of course a deformation map. The right-hand side of (4.41) measures how far $r : X \to A$ is from being an algebra map.

The next example shows that computing all deformation maps associated with a given matched pair is a highly non-trivial problem.

Examples 4.5.4 *Consider* $\mathcal{M}_n(K)$ *with the factorization given in Example 4.5.2 1). Then* $\mathcal{DM}\,(A, X \,|\, (\triangleright, \triangleleft, \leftharpoonup, \rightharpoonup))$ *is in bijection with the families of scalars* $\{(\alpha_{cd}^{ab})_{a,b,c,d \in \overline{1,n}} \mid \alpha_{cd}^{ab} \in K,\ c \leq d,\ a > b\}$ *subject to the compatibility condition:*

$$\sum_{q<t<k} \alpha_{ij}^{kt}\, \alpha_{rs}^{tq} = \delta_{jr}\, \alpha_{is}^{kq} + \sum_{r<u\leq s} \alpha_{ij}^{ur}\, \alpha_{us}^{kq} + \sum_{i \leq v < j} \alpha_{rs}^{jv}\, \alpha_{iv}^{kq} - \alpha_{ij}^{kr}\, \delta_{sq} - \alpha_{rs}^{jq}\, \delta_{ki}$$

for all $k > q$, $i \leq j$ and $r \leq s$, where δ_{jr} is the Kroneker symbol. The bijection is such that any deformation map $r : X \to A$ is implemented by a family of scalars via the following formula:

$$r(e_{ij}) = \sum_{v>t} \alpha_{ij}^{vt} \, e_{vt}, \quad \text{for all} \quad i \leq j$$

The following theorem is the key result in solving the (CCP) for associative algebras:

Theorem 4.5.5 *Let A be a subalgebra of E and X a given A-complement of E with the associated canonical matched pair $(A, X, \triangleright, \triangleleft, \leftharpoonup, \rightharpoonup)$.*

(1) *Let $r : X \to A$ be a deformation map of the above matched pair. Then $X_r := X$, as a vector space, with the new multiplication defined for any x, $y \in X$ by:*

$$x \cdot_r y := x\, y + r(x) \rightharpoonup y + x \triangleleft r(y) \tag{4.42}$$

is an associative algebra called the r-deformation of X. Furthermore, X_r is an A-complement of E.

(2) *$\overline{X}$ is an A-complement of E if and only if there exists an isomorphism of algebras $\overline{X} \cong X_r$, for some deformation map $r : X \to A$ of the matched pair $(A, X, \triangleright, \triangleleft, \leftharpoonup, \rightharpoonup)$.*

Proof: 1) The fact that the multiplication $\cdot_r$ defined by (4.42) is associative follows by a long but straightforward computation which relies on the axioms (MP1)-(MP6). However, we present here a different and more natural proof which will shed some light on the way we arrived at the multiplication given by (4.42). For a deformation map $r : X \to A$, consider $f_r : X \to E = A \bowtie X$ to be the K-linear map defined for any $x \in X$ by:

$$f_r(x) = (r(x),\, x)$$

We will prove that $\widetilde{X} := \mathrm{Im}(f_r)$ is an A-complement of E. We start by showing that $\widetilde{X}$ is a subalgebra of $A \bowtie X$. Indeed, for all x, $y \in X$ we have:

$$\begin{aligned}
\big(r(x),\, x\big)\big(r(y),\, y\big) &\overset{(4.35)}{=} \big(\underline{r(x)\, r(y) + r(x) \leftharpoonup y + x \triangleright r(y)},\\
&\qquad r(x) \rightharpoonup y + x \triangleleft r(y) + x\, y\big)\\
&\overset{(4.41)}{=} \Big(r\big(r(x) \rightharpoonup y + x \triangleleft r(y) + x\, y\big),\\
&\qquad r(x) \rightharpoonup y + x \triangleleft r(y) + x\, y\Big)
\end{aligned}$$

Therefore $\widetilde{X}$ is a subalgebra of $A \bowtie X$. We are left to prove that $A \cap \widetilde{X} = \{0\}$. To this end, consider $(a,\, x) \in A \cap \widetilde{X}$. Since in particular we have $(a,\, x) \in \widetilde{X}$, it follows that $a = r(x)$. As we also have $\big(r(x),\, x\big) \in A$ we obtain $x = 0$ and thus $A \cap \widetilde{X} = \{0\}$. Moreover, if $(b,\, y) \in E = A \bowtie X$ we can write $(b,\, y) = \big(b - r(y),\, 0\big) + \big(r(y),\, y\big) \in A + \widetilde{X}$. Hence, $\widetilde{X}$ is an A-complement of E.

The proof will be finished once we prove that X_r and $\widetilde{X}$ are isomorphic as algebras. Denote by $\widetilde{f}_r$ the linear isomorphism from X to $\widetilde{X}$ induced by f_r. We will prove that $\widetilde{f}_r$ is an algebra morphism if we consider X endowed with the multiplication given by (4.42). For all $x, y \in X$ we have:

$$\begin{aligned}
\widetilde{f}_r(x \cdot_r y) &\overset{(4.42)}{=} \widetilde{f}_r(x\,y + r(x) \rightharpoonup y + x \triangleleft r(y)) \\
&= \Big(\underline{r\big(x\,y + r(x) \rightharpoonup y + x \triangleleft r(y)\big)},\, x\,y + r(x) \rightharpoonup y + x \triangleleft r(y)\Big) \\
&\overset{(4.41)}{=} (r(x)\,r(y) + r(x) \leftharpoonup y + x \triangleright r(y),\, x\,y + r(x) \rightharpoonup y + x \triangleleft r(y)) \\
&= \big(r(x),\, x\big)\big(r(y),\, y\big) = \widetilde{f}_r(x)\widetilde{f}_r(y)
\end{aligned}$$

Therefore, X_r is an algebra and the proof is now finished.

2) Let $\overline{X}$ be an arbitrary A-complement of E. Since $E = A \oplus X = A \oplus \overline{X}$ we can find four K-linear maps:

$$u : X \to A, \quad v : X \to \overline{X}, \quad t : \overline{X} \to A, \quad w : \overline{X} \to X$$

such that for all $x \in X$ and $y \in \overline{X}$ we have:

$$x = u(x) \oplus v(x), \qquad y = t(y) \oplus w(y) \tag{4.43}$$

By an easy computation it follows that $v : X \to \overline{X}$ is a linear isomorphism of vector spaces. We denote by $\tilde{v} : X \to A \bowtie X$ the composition $\tilde{v} := i \circ v$ where $i : \overline{X} \to E = A \bowtie X$ is the canonical inclusion. Therefore, we have $\tilde{v}(x) \overset{(4.43)}{=} \big(-u(x),\, x\big)$, for all $x \in X$. In what follows we will prove that $r := -u$ is a deformation map and $\overline{X} \cong X_r$. Indeed, $\overline{X} = \mathrm{Im}(v) = \mathrm{Im}(\tilde{v})$ is a subalgebra of $E = A \bowtie X$ and we have:

$$\begin{aligned}
\big(r(x),\, x\big)\big(r(y),\, y\big) &\overset{(4.35)}{=} \big(r(x)\,r(y) + r(x) \leftharpoonup y + x \triangleright r(y), \\
&\qquad r(x) \rightharpoonup y + x \triangleleft r(y) + x\,y\big) \\
&= \big(r(z),\, z\big)
\end{aligned}$$

for some $z \in X$. Thus, we obtain:

$$r(z) = r(x)\,r(y) + r(x) \leftharpoonup y + x \triangleright r(y), \qquad z = r(x) \rightharpoonup y + x \triangleleft r(y) + x\,y \tag{4.44}$$

By applying r to the second part of (4.44) it follows that r is a deformation map of the matched pair $(A, X, \triangleright, \triangleleft, \leftharpoonup, \rightharpoonup)$. Furthermore, (4.42) and (4.44) show that $v : X_r \to \overline{X}$ is also an algebra map. The proof is now finished. □

Remark 4.5.6 *We should point out that in the context of associative algebras there exists another type of deformation in the literature, not related to the one we introduce in Theorem 4.5.5 (see for instance [106]).*

As we will see in Example 4.5.9, different deformation maps can give rise to isomorphic deformations. Therefore, in order to classify all complements we introduce the following:

Definition 4.5.7 *Let $(A, X, \triangleright, \triangleleft, \leftharpoonup, \rightharpoonup)$ be a matched pair of algebras. Two deformation maps r, $R : X \to A$ are called* equivalent *and we denote this by $r \sim R$ if there exists $\sigma : X \to X$, a K-linear automorphism of X such that for any x, $y \in X$ we have:*

$$\sigma(x\,y) - \sigma(x)\,\sigma(y) = \sigma(x) \triangleleft R\big(\sigma(y)\big) + R\big(\sigma(x)\big) \rightharpoonup \sigma(y) - \sigma\big(x \triangleleft r(y)\big) - \sigma\big(r(x) \rightharpoonup y\big)$$

The classification of complements now follows:

Theorem 4.5.8 *Let A be a subalgebra of E, X an A-complement of E and $(A, X, \triangleright, \triangleleft, \leftharpoonup, \rightharpoonup)$ the associated canonical matched pair. Then $\sim$ is an equivalence relation on the set $\mathcal{DM}\,(X, A \,|\, (\triangleright, \triangleleft, \leftharpoonup, \rightharpoonup))$ and the map*

$$\mathcal{HA}^2(X, A \,|\, (\triangleright, \triangleleft, \leftharpoonup, \rightharpoonup)) := \mathcal{DM}\,(X, A \,|\, (\triangleright, \triangleleft, \leftharpoonup, \rightharpoonup))/\sim \longrightarrow \mathcal{F}(A,\, E),$$

$$\overline{r} \mapsto X_r$$

is a bijection between $\mathcal{HA}^2(X, A \,|\, (\triangleright, \triangleleft, \leftharpoonup, \rightharpoonup))$ and the isomorphism classes of all A-complements of E. In particular, the factorization index of A in E is computed by the formula:

$$[E : A]^f = |\mathcal{HA}^2(X, A \,|\, (\triangleright, \triangleleft, \leftharpoonup, \rightharpoonup))|$$

Proof: Two deformation maps r and R are equivalent in the sense of Definition 4.5.7 if and only if the corresponding algebras X_r and X_R are isomorphic. The conclusion follows by Theorem 4.5.5. □

We end the chapter with a few examples illustrating our theory:

Examples 4.5.9 1) *Let A be a two-sided ideal of E. Then $[E : A]^f \leq 1$. Indeed, if an A-complement exists then it should be isomorphic to the factor algebra E/A. Therefore, the factorization index is at most 1.*

2) *Let $A = \begin{pmatrix} 0 & 0 \\ k & 0 \end{pmatrix}$ be a subalgebra of $E = \mathcal{M}_2(k)$. In this case $[E : A]^f = 2$. Indeed, using Example 4.5.2 1), it follows that $X = \begin{pmatrix} k & k \\ 0 & k \end{pmatrix}$ is an A-complement of E. The non-zero values of the canonical matched pair are given by:*

$$e_{21} \leftharpoonup e_{11} = e_{21}, \quad e_{21} \rightharpoonup e_{12} = e_{22}$$
$$e_{12} \triangleleft e_{21} = e_{11}, \quad e_{22} \triangleright e_{21} = e_{21}$$

By a straightforward computation it can be seen that the associated deformation maps are as follows:

$$r_a(e_{11}) = a\,e_{21}, \qquad r_a(e_{12}) = a^2\,e_{21}, \qquad r_a(e_{22}) = -a\,e_{21}, \qquad a \in k.$$

The multiplication on the r_a-deformation of X is described below:

$\cdot_{r_a}$	e_{11}	e_{12}	e_{22}
e_{11}	e_{11}	$e_{12} + a\, e_{22}$	0
e_{12}	$a\, e_{11}$	$a^2\, e_{22}$	e_{12}
e_{22}	0	$-a\, e_{22}$	e_{22}

If $a = 0$ then r_0 is the trivial map and $X_{r_0} = X$. On the other hand, for any $a \in k^*$ we have an isomorphism of algebras X_{r_a} and X_{r_1} given as follows:

$$\varphi : X_{r_a} \to X_{r_1}, \qquad \varphi(e_{11}) = e_{11}, \quad \varphi\left(a^{-1}\, e_{12}\right) = e_{12}, \quad \varphi(e_{22}) = e_{22}$$

To end with, it can be easily seen that X_{r_1} is not isomorphic to X. Indeed, it is enough to observe that $e_{11} + e_{22}$ is a unit for X while X_{r_1} is not unitary. Therefore the factorization index is equal to 2.

3) Let $M \in {}_k\mathcal{M}_k$, and consider $A = \begin{pmatrix} k & 0 \\ 0 & k \end{pmatrix}$ a subalgebra of $E = \begin{pmatrix} k & M \\ 0 & k \end{pmatrix}$. According to Example 4.5.2 3) for $R = S := k$ we obtain that $X = \begin{pmatrix} 0 & M \\ 0 & 0 \end{pmatrix}$ is an A-complement of E. Then any deformation map $r : \begin{pmatrix} 0 & M \\ 0 & 0 \end{pmatrix} \to \begin{pmatrix} k & 0 \\ 0 & k \end{pmatrix}$ is uniquely implemented by two K-linear maps α, $\beta : M \to k$ such that $r\begin{pmatrix} 0 & m \\ 0 & 0 \end{pmatrix} = \begin{pmatrix} \alpha(m) & 0 \\ 0 & \beta(m) \end{pmatrix}$. It can be easily seen that r satisfies (4.41) if and only if $\alpha(m)\beta(n) = 0$ for all $m, n \in M$. Therefore, we either have $\alpha(m) = 0$ for all $m \in M$ or $\beta(m) = 0$ for all $m \in M$. If $\alpha(m) = 0$ for all $m \in M$ we obtain a deformation map r_β defined by $r_\beta \begin{pmatrix} 0 & m \\ 0 & 0 \end{pmatrix} = \begin{pmatrix} 0 & 0 \\ 0 & \beta(m) \end{pmatrix}$ for all $m \in M$, where $\beta : M \to k$ is an arbitrary k-linear map. The multiplication induced by this deformation map is given as follows:

$$\begin{pmatrix} 0 & m \\ 0 & 0 \end{pmatrix} \cdot_{r_\beta} \begin{pmatrix} 0 & n \\ 0 & 0 \end{pmatrix} = \begin{pmatrix} 0 & m\beta(n) \\ 0 & 0 \end{pmatrix}$$

On the other hand, if $\beta(m) = 0$ for all $m \in M$ we obtain a deformation map r^α defined by $r^\alpha \begin{pmatrix} 0 & m \\ 0 & 0 \end{pmatrix} = \begin{pmatrix} \alpha(m) & 0 \\ 0 & 0 \end{pmatrix}$ for all $m \in M$, where $\alpha : M \to k$ is an arbitrary K-linear map. The multiplication induced by this deformation map is given as follows:

$$\begin{pmatrix} 0 & m \\ 0 & 0 \end{pmatrix} \cdot_{r^\alpha} \begin{pmatrix} 0 & n \\ 0 & 0 \end{pmatrix} = \begin{pmatrix} 0 & \alpha(m)n \\ 0 & 0 \end{pmatrix}$$

By a straightforward computation it can be proved that if the k-linear maps α, $\alpha' : M \to k$ are different from the trivial map then r^α is equivalent to $r^{\alpha'}$ in the sense of Definition 4.5.7. In the same manner, if the K-linear maps β, $\beta' : M \to K$ are different from the trivial map then r_β is equivalent to $r_{\beta'}$ in the sense of Definition 4.5.7. Finally, r^α is never equivalent to r_β in the *sense of Definition 4.5.7, except for the case when both α and β are equal to the trivial map. Therefore the factorization index $[E : A]^f = 3$.*

4) *Let* $A = \begin{pmatrix} k & k & k \\ k & k & k \\ 0 & 0 & 0 \end{pmatrix}$ *be a subalgebra of* $E = \mathcal{M}_3(k)$. *Then, by Example 4.5.2 2), it follows that* $X = \begin{pmatrix} 0 & 0 & 0 \\ 0 & 0 & 0 \\ k & k & k \end{pmatrix}$ *is an A-complement of E. The non-zero values of the canonical matched pair are given as follows:*

$$\begin{aligned} e_{31} \triangleleft e_{11} &= e_{31}, & e_{31} \triangleleft e_{12} &= e_{32}, & e_{31} \triangleleft e_{13} &= e_{33} \\ e_{32} \triangleleft e_{21} &= e_{31}, & e_{32} \triangleleft e_{22} &= e_{32}, & e_{32} \triangleleft e_{23} &= e_{33} \\ e_{13} \leftharpoonup e_{31} &= e_{11}, & e_{13} \leftharpoonup e_{32} &= e_{12}, & e_{13} \leftharpoonup e_{33} &= e_{13} \\ e_{23} \leftharpoonup e_{31} &= e_{21}, & e_{23} \leftharpoonup e_{32} &= e_{22}, & e_{23} \leftharpoonup e_{33} &= e_{23} \end{aligned}$$

In this case the computational complexity increases dramatically, making it very difficult to compute all associated deformation maps. However, we are still able to check by a straightforward computation that the following maps are deformations of the above canonical matched pair:

$$\begin{aligned} r_1 &: X \to A, & r_1(e_{33}) &= e_{22} \\ r_2 &: X \to A, & r_2(e_{33}) &= e_{11} + e_{22} \\ r_3 &: X \to A, & r_3(e_{31}) &= e_{12}, \quad r_3(e_{33}) = e_{11} + e_{22} \end{aligned}$$

We denote by X_i *the* r_i*-deformation of* X*, for all* $i = \overline{1,3}$*. The multiplication tables of the* X_i*'s,* $i = \overline{1,3}$*, are depicted below:*

$$\begin{aligned} X_1 &: \quad e_{33}e_{31} = e_{31}, \; e_{33}e_{32} = e_{32}e_{33} = e_{32}, \; e_{33}e_{33} = e_{33} \\ X_2 &: \quad e_{31}e_{33} = e_{33}e_{31} = e_{31}, \; e_{32}e_{33} = e_{33}e_{32} = e_{32}, \; e_{33}e_{33} = e_{33} \\ X_3 &: \quad e_{31}e_{31} = e_{32}, \; e_{31}e_{33} = e_{33}e_{31} = e_{31}, \; e_{32}e_{33} = e_{33}e_{32} = e_{32}, \\ &\quad\; e_{33}e_{33} = e_{33} \end{aligned}$$

It can be easily seen that the X_i*'s,* $i = \overline{1,3}$*, are isomorphic to the algebras* As_3^9, As_3^{10} *and respectively* As_3^{12} *listed in [207] (for a complete list of 3-dimensional associative algebras over* $\mathbb{C}$ *we refer to [105]). In each case the isomorphism sends* e_{31} *to* e_1, e_{32} *to* e_2 *and* e_{33} *to* e_3*, where according to the notations of [207],* $\{e_1, e_2, e_3\}$ *is a k-basis for the 3-dimensional algebras mentioned above. In particular, we obtain that the* X_i*'s are two-by-two non-isomorphic. Moreover, none of the three algebras listed above is isomorphic to* X*. Indeed, to start with, we should notice that* X *is not commutative and therefore it cannot be isomorphic to the commutative algebras* X_2 *or* X_3*. We prove now that* X *is not isomorphic to* X_1*. Assume that* $\varphi : X_1 \to X$ *is an isomorphism of algebras given by:*

$$\varphi(e_{31}) = \Sigma_{i=1}^3 a_i e_{3i}, \quad \varphi(e_{32}) = \Sigma_{i=1}^3 b_i e_{3i}, \quad \varphi(e_{33}) = \Sigma_{i=1}^3 c_i e_{3i}$$

where a_i, b_i, $c_i \in K$ *for all* $i = \overline{1,3}$. *Since* $\varphi(e_{31}) = \varphi(e_{33}e_{31}) = \varphi(e_{33})\varphi(e_{31}) = c_3\varphi(e_{31})$ *and* $\varphi(e_{31}) \neq 0$ *(as* φ *is an isomorphism) we obtain that* $c_3 = 1$. *Moreover, from* $0 = \varphi(e_{32}e_{31}) = \varphi(e_{32})\varphi(e_{31}) = b_3\varphi(e_{31})$ *it follows that* $b_3 = 0$. *Finally, as* $\varphi(e_{32}) = \varphi(e_{32}e_{33}) = \varphi(e_{32})\varphi(e_{33}) = b_3\varphi(e_{33}) = 0$ *we have reached a contradiction. Therefore* X_1 *is not isomorphic to* X *and the factorization index* $[E : A]^f \geq 4$.

Remark 4.5.10 *After a careful analysis of Example 4.5.9 we can easily conclude that the deformations of a given algebra* X *do not necessarily preserve the properties of* X. *For instance, in Example 4.5.9,* 2) *we obtained a non-unitary algebra as a deformation of a unitary one, while in Example 4.5.9,* 4) *we construct commutative deformations of a non-commutative algebra. This is not the case for the classical deformations studied in [106] as it is well known that finite-dimensional unitary algebras only deform to unitary algebras.*

Bibliographical Notes

The results of this chapter are part of the author's papers [25] and respectively [6].

Chapter 5

Jacobi and Poisson algebras

Noncommutative geometry as well as quantum group theory is based on the same idea: instead of working with points on a given *space* M (such as a compact topological group, a Lie group, an algebraic group, a manifold, etc.) we can work equivalently with the algebra $\mathrm{Fun}\,(M)$ of functions on M. The definition of the algebra of functions depends on the category of spaces that we are dealing with: for instance, if M is a compact topological group then $\mathrm{Fun}\,(M) := \mathcal{R}\,(M)$, the algebra of real-valued continuous representative functions on M, while if M is a manifold then $\mathrm{Fun}\,(M) := C^{\infty}(M)$, the algebra of all real smooth functions on M. Thus there exists a contravariant functor $\mathrm{Fun}\,(-)$, which very often is in fact a duality of categories, between a certain category of spaces and a certain category of algebras endowed with additional structures compatible with the algebra structure such as coalgebras (i.e., Hopf algebras), Lie brackets (i.e., Poisson algebras), etc. Adopting this categorical viewpoint, purely geometric concepts sat at the foundation of a research program started at the end of the 1980s which deals with their algebraic counterparts: the functor $\mathrm{Fun}\,(-)$ being the tool through which geometrical problems can be restated and approached in algebraic language by way of a certain category of algebras. Following this idea we deal with Jacobi algebras, which are abstract algebraic counterparts of Jacobi manifolds. Jacobi manifolds, as generalizations of symplectic or more generally Poisson manifolds, were introduced independently by Kirillov [150] and Lichnerowicz [161]. A Jacobi manifold is a smooth manifold endowed with a bivector field Λ and a vector field E satisfying some compatibility conditions. When the vector field E identically vanishes, the Jacobi manifold is just a Poisson manifold. Equivalently, a Jacobi manifold is a smooth manifold M such that the commutative algebra $A := C^{\infty}(M)$ of real smooth functions on M is endowed with a Lie bracket $[-,-]$ such that $[ab,\, c] = a\,[b,\, c] + [a,\, c]\,b - ab\,[1_A,\, c]$, for all $a,\, b,\, c \in A$. Such an algebraic object A is called a Jacobi algebra while a Poisson algebra is just a Jacobi algebra such that $[1_A,\, a] = 0$, for all $a \in A$. Poisson algebras appear in several areas of research situated at the border between mathematics and physics such as: Hamiltonian mechanics, differential geometry, Lie groups and representation theory, noncommutative algebraic/differential geometry, (super)integrable systems, quantum field theory, vertex operator algebras, quantum groups and so on; see the recent papers [75, 124, 159] and the references therein. If we look at Poisson algebras as the 'differential' version of Hopf algebras, then, mutatis mutandis, Jacobi algebras can be seen as generalizations of Poisson algebras in the same way as weak Hopf algebras [52] generalize Hopf algebras. It is therefore natural to expect that Jacobi algebras will play an important role in all fields enumerated above. For further details on the study of Jacobi algebras from geometric viewpoint we refer to [121, 123, 124, 177, 211].

The classification of finite-dimensional Poisson manifolds is a difficult task: the first steps towards the classification of low dimensional Poisson manifolds are taken in [122, 159] using mainly differential geometry tools. It is natural to ask a more general question, namely that of classifying Jacobi manifolds of a given dimension. Following the viewpoint of the aforementioned paper we look at its algebraic counterpart by considering the classification of all Jacobi algebras of a given dimension. The problem contains as subsequent questions the classical problems of classifying all associative (resp. Lie) algebras of a given dimension. For updates on the progress made so far and the geometrical significance we refer to [182] (resp. [201]) and their list of references—we just mention that the classification of all complex associative (resp. Lie) algebras is known only up to dimension 5 (resp. 7). As one of the main tools used intensively in the classification of finite 'objects' is Hölder's *extension problem*, we propose here a more general version of the extension problem. More precisely, this chapter is devoted to the extending structures (ES) problem in the context of Jacobi algebras which comes down to the following:

Let A be a Jacobi algebra and E a vector space containing A as a subspace. Describe and classify the set of all Jacobi algebra structures that can be defined on E such that A becomes a Jacobi subalgebra of E.

The chapter is organized as follows: in Section 5.1 we present basic properties of Jacobi algebras: in particular, we classify all complex Jacobi algebras of dimension 2 or 3. The *conformal deformation* of a Jacobi algebra A (Proposition 5.1.5) is the Jacobi version of the Drinfel'd twist deformation of a quantum group. We introduce the category $\mathcal{JM}_A^A$ of *Jacobi A-modules* which is an equivalent way of defining representations of a Jacobi algebra A and based on this concept we define a *Frobenius Jacobi* algebra as a Jacobi algebra A such that $A \cong A^*$, as Jacobi A-modules. A weaker version of this notion, at the level of Poisson algebras, was recently introduced in [232] where the term Frobenius Poisson algebra is used to denote Poisson algebras A such that $A \cong A^*$, an isomorphism of right A-modules. *Integrals* on a Jacobi algebra are introduced in Definition 5.1.12 having the Hopf algebra theory as a source of inspiration and are used in the characterization theorem of Frobenius Jacobi algebras (Theorem 5.1.15): a Jacobi algebra A is Frobenius if and only if there exists a non-degenerate integral on A; in the finite-dimensional case, this is also equivalent to the existence of a so-called *Jacobi-Frobenius pair* of A, which allows us to define the Euler-Casimir element associated to any finite-dimensional Frobenius Jacobi algebra.

In Section 5.2 we embark on the study of the ES problem for Jacobi algebras. If we fix V, a complement of the Jacobi algebra A in the vector space E, then the ES problem can be rephrased equivalently as follows:

Describe and classify all Jacobi algebras containing A as a subalgebra of codimension equal to $\dim(V)$.

The answer to the ES problem is given in three steps: the first one is taken in Theorem 5.2.2 where we perform the construction of the *unified product* $A \ltimes V$ associated to a Jacobi algebra A and a vector space V connected through four 'actions' and two 'cocycles'. The unified product is a very general construction containing as special cases the semi-direct product, the crossed product or the bicrossed product. The second step (Proposition 5.2.4) shows that a Jacobi algebra structure $(E, *_E, [-,\, -]_E)$ on E contains A as a Jacobi subalgebra if and only if there exists an isomorphism of Jacobi algebras $(E, *_E, [-,\, -]_E) \cong A \ltimes V$. Finally, the theoretical answer to the ES problem is given in Theorem 5.2.7: for a fixed complement V of A in E, a non-abelian cohomological type object $\mathcal{JH}^2(V,\, A)$ is explicitly constructed; it parameterizes and classifies all Jacobi algebras containing A as a subalgebra of codimension equal to $\dim(V)$. The classification obtained in this theorem follows the viewpoint of the extension problem: that is, up to an isomorphism of Jacobi algebras that stabilizes A and co-stabilizes V. For the sake of completeness we also write down the corresponding results for Poisson algebras. Computing the classifying object $\mathcal{JH}^2(V,\, A)$, for a given Jacobi algebra A and a vector space V is a highly nontrivial question. If $A := k$, the base field viewed as a Jacobi algebra with the trivial bracket, then $\mathcal{JH}^2(V,\, k)$ classifies in fact all Jacobi algebras of dimension $1 + \dim(V)$, which is of course a hopeless question for vector spaces of large dimension. For this reason we shall assume from now on that $A \neq k$. A very important step in computing $\mathcal{JH}^2(V,\, A)$ is taken in Corollary 5.2.9, where a decomposition of it as a coproduct over all Jacobi A-modules structures on V is given. Section 5.3 is devoted to computing $\mathcal{JH}^2(V,\, A)$ for what we have called *flag Jacobi* algebras over A: that is, a Jacobi algebra structure on E such that there exists a finite chain of Jacobi subalgebras $E_0 := A \subset E_1 \subset \cdots \subset E_m := E$, such that each E_i has codimension 1 in E_{i+1}. All flag Jacobi algebras over A can be completely described by a recursive reasoning where the key step is settled in Theorem 5.3.6: several applications and examples are given at the end of the section. In particular, if A is a Poisson algebra we indicate the difference between $\mathcal{JH}^2(V,\, A)$ and $\mathcal{PH}^2(V,\, A)$, the latter being the classifying object of the ES problem for Poisson algebras.

In Section 5.4 we recover the bicrossed product of Poisson algebras recently considered by Ni and Bai [189, Theorem 1] as a special case of the unified product. Furthermore, Proposition 5.4.2 shows that the bicrossed product is the construction responsible for the *factorization problem* at the level of Poisson algebras. The rest of the section is devoted to the converse of the factorization problem, called the *classifying complements* problem, which for Poisson algebras comes down to the following:

Let $P \subset R$ be an extension of Poisson algebras. If a P-complement in R exists (i.e., a Poisson subalgebra $Q \subset R$ such that $R = P + Q$ and $P \cap Q = \{0\}$), describe explicitly, classify all P-complements in R, and compute the cardinal *of the isomorphism classes of all P-complements in R (which will be called the factorization index $[R : P]^f$ of P in R).*

Let Q be a given P-complement and $\big(P, Q, \triangleleft, \triangleright, \leftharpoonup, \rightharpoonup\big)$ the associated canonical matched pair (Proposition 5.4.2). In Theorem 5.4.6 a general deformation of the Poisson algebra Q is introduced: it is associated to a deformation map $r : Q \to P$ in the sense of Definition 5.4.4. Theorem 5.4.6 proves that in order to find all complements of P in R it is enough to know only one P-complement: all the other P-complements are deformations of it. The answer to the bicrossed descent problem is given in Theorem 5.4.9: there exists a bijection between the isomorphism classes of all P-complements of R and a new cohomological type object $\mathcal{HA}^2\big(P, Q \mid (\triangleleft, \triangleright, \leftharpoonup, \rightharpoonup)\big)$ which is constructed and the formula for computing the factorization index $[R : P]^f$ is given. Examples are also provided: in particular, an extension $P \subseteq R$ of Poisson algebras such that P has infinitely many non-isomorphic complements in R is constructed in Proposition 5.4.10.

5.1 (Bi)modules, integrals and Frobenius Jacobi algebras

This section contains some basic notions about Jacobi algebras. Throughout we use the following convention: except for the situation when we deal with Poisson algebras (as in Section 5.4) by an algebra $A = (A, m_A)$ we will always mean a commutative and unitary algebra. We start with a few observations on the categorical behavior of Jacobi/Poisson algebras.

Remarks 5.1.1 (1) *The category ${}_k\mathrm{Poss}$ of unital Poisson algebras is a reflective subcategory in the category of Jacobi algebras, i.e., the inclusion functor $\iota : {}_k\mathrm{Poss} \hookrightarrow {}_k\mathrm{Jac}$ has a left adjoint which we will denote by $(-)_{\mathrm{poss}} : {}_k\mathrm{Jac} \to {}_k\mathrm{Poss}$ and whose construction goes as follows: for any Jacobi algebra A we define $A_{\mathrm{poss}} := A/I_{\mathrm{poss}}$, where I_{poss} is the Jacobi ideal of A generated by all brackets $[1_A, a]$, for all $a \in A$. Then, A_{poss} is a Poisson algebra and the quotient map $A \to A_{\mathrm{poss}}$, $a \mapsto \widehat{a}$ is universal among the maps from A to any unital Poisson algebra which are morphisms of Jacobi algebras. We mention that it is possible that $A_{\mathrm{poss}} = 0$; this happens when I_{poss} contains an invertible element of A. An example is the Jacobi algebra $J^{2,\,2}$ defined in Example 5.1.2 below.*

(2) *The category ${}_k\mathrm{Jac} = \big({}_k\mathrm{Jac}, - \otimes -, k, \tau_{-,-}\big)$ is braided monoidal: if A and B are Jacobi algebras, then $A \otimes B$ is a Jacobi algebra via*

$$(a \otimes b) \cdot (a' \otimes b') := aa' \otimes bb', \quad [a \otimes b, a' \otimes b'] := aa' \otimes [b, b'] + [a, a'] \otimes bb'$$

for all a, $a' \in A$ and b, $b' \in B$, the base field k viewed with the abelian Lie bracket is the unit and the braiding is the usual flip $\tau_{A,B} : A \otimes B \to B \otimes A$, $\tau_{A,B}(a \otimes b) := b \otimes a$.

The classification of Jacobi algebras of a given finite-dimension is the first non-trivial question which arises as the algebraic counterpart of the classification of finite-dimensional Jacobi manifolds. Of course, any 1-dimensional Jacobi algebra is isomorphic to the abelian Jacobi algebra k_0. The 2-dimensional case is covered below and it reveals an interesting fact, namely that two Jacobi algebras can be isomorphic both as algebras and Lie algebras (through different isomorphisms) but not isomorphic as Jacobi algebras. A similar result holds for Hopf algebras as well.

Example 5.1.2 *Let k be a field of characteristic $\neq 2$. If $k = k^2 := \{q^2 \mid q \in k\}$ then, up to an isomorphism, there exist four 2-dimensional Jacobi algebras over k. These are the Jacobi algebras denoted by $J^{2,1}$, $J^{2,2}$, $J^{2,3}$, $J^{2,4}$ having $\{1, x\}$ as a basis and the multiplication and the bracket defined by:*

$$J^{2,1} : x^2 = 0, \ [1, x] = 0; \qquad J^{2,2} : x^2 = 0, \ [1, x] = 1$$

$$J^{2,3} : x^2 = x, \ [1, x] = 0; \qquad J^{2,4} : x^2 = 0, \ [x, 1] = x$$

If $k \neq k^2$, besides the four Jacobi algebras listed above, there exists another one-parameter 2-dimensional Jacobi algebra J^2_d defined by:

$$J^2_d : \qquad x^2 = d, \qquad [1, x] = 0$$

for all $d \in S$, where $S \subseteq k \setminus k^2$ is a system of representatives for the following equivalence relation on $k \setminus k^2$: $d \equiv d'$ if and only if there exists $q \in k^*$ such that $d = q^2 d'$.

Indeed, we will fix $\{1, x\}$ as a basis in a 2-dimensional Jacobi algebra. The proof follows from the classical classification of 2-dimensional Lie algebras [162] and from the well-known classification of 2-dimensional associative algebras [215]. Indeed, the classification follows by a routine computation based on checking the compatibility condition (0.9). We only mention that the algebra defined by the multiplication $x^2 = x$ (or $x^2 = d$, for some $d \in S$, if $k \neq k^2$) together with the Lie bracket $[1, x] = 1$ or $[1, x] = x$ is not a Jacobi algebra since the compatibility condition (0.9) fails for $a = b = c := x$. We observe that $J^{2,2}$ and $J^{2,4}$ are Jacobi non-Poisson algebras, $(J^{2,2})_{\rm poss} = 0$ and $(J^{2,4})_{\rm poss} \cong k_0$. Moreover, the Jacobi algebras $J^{2,2}$ and $J^{2,4}$ are isomorphic as associative algebras as well as Lie algebras but they are not isomorphic as Jacobi algebras.

In particular, if $k = \mathbb{C}$ there are four isomorphism classes of 2-dimensional Jacobi algebras, while if $k = \mathbb{R}$ there exist five types of 2-dimensional Jacobi algebras, the fifth one being the Jacobi algebra J^2_{-1}.

Example 5.1.3 *Using the classical classification of 3-dimensional associative (resp. Lie) algebras from [215] (resp. [141, 162]) over the complex field $\mathbb{C}$ and*

Jacobi algebra	Algebra structure	Lie bracket
J_1^3	$x^2 = y^2 = xy = yx = 0$	abelian
J_2^3	$x^2 = y,\ y^2 = 0,\ xy = yx = 0$	abelian
J_3^3	$x^2 = x,\ y^2 = 0,\ xy = yx = 0$	abelian
J_4^3	$x^2 = x,\ y^2 = y,\ xy = yx = 0$	abelian
J_5^3	$x^2 = y^2 = xy = yx = 0$	$[x, 1] = x$
J_6^3	$x^2 = y^2 = xy = yx = 0$	$[x, y] = x$
J_7^3	$x^2 = y^2 = xy = yx = 0$	$[x, 1] = x+y,\ \ [y, 1] = y$
${}^uJ_8^3,\ u \in k^*$	$x^2 = y^2 = xy = yx = 0$	$[x, 1] = x,\ \ [y, 1] = u\,y$
J_9^3	$x^2 = y,\ y^2 = 0,\ xy = yx = 0$	$[x, 1] = x,\ \ [y, 1] = 2\,y$
J_{10}^3	$x^2 = y,\ y^2 = 0,\ xy = yx = 0$	$[x, 1] = 2^{-1}x,\ \ [y, 1] = y$
J_{11}^3	$x^2 = x,\ y^2 = 0,\ xy = yx = 0$	$[y, 1] = y$

Table 5.1: Jacobi algebras of dimension 3 over $\mathbb{C}$.

the same strategy as in Example 5.1.2 we can prove that, up to an isomorphism, there are exactly 11 *families of complex Jacobi algebras of dimension* 3*: they have* $\{1, x, y\}$ *as a basis and the multiplication and the bracket as listed in Table 5.1.*

Subsequent to the problem of classifying Jacobi algebras of a given dimension is the following question: *for a given algebra (resp. Lie algebra) A, describe and classify all Lie brackets $[-, -]$ (resp. all possible multiplications m_A) which endow A with a Jacobi algebra structure.* Some examples are given below:

Examples 5.1.4 (1) *Let C_n be a cyclic group of order $n \geq 2$ generated by c. Then the group algebra $k[C_n]$ can be made into a Jacobi algebra as follows:*

(1) *If* char $(k) = 0$ *or* char $(k) = p > 0$ *and* $(p, n-1) = 1$, *then the only Lie bracket which makes $k[C_n]$ into a Jacobi algebra is the trivial one, i.e., $k[C_n] = k[C_n]_0$;*

(2) *If* char $(k) = p \,|\, n-1$, *then any $y \in k[C_n]$ induces a unique Lie bracket given by $[c^i, c^j] = (j-i)\, c^{j+i-1}\, y$, for all $i,\ j = 1, 2, ..., n$ which makes $k[C_n]$ into a Jacobi algebra.*

Indeed, let $[-, -]$ be a Lie bracket that endows the group algebra $k[C_n]$ with a Jacobi algebra structure and we denote $y := [1, c]$. By using (0.9) and the induction we obtain that $[c^i, c^j] = (j-i)\, c^{j+i-1}\, y$, for all $i,\ j = 1, 2, ..., n$. Therefore, since $[c^n, c] = y$ we obtain $(n-1)\, y = 0$, which brings us to the two cases considered above. If char $(k) = p$ *and* $p|n-1$, *then it can be easily seen that Jacobi's identity is also fulfilled.*

(2) *Let C_∞ be the infinite cyclic group generated by c. Then, for any $y \in k[C_\infty]$, the group algebra $k[C_\infty]$ admits a Jacobi algebra structure with the bracket given by $[c^i, c^j] = (j-i)\, c^{j+i-1}\, y$, for all $i,\ j \in \mathbb{Z}$.*

(3) *Let* sl (2, ℂ) *be the complex special linear algebra of dimension* 3. *Since* sl (2, ℂ) *is perfect as a Lie algebra, a careful analysis of the Lie brackets on* 3*-dimensional Jacobi algebras given in table 5.1 brings us to the conclusion that the Lie algebra* sl (2, ℂ) *cannot be endowed with an algebra structure to make it into a Jacobi algebra.*

The following construction is the algebraic counterpart of conformal deformations of Jacobi manifolds [177]. Mutatis mutandis, it can be seen as the Jacobi version of the Drinfel'd twist deformation for quantum groups [97].

Proposition 5.1.5 *Let* $A = (A, m_A, [-, -])$ *be a Jacobi algebra and* $u \in U(A)$ *an invertible element of* A. *Then* $A_u := (A, m_A, [-, -]_u)$ *is a Jacobi algebra called the* u-conformal deformation *of* A, *where the bracket* $[-, -]_u$ *is given for any* $x, y \in A$ *by:*

$$[x, y]_u := u^{-1}\,[u\,x,\ u\,y] \tag{5.1}$$

Proof: The bilinear map $[-, -]_u$ is a Lie bracket on A since for any x, y, $z \in A$ we have:

$$\begin{aligned} & \big[x, [y, z]_u\big]_u + \big[y, [z, x]_u\big]_u + \big[z, [x, y]_u\big]_u = \\ = \;& u^{-1}\big[ux, [uy, uz]\big] + u^{-1}\big[uy, [uz, ux]\big] + u^{-1}\big[uz, [ux, uy]\big] = 0 \end{aligned}$$

Now, the compatibility condition (0.9) is equivalent to $[uxy, uz] = [ux, uz]\,y + x\,[uy, uz] - xy\,[u, uz]$, for all $x, y, z \in A$. The right-hand side gives:

$$\begin{aligned} & [ux, uz]\,y + x\,\underline{[uy, uz]} - xy\,[u, uz] = \\ \overset{(0.9)}{=}\; & [ux, uz]\,y + x\,\big(u\,[y, uz] + [u, uz]\,y - uy\,[1_A, uz]\big) - xy\,[u, uz] \\ = \; & [ux, uz]\,y + u\,x\,[y, uz] - u\,x\,y[1_A, uz] \overset{(0.9)}{=} [uxy, uz] \end{aligned}$$

as needed, where the last equality follows by applying (0.9) for the bracket $[-, -]$ in ux, y and respectively uz. □

Describing and classifying all u-conformal deformations of a given Jacobi algebra A is an interesting and difficult question that will not be addressed here. In what follows we provide an example which shows that a u-conformal deformation of a Jacobi algebra can be a Poisson algebra.

Example 5.1.6 *Let* A *be the* 3*-dimensional Jacobi non-Poisson algebra over of field of characteristic* $\neq 2$ *having* $\{1, x, y\}$ *as a basis and the multiplication and the non-trivial brackets given by* $x^2 := 0$, $\;xy = yx := x$, $\;y^2 := 2y - 1$, $\;[1, y] = [x, y] := x$ *(*A *is the Jacobi algebra denoted by* $J^{3,\,5}_{2,\,0,\,-1,-1}$ *in Proposition 5.3.10 below). The group* $U(A)$ *of units of* A *coincides with the set of elements of the form* $u = \alpha + \beta\,x + \gamma\,y$, *with* $\alpha + \gamma \neq 0$ *and the space* *of all u-conformal deformations of A is in bijection with the three-parameter Jacobi algebras having the bracket defined by:*

$$[1,\, x]_u = -\gamma\, x, \quad [1,\, y]_u = (\alpha + \beta + \gamma)\, x, \quad [x,\, y]_u = (\alpha + 2\gamma)\, x$$

for all $\alpha,\, \beta,\, \gamma \in k$ such that $\alpha + \gamma \neq 0$. In particular, the u-conformal deformation of A corresponding to $u := \alpha\,(1 - x)$, for any $\alpha \in k^*$ is a Poisson algebra.

(Bi)Modules, integrals and Frobenius Jacobi algebras

We introduce representations of a Jacobi algebra using the equivalent notion of modules.

Definition 5.1.7 *Let A be a Jacobi algebra. A right* Jacobi A-module *is a vector space V equipped with two bilinear maps $\triangleleft : V \times A \to V$ and $\leftharpoondown : V \times A \to V$ such that $(V, \triangleleft)$ is a right A-module, $(V, \leftharpoondown)$ is a right Lie A-module satisfying the following two compatibility conditions for any $a, b \in A$ and $x \in V$:*

$$x \leftharpoondown (ab) = (x \leftharpoondown a) \triangleleft b + (x \leftharpoondown b) \triangleleft a - (x \leftharpoondown 1_A) \triangleleft (ab) \qquad (5.2)$$

$$x \triangleleft [a,\, b] = (x \triangleleft a) \leftharpoondown b - (x \leftharpoondown b) \triangleleft a + (x \triangleleft a) \triangleleft [1_A,\, b] \qquad (5.3)$$

A right Jacobi A-bimodule *is a right Jacobi A-module which in addition satisfies the following compatibility condition for any $a, b \in A$ and $x \in V$:*

$$x \leftharpoondown (ab) = (x \triangleleft a) \leftharpoondown b + (x \triangleleft b) \leftharpoondown a - (x \triangleleft (ab)) \leftharpoondown 1_A \qquad (5.4)$$

Let $\mathcal{JM}_A^A$ (resp. $\mathcal{JB}_A^A$) be the category of right Jacobi A-modules (resp. Jacobi A-bimodules) having as morphisms the linear maps which are compatible with both actions.

The categories ${}_A^A\mathcal{JM}$ (resp. ${}_A^A\mathcal{JB}$) of left Jacobi A-(bi)modules are defined analogously and there exists an isomorphism of categories ${}_A^A\mathcal{JM} \cong \mathcal{JM}_A^A$ and ${}_A^A\mathcal{JB} \cong \mathcal{JB}_A^A$.

Remarks 5.1.8 (1) *The compatibility conditions (5.2)-(5.3) defining the category $\mathcal{JM}_A^A$ are the Jacobi version of the compatibility conditions (0.8) defining Poisson modules over a Poisson algebra: they are precisely axioms (J2) and (J4) from the construction of the unified product (Theorem 5.2.2). On the other hand, axiom (5.4) defining Jacobi bimodules has another explanation which will be given in Lemma 5.1.10 below.*

(2) *Rephrasing Definition 5.1.7 in terms of representations can be done as follows: a* representation *of a Jacobi algebra A on a vector space V is a pair (ψ, φ) consisting of a representation ψ of the associative algebra A on V, that is an algebra map $\psi : A \to \mathrm{End}_k(A)$ and a representation φ of a Lie algebra*

A on V, i.e., a Lie algebra map $\varphi : A \to \mathrm{gl}(V)$*, satisfying the following two compatibilities for any* $a, b \in A$*:*

$$\begin{aligned} \varphi(ab) &= \psi(b) \circ \varphi(a) - \psi(a) \circ \varphi(b) + \psi(ab) \circ \varphi(1_A) && (5.5) \\ \psi\big([a,\, b]\big) &= \psi(a) \circ \varphi(b) - \varphi(b) \circ \psi(a) + \psi\big([1_A,\, b]\, a\big) && (5.6) \end{aligned}$$

Representations of a Jacobi algebra A and right Jacobi A-modules are two different ways of describing the same structure: more precisely, there exists an equivalence of categories $\mathcal{JM}_A^A \cong \mathrm{Rep}(A)$*, where* $\mathrm{Rep}(A)$ *is the category of representations of A with the obvious morphisms. The one-to-one correspondence between right Jacobi A-modules* $(\triangleleft, \leftharpoondown)$ *on V and representations* (ψ, φ) *of A on V is given by the two-sided formulas:* $\psi(a)(x) :=: x \triangleleft a$ *and* $\varphi(a)(x) :=: -x \leftharpoondown a$*, for all* $a \in A$ *and* $x \in V$*.*

Examples 5.1.9 (1) *Any Jacobi algebra A has a canonical structure of a Jacobi A-bimodule via:* $x \triangleleft a := xa$ *and* $x \leftharpoondown a := [x,\, a]$*, for all* $x,\, a \in A$*. Indeed, for these actions, axioms (5.2) and (5.3) are both equivalent to the compatibility condition (0.9) defining a Jacobi algebra. On the other hand, axiom (5.4) is equivalent to*

$$[x,\, ab] = [xa,\, b] + [xb,\, a] - [xab,\, 1_A]$$

which follows trivially by applying several times the compatibility condition (0.9).

(2) *Any vector space V can be viewed as a Jacobi A-bimodule via the trivial actions:* $x \triangleleft a := x$ *and* $x \leftharpoondown a := 0$*, for all* $x \in V$*,* $a \in A$*. We shall denote by* V_0 *the vector space V equipped with these trivial actions.*

(3) *There exists a bijection between the set of all right Jacobi A-module structures* $(\triangleleft, \leftharpoondown)$ *that can be defined on k and the set of all pairs* $(\Lambda, \lambda) \in A^* \times A^*$*, consisting of an algebra map* $\Lambda : A \to k$ *and a Lie algebra map* $\lambda : A \to k$ *satisfying the following two compatibility conditions for any* $a,\, b \in A$*:*

$$\begin{aligned} \lambda(ab) &= \lambda(a)\, \Lambda(b) + \lambda(b)\, \Lambda(a) - \lambda(1_A)\, \Lambda(ab), \\ \Lambda([a,\, b]) &= \Lambda(a)\, \Lambda([1_A,\, b]) && (5.7) \end{aligned}$$

The bijection is given such that the Jacobi A-module structure $(\triangleleft, \leftharpoondown)$ *on k associated to* $(\Lambda, \lambda) \in A^* \times A^*$ *is defined by* $x \triangleleft a := x\, \Lambda(a)$ *and* $x \leftharpoondown a := x\, \lambda(a)$*, for all* $a \in A$*,* $x \in k$*. The actions associated to such a pair* (Λ, λ) *endow k with a Jacobi A-bimodule structure since (5.4) also holds thanks to the first compatibility condition of (5.7).*

The proof of the next lemma provides a motivation for introducing axiom (5.4) in Definition 5.1.7: without it the linear dual of a right Jacobi A-module is not necessarily a right Jacobi A-module.

Lemma 5.1.10 *Let A be a Jacobi algebra and $(V, \triangleleft, \leftharpoonup) \in \mathcal{JB}_A^A$ a Jacobi A-bimodule. Then the k-linear dual $V^* = (V^*, \blacktriangleleft, \curvearrowleft) \in \mathcal{JB}_A^A$ is a Jacobi A-bimodule, where the actions $\blacktriangleleft$, $\curvearrowleft$ are defined for any $v^* \in V^*$, $a \in A$ and $x \in V$ by:*

$$(v^* \blacktriangleleft a)(x) := v^*(x \triangleleft a), \qquad (v^* \curvearrowleft a)(x) := -v^*(x \leftharpoonup a) \tag{5.8}$$

In particular, there exists a well-defined contravariant functor $(-)^* : \mathcal{JB}_A^A \to \mathcal{JB}_A^A$ which for finite-dimensional Jacobi A-bimodules provides a duality of categories.

Proof: $(V^*, \curvearrowleft)$ is a right Lie A-module and $(V^*, \blacktriangleleft)$ is a right A-module since A is a commutative algebra. A straightforward computation shows that the compatibility condition (5.2) holds for $(V^*, \blacktriangleleft, \curvearrowleft)$ if and only if (5.4) holds for $(V, \triangleleft, \leftharpoonup)$ and similarly, (5.4) holds for $(V^*, \blacktriangleleft, \curvearrowleft)$ if and only if (5.2) holds for $(V, \triangleleft, \leftharpoonup)$. Finally, (5.3) for $(V^*, \blacktriangleleft, \curvearrowleft)$ is equivalent to (5.3) for $(V, \triangleleft, \leftharpoonup)$. □

We shall view the Jacobi algebra A as an object in $\mathcal{JB}_A^A$ via the actions $x \triangleleft a := xa$ and $x \leftharpoonup a := [x, a]$, for all $x, a \in A$. It follows from Lemma 5.1.10 that the k-linear dual A^* is a Jacobi A-bimodule whose actions take the following form:

$$(a^* \blacktriangleleft a)(x) := a^*(ax), \qquad (a^* \curvearrowleft a)(x) := a^*([a, x]) \tag{5.9}$$

From now on we will see A and A^* as objects in $\mathcal{JB}_A^A$ via the above structures.

Definition 5.1.11 *A Jacobi algebra A is called* Frobenius *if there exists an isomorphism $A \cong A^*$ in $\mathcal{JB}_A^A$, i.e., an isomorphism of right A-modules and right Lie A-modules.*

Any Frobenius Jacobi algebra is Frobenius as an associative algebra and as a Lie algebra. Now we introduce integrals on a Jacobi algebra: as in the case of Hopf algebras [204], integrals will be intimately related to the property of being Frobenius.

Definition 5.1.12 *An* integral *on a Jacobi algebra A is an element $\nu \in A^*$ such that*

$$\nu\big([a, b]\, c\big) = \nu\big(a\, [b, c]\big) \tag{5.10}$$

for all $a, b, c \in A$. We denote by $\int_A$ the space of integrals on a Jacobi algebra A. An integral ν is called non-degenerate *if: $\nu(ax) = 0$, for all $x \in A$ implies $a = 0$.*

If ν is an integral on A then by taking $b = c = 1_A$ in (5.10) we obtain that $\nu\big([a, 1]\big) = 0$.

Examples 5.1.13 (1) *If A is an abelian Jacobi algebra, then* $\int_A = A^*$.

(2) *Let A be a unital Poisson algebra and $\nu \in \int_A$. Then by taking $c = 1_A$ in (5.10) we obtain that $\nu(x) = 0$, for any $x \in A' := [A, A]$. In particular, it follows that $\int_A = 0$, for any Poisson algebra A which is perfect as a Lie algebra.*

(3) *The trace map* $\mathrm{Tr} : \mathrm{M}_n(k) \to k$ *satisfies (5.10) since* $\mathrm{Tr}\big([A, B]\, C\big) = \mathrm{Tr}\big(A\, [B, C]\big)$, *for all $n \times n$-matrices A, B, C. Thus, for any finite-dimensional Jacobi algebra A which is also a subalgebra of* $\mathrm{M}_n(k)$ *and a Lie subalgebra of* $\mathrm{gl}(n, k)$, *the restriction of the trace map* $\mathrm{Tr}_{|\, A}$ *is an integral on A.*

(4) *Let $u \in U(A)$ be an invertible element of a Jacobi algebra A. Then the map*

$$\int_A \longrightarrow \int_{A_u}, \qquad \nu \mapsto \nu_u := \nu(u^2\, \cdot)$$

is a bijection between the integrals on A and those on the u-conformal deformation A_u. Indeed, we can easily prove that $\nu_u(a) := \nu(u^2 a)$ is an integral on A_u for any integral ν on A and the inverse of the map $\nu \mapsto \nu_u$ is given by $\mu \mapsto \mu(u^{-2}\, \cdot)$—we note that $(A_u)_{u^{-1}} = A$. We can also prove that an integral ν on A remains an integral on A_u if and only if $2\,[u, x] = 2u\,[1_A, x]$, for all $x \in A$.

(5) *Let $J^{2,4}$ be the Jacobi algebra of Example 5.1.2. Then ν is an integral on $J^{2,4}$ if and only if $\nu(1) = \alpha$ and $\nu(x) = 0$, for some $\alpha \in k$, i.e.,* $\int_{J^{2,4}} \cong k$.

(6) *Let J^3_{11} be the Jacobi algebra from Table 5.1. Then ν is an integral on J^3_{11} if and only if $\nu(1) = \alpha$, $\nu(x) = \nu(y) = 0$, for some $\alpha \in k$, i.e.,* $\int_{J^3_{11}} \cong k$.

Lemma 5.1.14 *Let A be a Jacobi algebra. There exists a bijection between $\int_A$ and the space of all (symmetric) associative, invariant bilinear forms on A, i.e., bilinear maps $B : A \times A \to k$ satisfying the following compatibility conditions for any a, b, $c \in A$:*

$$B(ab,\, c) = B(a,\, bc), \qquad B([a,\, b],\, c) = B(a,\, [b,\, c]) \tag{5.11}$$

Proof: Any bilinear form B satisfying (5.11) is symmetric since A is a commutative algebra: $B(a, b) = B(1_A, ab) = B(1_A, ba) = B(b, a)$, for all a, $b \in A$. If $\nu \in \int_A$ is an integral on A, then $B_\nu(a, b) := \nu(ab)$ is an associative, invariant, bilinear form on A; conversely, if B is such a form then $\nu_B : A \to k$, $\nu_B(a) := B(a, 1_A) = B(1_A, a)$ is an integral on A and the correspondence $(\nu \mapsto B_\nu,\, B \mapsto \nu_B)$ is clearly bijective. □

The equivalences (1) – (2) – (4) in the theorem below can be seen as the Jacobi versions of the classical characterization of Frobenius algebras ([65]). The equivalence (1) – (3) is the Jacobi counterpart of Sullivan's theorem [204] for Hopf algebras: a Hopf algebra H is co-Frobenius if and only if there exists a non-zero integral on H.

Theorem 5.1.15 *Let A be a Jacobi algebra. The following are equivalent:*

(1) *A is a Frobenius Jacobi algebra;*

(2) *there exists a nondegenerate associative, invariant, bilinear form on A;*

(3) *there exists a nondegenerate integral on the Jacobi algebra A.*

Furthermore, if A is finite-dimensional then the above statements are also equivalent to:

(4) *there exists a pair $(\nu,\, e = \sum e^1 \otimes e^2)$, consisting of an integral $\nu \in \int_A$ on A and an element $e = \sum e^1 \otimes e^2 \in A \otimes A$ such that for any $a \in A$ we have:*

$$\sum ae^1 \otimes e^2 = \sum e^1 \otimes e^2 a, \qquad \sum \nu(e^1)e^2 = \sum e^1 \nu(e^2) = 1_A \qquad (5.12)$$

We call such a pair $(\nu,\, e = \sum e^1 \otimes e^2) \in \int_A \times (A \otimes A)$ a Jacobi-Frobenius pair *and $\omega_A := \sum e^1 e^2 \in A$ the* Euler-Casimir *element of A.*

Proof: (1) $\Leftrightarrow$ (2) Follows from the one-to-one correspondence between the set of all k-linear isomorphisms $f : A \to A^*$ and the set of all nondegenerate bilinear forms $B : A \times A \to k$ given by the two-sided formula $f(a)(b) =: B(a, b)$, for all a, $b \in A$. Under this bijection, the right A-module (resp. right Lie A-module) maps $f : A \to A^*$ correspond to those bilinear forms $B : A \times A \to k$ that satisfy the left (resp. right) hand part of (5.11).

(2) $\Leftrightarrow$ (3) Follows from Lemma 5.1.14 since under the bijective correspondence ($\nu \mapsto B_\nu$, $B \mapsto \nu_B$) nondegenerate forms on A correspond to nondegenerate integrals on A. We note that the isomorphism of right Jacobi A-modules $f = f_\nu : A \to A^*$ associated to a nondegenerate integral ν is defined by $f_\nu(a) := \nu \blacktriangleleft a$, i.e., $f_\nu(a)(x) = \nu(ax)$, for all a, $x \in A$. Conversely any nondegenerate integral $\nu = \nu_f$ on A arises from an isomorphism of right Jacobi A-modules $f : A \to A^*$ via $\nu_f := f(1_A)$.

(1) $\Leftrightarrow$ (4) Let $\{e_i,\, e_i^* \,|\, i = 1, \cdots, n\}$ be a dual basis of A and $f : A \to A^*$ and isomorphism of Jacobi A-bimodules. Then, $\nu := f(1_A)$ is a Jacobi integral and by a straightforward computation we can see that $(f(1_A),\, e := \sum_{i=1}^n e_i \otimes f^{-1}(e_i^*)$ is a Jacobi-Frobenius pair. Conversely, if $(\nu,\, e = \sum e^1 \otimes e^2)$ is a Jacobi-Frobenius pair then the map $f = f_\nu : A \to A^*$, $f(a)(b) := \nu(ab)$, for all a, $b \in A$ is an isomorphism of Jacobi A-bimodules with the inverse $f^{-1} : A^* \to A$ given by $f^{-1}(a^*) := \sum a^*(e^1)\, e^2$, for all $a^* \in A^*$. □

Remarks 5.1.16 (1) *Integrals on a unital Poisson algebra A are defined exactly as in Definition 5.1.12 and a Poisson algebra A is called Frobenius if there exists an isomorphism of Poisson modules $A \cong A^*$. The Poisson version of Theorem 5.1.15 has the same statement and we can rephrase this characterization by saying that a Poisson algebra is Frobenius if and only if it is Frobenius when viewed as a Jacobi algebra.*

(2) *Let $\nu \in \int_A$ be a nondegenerate integral on a finite-dimensional Jacobi algebra A. It follows from Theorem 5.1.15 that $(A^*, \blacktriangleleft)$ is free as a right A-module having $\{\nu\}$ as a basis, i.e., $A^* = \nu \blacktriangleleft A$. This is the Jacobi version of a well-known result for Hopf algebras ([204, Corollary 10.6.6]).*

(3) *Using* (1) $\Leftrightarrow$ (2) *of Theorem 5.1.15 we obtain that the abelian Jacobi algebra A_0 is Frobenius as a Jacobi algebra if and only if A is Frobenius as an associative algebra.*

(4) *Let A be a finite-dimensional Frobenius Jacobi algebra with the Jacobi-Frobenius pair $(\nu,\, e = \sum e^1 \otimes e^2)$. In the case of associative Frobenius algebras the element $\omega_A := \sum e^1 e^2 \in A$ does not depend on the choice of the Jacobi-Frobenius pair and it is called in [1] the* Euler (characteristic) *element of A since it is the algebraic counterpart of the Euler class of a connected, oriented, finite-dimensional and compact manifold. In the context of finite-dimensional Lie algebras equipped with an invariant bilinear form, the same element viewed in the enveloping algebra is called the Casimir element [55, Proposition 11]. These explain the terminology used in* (4) *of Theorem 5.1.15.*

Examples 5.1.17 (1) *Let C_n be the cyclic group of order n and k a field such that* $\mathrm{char}\,(k) = p \,|\, n-1$. *Let $A := k[C_n]$ be the Jacobi algebra with the bracket $[c^i,\, c^j] := (j-i)c^{j+i}$, for all $i,\, j = 0, \cdots, n-1$. Then $\int_{k[C_n]} = 0$. Indeed, let $\nu \in \int_{k[C_n]}$. By applying (5.10) in $a = b := c^0 = 1$ and $c := c^i$, we obtain $\nu(c^i) = 0$, for any $i = 0, \cdots, n-1$. Since any group algebra $k[G]$ of a finite group G is Frobenius [145], we obtain that $k[C_n]$ is Frobenius as an associative algebra and using Theorem 5.1.15 we obtain that it is not Frobenius as a Jacobi algebra.*

(2) *Let $u \in U(A)$ be an invertible element of a Jacobi algebra A. Then, the u-conformal deformation A_u is a Frobenius Jacobi algebra if and only if A is a Frobenius Jacobi algebra. The result follows from Theorem 5.1.15 and* (4) *of Example 5.1.13 since the bijection there preserves the non-degeneration of integrals.*

(3) *The equivalence* (1) – (3) *of Theorem 5.1.15 provides a very efficient criterion for deciding when a given Jacobi algebra is Frobenius. For instance, the only 2-dimensional Frobenius Jacobi algebras are $J^{2,1}$, $J^{2,3}$ and $J^2_{\rm d}$: for each of them the linear map ν defined by $\nu(1) := 0$ and $\nu(x) := 1$ is a non-degenerate integral. In the same manner, we can easily prove that among the eleven types of* 3*-dimensional complex Jacobi algebras given in Example 5.1.3 there are only three Frobenius Jacobi algebras, namely J^3_2, J^3_3 and J^3_4. For each of them the linear map ν defined by $\nu(1) := a_1$, $\nu(x) := a_2$ and $\nu(y) := a_3$ is a non-degenerate integral, for all $a_1,\, a_2,\, a_3 \in k$ such that $a_2 a_3 \neq 0$ and $a_1 \neq a_2 + a_3$.*

5.2 Unified products for Jacobi algebras

In this section we shall answer the extending structures ES problem for Jacobi algebras: i.e., we shall describe and classify all Jacobi algebras containing a given Jacobi algebra A as a subalgebra of a fixed codimension. We start by explaining what we mean by classification within the ES problem. Let A be a Jacobi algebra, E a vector space such that A is a subspace of E and let V be a fixed complement of A in E, i.e., V is a subspace of E

such that $E = A + V$ and $A \cap V = \{0\}$. Let $\mathcal{J}(A, E)$ be the category whose objects are all Jacobi algebra structures $(\cdot_E, [-,-]_E)$ that can be defined on E such that A becomes a Jacobi subalgebra of $(E, \cdot_E, [-,-]_E)$. A morphism $\varphi : (\cdot_E, [-,-]_E) \to (\cdot'_E, [-,-]'_E)$ in $\mathcal{J}(A, E)$ is a morphism of Jacobi algebras $\varphi : (E, \cdot_E, \{-, -\}_E) \to (E, \cdot'_E, \{-, -\}'_E)$ which stabilizes A and co-stabilizes V, i.e., the diagram

$$\begin{array}{ccccc} A & \xrightarrow{i} & E & \xrightarrow{\pi} & V \\ {\scriptstyle Id}\downarrow & & \downarrow{\scriptstyle \varphi} & & \downarrow{\scriptstyle Id} \\ A & \xrightarrow{i} & E & \xrightarrow{\pi} & V \end{array} \tag{5.13}$$

is commutative, where $\pi : E \to V$ is the canonical projection of $E = A + V$ on V and $i : A \to E$ is the inclusion map. In this case we say that the Jacobi algebra structures $(\cdot_E, [-,-]_E)$ and $(\cdot'_E, [-,-]'_E)$ on E are *cohomologous* and we denote this by $(\cdot_E, \{-, -\}_E) \approx (\cdot'_E, \{-, -\}'_E)$. Any linear map φ which makes diagram (5.13) commutative is bijective, thus the category $\mathcal{J}(A, E)$ is a groupoid, i.e., any morphism is an isomorphism. In particular, we obtain that $\approx$ is an equivalence relation on the set of objects of $\mathcal{J}(A, E)$ and we denote by $\mathrm{Extd}_{\mathcal{J}}(E, A)$ the set of all equivalence classes, i.e., $\mathrm{Extd}_{\mathcal{J}}(E, A) := \mathcal{J}(A, E)/\approx$. $\mathrm{Extd}_{\mathcal{J}}(E, A)$ is the classifying object for the ES problem: it classifies all Jacobi algebra structures that can be defined on E containing A as a Jacobi subalgebra up to an isomorphism that stabilizes A and co-stabilizes V. The answer to the ES problem will be provided by explicitly computing $\mathrm{Extd}_{\mathcal{J}}(E, A)$ for a given Jacobi algebra A and a vector space E. From a geometrical point of view this means to decompose the groupoid $\mathcal{J}(A, E)$ into connected components and to indicate a 'point' in each such component. The main result of this section proves that $\mathrm{Extd}_{\mathcal{J}}(E, A)$ is parameterized by a non-abelian cohomological type object $\mathcal{JH}^2(V, A)$ that will be explicitly constructed and the bijection between $\mathcal{JH}^2(V, A)$ and $\mathrm{Extd}_{\mathcal{J}}(E, A)$ will be indicated.

Definition 5.2.1 *Let A be a Jacobi algebra and V a vector space. An extending datum of A through V is a system $\Upsilon(A,V) = (\triangleleft, \triangleright, f, \cdot, \leftharpoonup, \rightharpoonup, \theta, \{-, -\})$ consisting of eight bilinear maps*

$$\triangleleft : V \times A \to V, \quad \triangleright : V \times A \to A, \quad f : V \times V \to A, \quad \cdot : V \times V \to V$$
$$\leftharpoonup : V \times A \to V, \quad \rightharpoonup : V \times A \to A, \quad \theta : V \times V \to A, \quad \{-, -\} : V \times V \to V$$

Let $\Upsilon(A,V) = (\triangleleft, \triangleright, f, \cdot, \leftharpoonup, \rightharpoonup, \theta, \{-, -\})$ be an extending datum of a Jacobi algebra A through a vector space V. We denote by $A \ltimes_{\Upsilon(A,V)} V = A \ltimes V$ the vector space $A \times V$ together with the multiplication $\bullet$ and the bracket $[-, -]$ defined by:

$$(a, x) \bullet (b, y) := \big(ab + x \triangleright b + y \triangleright a + f(x,y),\ x \triangleleft b + y \triangleleft a + x \cdot y\big) \tag{5.14}$$

$$[(a,x), (b,y)] := \big([a, b] + x \rightharpoonup b - y \rightharpoonup a + \theta(x,y),\ x \leftharpoonup b - y \leftharpoonup a + \{x, y\}\big) \tag{5.15}$$

for all $a, b \in A$ and $x, y \in V$. The object $A \ltimes V$ is called the unified product of A and $\Upsilon(A, V)$ if it is a Jacobi algebra with the multiplication defined by (5.14), the unit $(1_A, 0_V)$ and the bracket given by (5.15). In this case the extending datum $\Upsilon(A, V)$ is called a Jacobi extending structure of A through V.

The next theorem provides the necessary and sufficient conditions that need to be fulfilled by an extending datum $\Upsilon(A, V)$ such that $A \ltimes V$ is a unified product.

Theorem 5.2.2 *Let $A = (A, m_A, [-, -])$ be a Jacobi algebra, V a vector space and $\Upsilon(A, V) = (\triangleleft, \triangleright, f, \cdot, \leftharpoondown, \rightharpoonup, \theta, \{-, -\})$ an extending datum of A through V. Then $A \ltimes V$ is a unified product if and only if the following compatibilities hold:*

(J0) *$(\triangleleft, \triangleright, f, \cdot)$ is an algebra extending system of the associative algebra A through V and $(\leftharpoondown, \rightharpoonup, \theta, \{-, -\})$ is a Lie extending system of the Lie algebra A through V;*

(J1) $x \rightharpoonup (ab) = (x \rightharpoonup a)\, b + (x \leftharpoondown a) \triangleright b + a\, (x \rightharpoonup b) + (x \leftharpoondown b) \triangleright a - ab\, (x \rightharpoonup 1_A) -$
$$- (x \leftharpoondown 1_A) \triangleright (ab)$$

(J2) $x \leftharpoondown (ab) = (x \leftharpoondown a) \triangleleft b + (x \leftharpoondown b) \triangleleft a - (x \leftharpoondown 1_A) \triangleleft (ab)$

(J3) $x \triangleright [a, b] = [x \triangleright a, b] + (x \triangleleft a) \rightharpoonup b - a\, (x \rightharpoonup b) - (x \leftharpoondown b) \triangleright a + (x \triangleright a)[1_A, b] + (x \triangleleft a) \triangleright [1_A, b]$

(J4) $x \triangleleft [a, b] = (x \triangleleft a) \leftharpoondown b - (x \leftharpoondown b) \triangleleft a + (x \triangleleft a) \triangleleft [1_A, b]$

(J5) $\{x, y\} \triangleright a = \theta(x \triangleleft a, y) - a\, \theta(x, y) + f(y \leftharpoondown a, x) - f(x \triangleleft a, y \leftharpoondown 1_A) - y \rightharpoonup (x \triangleright a) +$
$$+ x \triangleright (y \rightharpoonup a) - (x \triangleright a)(y \rightharpoonup 1_A) - (x \triangleleft a) \triangleright (y \rightharpoonup 1_A) - (y \leftharpoondown 1_A) \triangleright (x \triangleright a)$$

(J6) $\{x, y\} \triangleleft a = \{x \triangleleft a, y\} - y \leftharpoondown (x \triangleright a) + x \triangleleft (y \rightharpoonup a) + (y \leftharpoondown a) \cdot x -$
$$- (x \triangleleft a) \triangleleft (y \rightharpoonup 1_A) - (y \leftharpoondown 1_A) \triangleleft (x \triangleright a) - (x \triangleleft a) \cdot (y \leftharpoondown 1_A)$$

(J7) $(x \cdot y) \rightharpoonup a = x \triangleright (y \rightharpoonup a) + y \triangleright (x \rightharpoonup a) + f(x \leftharpoondown a, y) + f(x, y \leftharpoondown a) - [f(x, y), a] -$
$$- f(x, y)[1_A, a] - (x \cdot y) \triangleright [1_A, a]$$

(J8) $(x \cdot y) \leftharpoondown a = x \cdot (y \leftharpoondown a) + (x \leftharpoondown a) \cdot y + x \triangleleft (y \rightharpoonup a) + y \triangleleft (x \rightharpoonup a) - (x \cdot y) \triangleleft [1_A, a]$

(J9) $\theta(x \cdot y, z) = x \triangleright \theta(y, z) + y \triangleright \theta(x, z) + z \rightharpoonup f(x, y) + f(\{x, z\}, y) + f(x, \{y, z\}) +$
$$+ f(x, y)(z \rightharpoonup 1_A) + (x \cdot y) \triangleright (z \rightharpoonup 1_A) + (z \leftharpoondown 1_A) \triangleright f(x, y) + f(x \cdot y, z \leftharpoondown 1_A)$$

(J10) $\{x \cdot y, z\} = x \cdot \{y, z\} + \{x, z\} \cdot y + z \leftharpoondown f(x, y) + x \triangleleft \theta(y, z) + y \triangleleft \theta(x, z) +$
$$+ (x \cdot y) \triangleleft (z \rightharpoonup 1_A) + (z \leftharpoondown 1_A) \triangleleft f(x, y) + (x \cdot y) \cdot (z \leftharpoondown 1_A)$$

for all $a, b \in A$, $x, y, z \in V$.

Proof: We have already noticed in Preliminaries that $(A \ltimes V, \bullet)$ is a commutative algebra with unit $(1_A, 0)$ if and only if $(\triangleleft, \triangleright, f, \cdot)$ is an algebra extending system of the algebra A through V and $(A \ltimes V, [-, -])$ is a Lie algebra if and only $(\leftharpoonup, \rightharpoonup, \theta, \{-, -\})$ is a Lie extending system of the Lie algebra A through V. These are the assumptions from (J0) which from now on we assume to be fulfilled. Then, $(A \ltimes V, \bullet, [-, -])$ is a Jacobi algebra if and only if the following compatibility holds for any $a, b, c \in A$ and $x, y, z \in V$

$$\begin{aligned}[(a,x) \bullet (b,y), (c,z)] &= (a,x) \bullet [(b,y), (c,z)] + [(a,x), (c,z)] \bullet (b,y) - \\ &- (a,x) \bullet (b,y) \bullet \big([1_A, c] - z \rightharpoonup 1_A, -z \leftharpoonup 1_A\big)\end{aligned}$$

If we denote the last equation by (J), the proof relies on a detailed analysis of this identity. Since in $A \times V$ we have $(a, x) = (a, 0) + (0, x)$ it follows that (J) holds if and only if it holds for all generators of $A \times V$, i.e., for the set $\{(a, 0) \mid a \in A\} \cup \{(0, x) \mid x \in V\}$. However, since the computations are rather long but straightforward we will only indicate the main steps of the proof, the details being available upon request. First, we notice that (J) holds for the triple $(a, 0)$, $(b, 0)$, $(c, 0)$ since A is a Jacobi algebra. The left-hand side of (J) evaluated in $(a, 0)$, $(b, 0)$, $(0, x)$ is equal to $\big(-x \rightharpoonup (ab), -x \leftharpoonup (ab)\big)$ while the right-hand side of (J) evaluated in the same triple comes down to:

$$\begin{aligned}&-\big((x \rightharpoonup a)\, b + (x \leftharpoonup a) \triangleright b,\ (x \leftharpoonup a) \triangleleft b\big) - \big(a\,(x \rightharpoonup b) + (x \leftharpoonup b) \triangleright a,\\ &(x \leftharpoonup b) \triangleleft a\big) + \big(ab\,(x \rightharpoonup 1_A) + (x \leftharpoonup 1_A) \triangleright (ab),\ (x \leftharpoonup 1_A) \triangleleft (ab)\big)\end{aligned}$$

We obtain that (J) holds for the triple $(a, 0)$, $(b, 0)$, $(0, x)$ if and only if (J1) and (J2) hold. Similar computations show the following: (J) holds for the triple $(a, 0)$, $(0, x)$, $(b, 0)$ if and only if (J3) and (J4) hold; (J) holds for the triple $(a, 0)$, $(0, x)$, $(0, y)$ if and only if (J5) and (J6) hold; (J) holds for the triple $(0, x)$, $(0, y)$, $(a, 0)$ if and only if (J7) and (J8) hold and finally, (J) holds for the triple $(0, x)$, $(0, y)$, $(0, z)$ if and only if (J9) and (J10) hold. Moreover, since A is commutative, we observe that the Jacobi compatibility (0.9) holds for the triple (a, b, c) if and only if it holds for the triple (b, a, c). Based on this remark we obtain that (J) holds for the triple $(0, x)$, $(a, 0)$, $(b, 0)$ whenever it holds for $(a, 0)$, $(0, x)$, $(b, 0)$ and (J) holds for the triple $(0, x)$, $(a, 0)$, $(0, y)$ whenever it holds for $(a, 0)$, $(0, x)$, $(0, y)$. The proof is now finished. □

From now on, a Jacobi extending structure of a Jacobi algebra A through a vector space V will be viewed as a system $\Upsilon(A, V) = \big(\triangleleft, \triangleright, f, \cdot, \leftharpoonup, \rightharpoonup, \theta, \{-, -\}\big)$ satisfying the compatibility conditions (J0)-(J10). We denote by $\mathcal{JE}(A, V)$ the set of all Jacobi algebra extending structures of A through V. Theorem 5.2.2 takes a simplified form at the level of Poisson algebras:

Corollary 5.2.3 *Let $A = (A, m_A, [-, -])$ be a unital Poisson algebra, V a vector space and $\Upsilon(A, V) = \big(\triangleleft, \triangleright, f, \cdot, \leftharpoonup, \rightharpoonup, \theta, \{-, -\}\big)$ an extending datum of A through V. Then $A \ltimes V = (A \ltimes V, \bullet, [-, -])$ is a Poisson algebra with $(1_A, 0_V)$ as a unit if and only if the following compatibilities hold for any a, $b \in A$, x, y, $z \in V$:*

(*P*0) $(\triangleleft, \triangleright, f, \cdot)$ *is an algebra extending system of the associative algebra A through V and* $(\leftharpoonup, \rightharpoonup, \theta, \{-, -\})$ *is a Lie extending system of the Lie algebra A through V*

(*P*1) $x \rightharpoonup (ab) = (x \rightharpoonup a)\, b + (x \leftharpoonup a) \triangleright b + a\, (x \rightharpoonup b) + (x \leftharpoonup b) \triangleright a$

(*P*2) $x \leftharpoonup (ab) = (x \leftharpoonup a) \triangleleft b + (x \leftharpoonup b) \triangleleft a$

(*P*3) $x \triangleright [a, b] = [x \triangleright a, b] + (x \triangleleft a) \rightharpoonup b - a\,(x \rightharpoonup b) - (x \leftharpoonup b) \triangleright a$

(*P*4) $x \triangleleft [a, b] = (x \triangleleft a) \leftharpoonup b - (x \leftharpoonup b) \triangleleft a$

(*P*5) $\{x, y\} \triangleright a = \theta(x \triangleleft a, y) - a\, \theta(x, y) + f(y \leftharpoonup a, x) - y \rightharpoonup (x \triangleright a) + x \triangleright (y \rightharpoonup a)$

(*P*6) $\{x, y\} \triangleleft a = \{x \triangleleft a, y\} - y \leftharpoonup (x \triangleright a) + x \triangleleft (y \rightharpoonup a) + (y \leftharpoonup a) \cdot x$

(*P*7) $(x \cdot y) \rightharpoonup a = x \triangleright (y \rightharpoonup a) + y \triangleright (x \rightharpoonup a) + f(x \leftharpoonup a, y) + f(x, y \leftharpoonup a) - [f(x, y), a]$

(*P*8) $(x \cdot y) \leftharpoonup a = x \cdot (y \leftharpoonup a) + (x \leftharpoonup a) \cdot y + x \triangleleft (y \rightharpoonup a) + y \triangleleft (x \rightharpoonup a)$

(*P*9) $\theta(x \cdot y, z) = x \triangleright \theta(y, z) + y \triangleright \theta(x, z) + z \rightharpoonup f(x, y) + f(\{x, z\}, y) + f(x, \{y, z\})$

(*P*10) $\{x \cdot y, z\} = x \cdot \{y, z\} + \{x, z\} \cdot y + z \leftharpoonup f(x, y) + x \triangleleft \theta(y, z) + y \triangleleft \theta(x, z)$

An extending datum $\Upsilon(A, V) = (\triangleleft, \triangleright, f, \cdot, \leftharpoonup, \rightharpoonup, \theta, \{-, -\})$ of a unital Poisson algebra A through a vector space V satisfying the axioms (P0)-(P10) is called a *Poisson extending structure of A through V* and we denote by $\mathcal{PE}(A, V)$ the set of all Poisson extending structures of A through V.

Let $\Upsilon(A, V) = (\triangleleft, \triangleright, f, \cdot, \leftharpoonup, \rightharpoonup, \theta, \{-, -\}) \in \mathcal{JE}(A, V)$ be a Jacobi extending structure of a Jacobi algebra A through a vector space V. Then A is a Jacobi subalgebra of the unified product $A \ltimes V$ through the identification $A \cong i_A(A) = A \times \{0\}$, where $i_A : A \to A \ltimes V$, $i_A(a) = (a, 0)$ is the canonical injection. Conversely, the following result provides the answer to the description part of the ES problem:

Proposition 5.2.4 *Let A be a Jacobi algebra, E a vector space containing A as a subspace and* $(*_E, [-, -]_E)$ *a Jacobi algebra structure on E such that A is a subalgebra of* $(E, *_E, [-, -]_E)$. *Then there exists a Jacobi extending structure* $\Upsilon(A, V) = (\triangleleft, \triangleright, f, \cdot, \leftharpoonup, \rightharpoonup, \theta, \{-, -\})$ *of A through a subspace V of E and an isomorphism of Jacobi algebras* $(E, *_E, [-, -]_E) \cong A \ltimes V$ *that stabilizes A and co-stabilizes V.*

Proof: Since k is a field, there exists a linear map $p : E \to A$ such that $p(a) = a$, for all $a \in A$. Then $V := \mathrm{Ker(p)}$ is a complement of A in E. Using the retraction p, we define the extending datum $\Upsilon(A, V) = (\triangleleft, \triangleright, f, \cdot, \leftharpoonup, \rightharpoonup, \theta, \{-, -\})$ of A through V by the following formulas for any $a \in A$ and x, $y \in V$:

$$\begin{aligned} x \triangleright a &:= p(x *_E a), & x \triangleleft a &:= x *_E a - p(x *_E a) \\ f(x, y) &:= p(x *_E y), & x \cdot y &:= x *_E y - p(x *_E y) \\ x \rightharpoonup a &:= p\big([x, a]_E\big), & x \leftharpoonup a &:= [x, a]_E - p\big([x, a]_E\big) \\ \theta(x, y) &:= p\big([x, y]_E\big), & \{x, y\} &:= [x, y]_E - p\big([x, y]_E\big) \end{aligned}$$

Then by arguments similar to those used for Lie algebras in Theorem 3.1.4 and associative algebras in Theorem 4.1.4 we can prove that $\Upsilon(A, V) = \big(\triangleleft,\, \triangleright,\, f,\, \cdot,\, \leftharpoonup,\, \rightharpoonup,\, \theta,\, \{-,\, -\}\big)$ is a Jacobi extending structure of A through V and the linear map $\varphi : A \ltimes V \to (E, *_E, [-,\, -]_E)$, $\varphi(a, x) := a + x$, is an isomorphism of Jacobi algebras that stabilizes A and co-stabilizes V, i.e., the following diagram is commutative:

$$\begin{array}{ccccc} A & \xrightarrow{\ i\ } & A \ltimes V & \xrightarrow{\ q\ } & V \\ {\scriptstyle Id}\downarrow & & \downarrow{\scriptstyle \varphi} & & \downarrow{\scriptstyle Id} \\ A & \xrightarrow{\ i\ } & E & \xrightarrow{\ \pi\ } & V \end{array}$$

where $q : A \ltimes V \to V$, $q(a, x) := x$ is the canonical projection. □

Proposition 5.2.4 reduces the classification of all Jacobi algebra structures on E that contain A as a subalgebra to the classification of all unified products $A \ltimes V$, associated to all Jacobi extending structures $\Upsilon(A, V) = \big(\triangleleft,\, \triangleright,\, f,\, \cdot,\, \leftharpoonup,\, \rightharpoonup,\, \theta,\, \{-,\, -\}\big)$, for a fixed complement V of A in E. First we need the following technical result:

Proposition 5.2.5 *Let A be a Jacobi algebra, $\Upsilon(A, V) = \big(\triangleleft,\, \triangleright,\, f,\, \cdot,\, \leftharpoonup,\, \rightharpoonup,\, \theta,\, \{-,\, -\}\big)$ and $\Upsilon'(A, V) = \big(\triangleleft',\, \triangleright',\, f',\, \cdot',\, \leftharpoonup',\, \rightharpoonup',\, \theta',\, \{-,\, -\}'\big)$ two Jacobi extending structures of A through V. Let $A \ltimes V$ and $A \ltimes' V$ be the unified products associated to the Jacobi extending structures $\Upsilon(A, V)$ and respectively $\Upsilon'(A, V)$. The following are equivalent:*

(1) There exists $\psi : A \ltimes V \to A \ltimes' V$, a morphism of Jacobi algebras which stabilizes A and co-stabilizes V;

(2) $\triangleleft' = \triangleleft$, $\leftharpoonup' = \leftharpoonup$ and there exists a linear map $r : V \to A$ such that $\triangleright'$, f', $\cdot'$, $\rightharpoonup'$, θ' and $\{-,\, -\}'$ are implemented by r via the following formulas for any $a \in A$, x, $y \in V$:

$$x \triangleright' a = x \triangleright a + r(x \triangleleft a) - r(x)\, a, \qquad x \cdot' y = x \cdot y - x \triangleleft r(y) - y \triangleleft r(x)$$

$$f'(x, y) = f(x,\, y) + r(x \cdot y) + r(x) r(y) - x \triangleright r(y) - r\big(x \triangleleft r(y)\big) - y \triangleright r(x) - r\big(y \triangleleft r(x)\big)$$

$$x \rightharpoonup' a = x \rightharpoonup a + r(x \leftharpoonup a) - [r(x),\, a], \qquad \{x, y\}' = \{x, y\} - x \leftharpoonup r(y) + y \leftharpoonup r(x)$$

$$\theta'(x, y) = \theta(x, y) + r\big(\{x,\, y\}\big) + [r(x),\, r(y)] + y \rightharpoonup r(x) - x \rightharpoonup r(y) + r\big(y \leftharpoonup r(x)\big) - r\big(x \leftharpoonup r(y)\big)$$

Proof: There exists a bijection between the set of all linear maps $\psi : A \times V \to A \times V$ which stabilize A and co-stabilize V and the set of all linear map $r : V \to A$ given as follows: $\psi \mapsto r_\psi$, where $r_\psi(x) := \psi(0, x)$ respectively $r \mapsto \psi_r$, where $\psi_r(a,\, x) := (a + r(x),\, x)$, for all $a \in A$ and $x \in V$. We denote by $\psi_r = \psi$, the linear map associated to $r : V \to A$. We prove that $\psi = \psi_r : A \ltimes V \to A \ltimes' V$ is a morphism of Jacobi algebras if and only if the compatibility conditions from (2) hold. Indeed, first we can see that $\psi\big((a,0)\bullet(0,x)\big) = \psi(a,0)\bullet'\psi(0,x)$ if and only if $\triangleleft' = \triangleleft$ and $x\triangleright' a = x\triangleright a + r(x\triangleleft a)$. Taking these two compatibilities into account, we can easily see that $\psi\big((0,x)\bullet(0,y)\big) = \psi(0,x)\bullet'\psi(0,y)$ if and only if $x\cdot' y = x\cdot y - x\triangleleft r(y) - y\triangleleft r(x)$ and $f'(x,y) = f(x,\,y) + r(x\cdot y) + r(x)r(y) - x\triangleright r(y) - r\big(x\triangleleft r(y)\big) - y\triangleright r(x) - r\big(y\triangleleft r(x)\big)$, for all $x,\,y\in V$. This shows that $\psi = \psi_r : A\ltimes V \to A\ltimes' V$ is a morphism of associative algebras if and only if $\triangleleft' = \triangleleft$ and the first three compatibility conditions of (2) hold. In a similar fashion, we can prove that $\psi = \psi_r : A\ltimes V \to A\ltimes' V$ is a morphism of Lie algebras if and only if $\leftharpoonup' = \leftharpoonup$ and the last three compatibility conditions of (2) hold. □

The Jacobi algebra morphism $\psi_r : A\times V \to A\times V$ defined in the proof of Proposition 5.2.5 is bijective. This allows us to introduce the following equivalence relation:

Definition 5.2.6 *Let A be a Jacobi algebra and V a vector space. We call two Jacobi extending structures $\Upsilon(A,V) = \big(\triangleleft,\,\triangleright,\,f,\,\cdot,\,\leftharpoonup,\,\rightharpoonup,\,\theta,\,\{-,\,-\}\big)$ and $\Upsilon'(A,V) = \big(\triangleleft',\,\triangleright',\,f',\,\cdot',\,\leftharpoonup',\,\rightharpoonup',\,\theta',\,\{-,\,-\}'\big)$* cohomologous, *and we denote this by $\Upsilon(A,V)\approx\Upsilon'(A,V)$, if $\triangleleft' = \triangleleft$, $\leftharpoonup' = \leftharpoonup$ and there is a linear map $r : V\to A$ such that $\triangleright'$, f', $\cdot'$, $\rightharpoonup'$, θ' and $\{-,\,-\}'$ are implemented by r via the formulas given in* (2) *of Proposition 5.2.5.*

The theoretical answer to the ES problem for Jacobi algebras now follows:

Theorem 5.2.7 *Let A be a Jacobi algebra, E a vector space which contains A as a subspace and V a complement of A in E. Then $\approx$ is an equivalence relation on the set $\mathcal{JE}(A,V)$ of all Jacobi algebra extending structures of A through V. If we denote by $\mathcal{JH}^2(V,\,A) := \mathcal{JE}(A,V)/\approx$, then the map*[1]

$$\mathcal{JH}^2(V,\,A) \to \mathrm{Extd}_{\mathcal{J}}(E,\,A), \qquad \overline{\big(\triangleleft,\,\triangleright,\,f,\,\cdot,\,\leftharpoonup,\,\rightharpoonup,\,\theta,\,\{-,\,-\}\big)} \longmapsto A\ltimes V$$

is bijective, where $A\ltimes V$ is the unified product associated to A and $\big(\triangleleft,\,\triangleright,\,f,\,\cdot,\,\leftharpoonup,\,\rightharpoonup,\,\theta,\,\{-,\,-\}\big)$.

Proof: Proposition 5.2.5 proves that $\Upsilon(A,V)\approx\Upsilon'(A,V)$ if and only if there exists an isomorphism of Jacobi algebras $\psi : A\ltimes V \to A\ltimes' V$ which stabilizes A and co-stabilizes V. This shows that $\approx$ is an equivalence relation on $\mathcal{JE}(A,V)$. The last part follows from this observation together with Theorem 5.2.2 and Proposition 5.2.4. □

Remark 5.2.8 *The Poisson version of Theorem 5.2.7 has the following form. Let A be a unital Poisson algebra, E a vector space which contains A as a subspace, V a complement of A in E and let $\mathcal{PE}(A,V)$ be the set of all*

[1] $\overline{\big(\triangleleft,\,\triangleright,\,f,\,\cdot,\,\leftharpoonup,\,\rightharpoonup,\,\theta,\,\{-,\,-\}\big)}$ denotes the equivalence class of $\big(\triangleleft,\,\triangleright,\,f,\,\cdot,\,\leftharpoonup,\,\rightharpoonup,\,\theta,\,\{-,\,-\}\big)$ via $\approx$.

Poisson extending structures of A through V in the sense of Corollary 5.2.3. We denote by $\mathcal{PH}^2(V, A) := \mathcal{PE}(A, V)/\approx$, where $\approx$ is the equivalence relation on $\mathcal{PE}(A, V)$ defined exactly as in Definition 5.2.6 and by $\mathrm{Extd}_{\mathcal{P}}(E, A)$ *the set of all equivalence classes of isomorphism of all Poisson algebra structures which can be defined on E that contain and stabilize A as a Poisson subalgebra and co-stabilize V. Then the map*

$$\mathcal{PH}^2(V, A) \to \mathrm{Extd}_{\mathcal{P}}(E, A), \qquad \overline{\overline{(\triangleleft, \triangleright, f, \cdot, \leftharpoonup, \rightharpoonup, \theta, \{-, -\})}} \longmapsto A \ltimes V$$

is bijective.

Computing the classifying object $\mathcal{JH}^2(V, A)$ for a given Jacobi algebra A and a given vector space V is a highly nontrivial problem. However, the first step in computing $\mathcal{JH}^2(V, A)$ is suggested by the first part of Definition 5.2.6: if two Jacobi extending structures $\Upsilon(A, V) = \big(\triangleleft, \triangleright, f, \cdot, \leftharpoonup, \rightharpoonup, \theta, \{-, -\}\big)$ and $\Upsilon'(A, V) = \big(\triangleleft', \triangleright', f', \cdot', \leftharpoonup', \rightharpoonup', \theta', \{-, -\}'\big)$ are cohomologous, then we must have $\triangleleft' = \triangleleft$, $\leftharpoonup' = \leftharpoonup$. Thus, in order to compute $\mathcal{JH}^2(V, A)$ we can fix the right A-module action $\triangleleft$ and the right Lie A-module action $\leftharpoonup$ such that $(V, \triangleleft, \leftharpoonup)$ is a right Jacobi module as defined in Definition 5.1.7; we observe that the compatibility conditions (5.2) and (5.3) of Definition 5.1.7 coincide with axioms (J2) and (J4) of Theorem 5.2.2 defining the Jacobi extending structures. Hence, we can decompose the object $\mathcal{JH}^2(V, A)$ as follows. Let A be a Jacobi algebra, V a vector space and let $(V, \triangleleft, \leftharpoonup)$ be a fixed right Jacobi A-module. Let $\mathcal{JE}_{(\triangleleft, \leftharpoonup)}(A, V)$ be the set of all $(\triangleleft, \leftharpoonup)$-*Jacobi extending structures* of A through $(V, \triangleleft, \leftharpoonup)$ which is the set of all 6-tuples $\big(\triangleright, f, \cdot, \rightharpoonup, \theta, \{-, -\}\big)$ consisting of six bilinear maps such that $\big(\triangleleft, \triangleright, f, \cdot, \leftharpoonup, \rightharpoonup, \theta, \{-, -\}\big)$ is a Jacobi extending structure of A through V. Two elements $\big(\triangleright, f, \cdot, \rightharpoonup, \theta, \{-, -\}\big)$ and $\big(\triangleright', f', \cdot', \rightharpoonup', \theta', \{-, -\}'\big)$ of $\mathcal{JE}_{(\triangleleft, \leftharpoonup)}(A, V)$ are $(\triangleleft, \leftharpoonup)$-*cohomologous* and we denote this by $\big(\triangleright, f, \cdot, \rightharpoonup, \theta, \{-, -\}\big) \approx_l \big(\triangleright', f', \cdot', \rightharpoonup', \theta', \{-, -\}'\big)$ if $\big(\triangleleft, \triangleright, f, \cdot, \leftharpoonup, \rightharpoonup,' \theta, \{-, -\}\big) \approx \big(\triangleleft, \triangleright', f', \cdot', \leftharpoonup, \rightharpoonup', \theta', \{-, -\}'\big)$. Then $\approx_l$ is an equivalence relation on $\mathcal{JE}_{(\triangleleft, \leftharpoonup)}(A, V)$ and we denote by $\mathcal{JH}^2\big((V, \triangleleft, \leftharpoonup), A\big)$ the quotient set $\mathcal{JE}_{(\triangleleft, \leftharpoonup)}(A, V)/\approx_l$. Theorem 5.2.7 and the above considerations provide the following decomposition of $\mathcal{JH}^2(V, A)$:

Corollary 5.2.9 *Let A be a Jacobi algebra and V a vector space. Then*

$$\mathcal{JH}^2(V, A) = \sqcup_{(\triangleleft, \leftharpoonup)} \mathcal{JH}^2\big((V, \triangleleft, \leftharpoonup), A\big) \tag{5.16}$$

where the coproduct in the right-hand side is in the category of sets over all possible right Jacobi A-module structures $(\triangleleft, \leftharpoonup)$ on V.

The decomposition given by (5.16) is a very important step in computing the classifying object $\mathcal{JH}^2(V, A)$. However, even computing every object $\mathcal{JH}^2\big((V, \triangleleft, \leftharpoonup), A\big)$, for a fixed right Jacobi A-module $(V, \triangleleft, \leftharpoonup)$ is a problem far from being trivial. However, the decomposition is important as some of the components in the right-hand side of (5.16) might be equal to the empty set as the following example shows; several explicit examples of computing $\mathcal{JH}^2(V, A)$ are provided in Section 5.3.

Example 5.2.10 *We consider the trivial right Jacobi A-module structure on V, that is* $x \triangleleft a := x$ *and* $x \leftharpoonup a := 0$, *for all* $x \in V$, $a \in A$. *This right Jacobi A-module structure on V was denoted by* V_0. *If* $V \neq \{0\}$, *then* $\mathcal{JH}^2(V_0, A) = \emptyset$. *We will prove that in fact the corresponding set* $\mathcal{JE}_{(\triangleleft, \leftharpoonup := 0)}(A, V)$ *is empty. Indeed, if* $(\triangleright, f, \cdot, \rightharpoonup, \theta, \{-, -\}) \in \mathcal{JE}_{(\triangleleft, \leftharpoonup := 0)}(A, V)$, *then taking into account that* $\triangleleft$ *acts trivially on V, if follows from axiom (A3) that* $x \cdot y = x + x \cdot y$, *for all* $x, y \in V$, *thus* $V = 0$ *and we have reached a contraction.*

5.3 Flag Jacobi algebras: Examples

In this section we will test the efficiency of Theorem 5.2.7 and the decomposition given by (5.16) in order to compute $\mathcal{JH}^2(V, A)$ for the class of Jacobi algebras defined below:

Definition 5.3.1 *Let A be a Jacobi algebra and E a vector space containing A as a subspace. A Jacobi algebra structure on E is called a* flag extending structure *of A to E if there exists a finite chain of Jacobi subalgebras of E*

$$E_0 := A \subset E_1 \subset \cdots \subset E_m := E \tag{5.17}$$

such that E_i *has codimension* 1 *in* E_{i+1}, *for all* $i = 0, \cdots, m-1$. *A Jacobi algebra that is a flag extending structure of the Jacobi algebra* $k = k_0$ *is called a* flag Jacobi algebra.

All flag extending structures of a given Jacobi algebra A to a vector space E can be completely described by a recursive reasoning where the key step is the case when A has codimension 1 in E. This step will provide the description and classification of all unified products $A \ltimes k$ of Jacobi algebras. Then, by replacing A with each of these unified products $A \ltimes k$, we go on with the recursive process in m steps, where m is the codimension of A in E. The tool that will play the key role in the description of flag extending structures is the following:

Definition 5.3.2 *A* Jacobi flag datum *of a Jacobi algebra A is a 6-tuple* $(\Lambda, \Delta, f_0, u, \lambda, D)$, *consisting of four k-linear maps* $\Lambda, \lambda : A \to k$, $\Delta, D : A \to A$ *and two elements* $f_0 \in A$ *and* $u \in k$ *satisfying the following compatibilities for any* $a, b \in A$:

(JF0) $(\Lambda, \Delta, f_0, u)$ *is a flag datum of the associative algebra A and* (λ, D) *is a twisted derivation of the Lie algebra A*

(JF1) $D(ab) = D(a)\, b + a\, D(b) + \lambda(a)\, \Delta(b) + \lambda(b)\, \Delta(a) - \lambda(1_A)\Delta(ab) - ab\, D(1_A)$

(JF2) $\lambda(ab) = \lambda(a)\, \Lambda(b) + \lambda(b)\, \Lambda(a) - \lambda(1_A)\, \Lambda(ab)$

(JF3) $\Delta([a,\, b]) = [\Delta(a),\, b] + \Lambda(a)\, D(b) - a\, D(b) - \lambda(b)\, \Delta(a) + \Delta(a)[1_A,\, b] + \Lambda(a)\, \Delta([1_A,\, b])$

(JF4) $\Lambda([a,\, b]) = \Lambda(a)\, \Lambda([1_A,\, b])$

(JF5) $\Delta\big(D(a)\big) - D\big(\Delta(a)\big) = \Delta(a) D(1_A) + \Lambda(a)\, \lambda(1_A)\, f_0 - \lambda(a)\, f_0 + \Lambda(a)\, \Delta\big(D(1_A)\big) +$

$$+ \lambda(1_A)\, \Delta^2(a)$$

(JF6) $\Lambda\big(D(a)\big) - \lambda\big(\Delta(a)\big) = \Lambda(a)\, \Lambda\big(D(1_A)\big) + \lambda(1_A)\, \Lambda(a)\, u - \lambda(a)\, u$

(JF7) $2\, \Delta\big(D(a)\big) + 2\, \lambda(a)\, f_0 = u\, D(a) + [f_0,\, a] + f_0\, [1_A,\, a] + u\, \Delta([1_A,\, a])$

(JF8) $2\, \Lambda\big(D(a)\big) = -\lambda(a)\, u + u\, \Lambda([1_A,\, a])$

(JF9) $D(f_0) + f_0\, D(1_A) + u\, \Delta\big(D(1_A)\big) + \lambda(1_A)\, \Delta(f_0) + u\, \lambda(1_A)\, f_0 = 0$

(JF10) $\lambda(f_0) + u\, \Lambda\big(D(1_A)\big) + \lambda(1_A)\, \Lambda(f_0) + u^2\, \lambda(1_A) = 0$

We denote by $\mathcal{JF}\,(A)$ the set of all Jacobi flag datums of A.

For further computations we point out that for any $(\Lambda,\, \Delta,\, f_0,\, u,\, \lambda,\, D) \in \mathcal{FJ}\,(A)$ we have $\Delta(1_A) = 0$ and $\Lambda(1_A) = 1$.

Proposition 5.3.3 *Let A be a Jacobi algebra. Then there exists a bijection between the set $\mathcal{JE}\,(A,\, k)$ of all Jacobi extending structures of A through k and $\mathcal{JF}\,(A)$ given such that the unified product $A \ltimes_{(\Lambda,\, \Delta,\, f_0,\, u,\, \lambda,\, D)} k$ corresponding to the Jacobi flag datum $(\Lambda,\, \Delta,\, f_0,\, u,\, \lambda,\, D) \in \mathcal{JF}\,(A)$, denoted by $A_{(\Lambda,\, \Delta,\, f_0,\, u,\, \lambda,\, D)}$, is the vector space $A \times k$ with the Jacobi algebra structure given for any $a,\, b \in A$ and $x,\, y \in k$ by:*

$$(a, x) \bullet (b, y) = \big(ab + x\Delta(b) + y\Delta(a) + xy\, f_0,\ x\Lambda(b) + y\Lambda(a) + xy\, u\big) \quad (5.18)$$

$$[(a, x),\, (b, y)] = \big([a,\, b] + xD(b) - yD(a),\ x\lambda(b) - y\lambda(a)\big) \quad (5.19)$$

Furthermore, a Jacobi algebra B contains A as a Jacobi subalgebra of codimension 1 if and only if $B \cong A_{(\Lambda,\, \Delta,\, f_0,\, u,\, \lambda,\, D)}$, for some Jacobi flag datum $(\Lambda,\, \Delta,\, f_0,\, u,\, \lambda,\, D)$ of A.

Proof: We have to compute the set of all bilinear maps

$$\triangleleft : k \times A \to k, \quad \triangleright : k \times A \to A, \quad f : k \times k \to A, \quad \cdot : k \times k \to k$$

$$\leftharpoonup : k \times A \to k, \quad \rightharpoonup : k \times A \to A, \quad \theta : k \times k \to A, \quad \{-,\, -\} : k \times k \to k$$

satisfying the compatibility conditions (J0)-(J10) in Theorem 5.2.2. To start with, the first part of axiom $(J0)$ tells us that $(\triangleleft,\, \triangleright,\, f,\, \cdot)$ is an algebra extending system of A through k, and Example 4.2.4 proves that there is a bijection between the set of all such maps $(\triangleleft,\, \triangleright,\, f,\, \cdot)$ and the set of all 4-tuples $(\Lambda,\, \Delta,\, f_0,\, u) \in A^* \times \mathrm{End}_k(A) \times A \times k$ that are flag datums of the associative algebra A: the bijection is given such that the algebra extending system $\big(\triangleleft,\, \triangleright,\, f,\, \cdot\big)$ associated to $(\Lambda,\, \Delta,\, f_0,\, u) \in \mathcal{F}(A)$ is defined by the formulas (4.22). Secondly, the last assertion of $(J0)$ tells us that $(\leftharpoonup,\, \rightharpoonup,\, \theta,\, \{-,\, -\})$

is a Lie extending system of the Lie algebra A through the vector space k and Example 3.2.6 shows that there is a bijection between this set and the space $\mathrm{TwDer}(A)$ of all twisted derivations of A; the bijection is given such that the Lie extending system $(\leftharpoonup, \rightharpoonup, \theta, \{-, -\})$ associated to a twisted derivation $(\lambda, D) \in \mathrm{TwDer}(A)$ is defined by the formulas (3.13). Hence, there exists a bijection between the set of all bilinear maps $(\triangleleft, \triangleright, f, \cdot, \leftharpoonup, \rightharpoonup, \theta, \{-, -\})$ satisfying (J0) and the set of all 6-tuples $(\Lambda, \Delta, f_0, u, \lambda, D)$ as defined in Definition 5.3.2 satisfying the compatibility conditions (JF0) and the bijection is given by the formulas (4.22) and (3.13). The rest of the proof is a long but straightforward computation which shows that, under this bijection, the compatibility conditions (J1)-(J10) of Theorem 5.2.2 take the equivalent forms given by (JF1)-(JF10). Finally, the Jacobi algebra defined by (5.18)-(5.19) is exactly the associated unified product $A \ltimes k$ defined by (5.14)-(5.15) written in this context. Finally, the last statement follows from the first part and Proposition 5.2.4. □

Let $\{e_i \,|\, i \in I\}$ be a basis of a Jacobi algebra A and $(\Lambda, \Delta, f_0, u, \lambda, D) \in \mathcal{JF}(A)$ a Jacobi flag datum. Then $A_{(\Lambda, \Delta, f_0, u, \lambda, D)}$ is the Jacobi algebra having $\{E, e_i \,|\, i \in I\}$ as a basis with the multiplication and the bracket defined for any $i \in I$ by:

$$e_i \bullet e_j := e_i \cdot_A e_j, \quad E \bullet e_i = e_i \bullet E := \Delta(e_i) + \Lambda(e_i)\, E, \quad E^2 := f_0 + u\, E \tag{5.20}$$

$$[e_i, e_j] := [e_i, e_j]_A, \quad [E, e_i] := D(e_i) + \lambda(e_i)\, E \tag{5.21}$$

where $\cdot_A$ (resp. $[-, -]_A$) is the multiplication (resp. the bracket) on A. The existence of these Jacobi algebras depends on the Jacobi algebra A. An interesting fact is the following:

Corollary 5.3.4 *Let A be a perfect Poisson algebra, i.e., $[A, A] = A$. Then, there is no Jacobi algebra which contains A as a subalgebra of codimension* 1.

Proof: It follows from Proposition 5.3.3 if we prove that the set $\mathcal{JF}(A)$ is empty. Indeed, let $(\Lambda, \Delta, f_0, u, \lambda, D) \in \mathcal{JF}(A)$. Since $[1_A, a] = 0$, for all $a \in A$, it follows from axiom (JF4) that $\Lambda([a, b]) = 0$, for all $a, b \in A$. As A is perfect as a Lie algebra, we obtain that $\Lambda(x) = 0$, for any $x \in A$, contrary to axiom (FA1) which ensures that $\Lambda(1_A) = 1$. □

Remark 5.3.5 *A basic invariant of a finite-dimensional Lie algebra $\mathfrak{g}$ is the Schur invariant defined by $\alpha(\mathfrak{g}) :=$ the maximal dimension of an abelian subalgebra of $\mathfrak{g}$. There is a vast literature devoted to computing this number for several classes of Lie algebras such as (semi)simple, (super)solvable, etc.; see [62, 74] and their references. Lie algebras for which $\alpha(\mathfrak{g}) = \dim(\mathfrak{g}) - 1$ are fully described in [118]: below we give the Jacobi algebra version of this result. For a finite-dimensional Jacobi algebra J we define the Schur invariant by the formula:*

$$\alpha(J) := \max\{\dim(A) \mid A \text{ is an abelian Jacobi subalgebra of } J\}$$

Using Proposition 5.3.3 we obtain that an $(n+1)$*-dimensional Jacobi algebra* J *has* $\alpha(J) = \dim(J) - 1$ *if and only if* $J \cong A_{(\Lambda,\, \Delta,\, f_0,\, u,\, \lambda,\, D)}$*, where* A *is an* n*-dimensional algebra with basis* $\{e_i \,|\, i = 1, \cdots, n\}$ *and* $A_{(\Lambda,\, \Delta,\, f_0,\, u,\, \lambda,\, D)}$ *is the Jacobi algebra having* $\{E,\, e_i \,|\, i = 1, \cdots, n\}$ *as a basis and the multiplication and the bracket is given for any* $i,\, j = 1, \cdots n$ *by:*

$$\begin{aligned} e_i \bullet e_j &:= e_i \cdot_A e_j, \quad E \bullet e_i = e_i \bullet E := \Delta(e_i) + \Lambda(e_i)\, E, \quad E^2 := f_0 + u\, E \\ [E,\, e_i] &:= D(e_i) + \lambda(e_i)\, E \end{aligned}$$

for some 6*-tuple* $(\Lambda,\, \Delta,\, f_0,\, u,\, \lambda,\, D) \in \mathcal{JF}(A)$ *with* $(D,\, \lambda) \neq (0, 0)$*. We mention that in this case the axioms (JF0)–(JF10), which need to be fulfilled by the* 6*-tuples* $(\Lambda,\, \Delta,\, f_0,\, u,\, \lambda,\, D)$*, take a simplified form as the Lie bracket on* A *is trivial.*

Now we will classify the algebras $A_{(\Lambda,\, \Delta,\, f_0,\, u,\, \lambda,\, D)}$ by providing the first explicit classification result of the ES problem for Jacobi algebras:

Theorem 5.3.6 *Let* A *be a Jacobi algebra of codimension* 1 *in the vector space* E*. Then there exists a bijection*

$$\mathrm{Extd}_{\mathcal{J}}\,(E,\, A) \cong \mathcal{JH}^2\,(k,\, A) \cong \mathcal{JF}\,(A)/\equiv \tag{5.22}$$

where $\equiv$ *is the equivalence relation on the set* $\mathcal{JF}\,(A)$ *of all Jacobi flag datums of* A *defined as follows:* $(\Lambda,\, \Delta,\, f_0,\, u,\, \lambda,\, D) \equiv (\Lambda',\, \Delta',\, f_0',\, u',\, \lambda',\, D')$ *if and only if* $\Lambda' = \Lambda$*,* $\lambda' = \lambda$*,* $u' = u$ *and there exists* $\alpha \in A$ *such that for any* $a \in A$ *we have:*

$$\Delta'(a) = \Delta(a) + \Lambda(a)\alpha - a\alpha \tag{5.23}$$

$$f_0' = f_0 + \alpha^2 + u\alpha - 2\Lambda(\alpha)\alpha - 2\Delta(\alpha) \tag{5.24}$$

$$D'(a) = D(a) + \lambda(a)\alpha - [\alpha,\, a] \tag{5.25}$$

The bijection between $\mathcal{JF}\,(A)/\equiv$ *and* $\mathrm{Extd}_{\mathcal{J}}\,(E,\, A)$ *is given by* $\overline{(\Lambda,\, \Delta,\, f_0,\, u,\, \lambda,\, D)} \mapsto A_{(\Lambda,\, \Delta,\, f_0,\, u,\, \lambda,\, D)}$*, where* $\overline{(\Lambda,\, \Delta,\, f_0,\, u,\, \lambda,\, D)}$ *is the equivalence class of* $(\Lambda,\, \Delta,\, f_0,\, u,\, \lambda,\, D)$ *via the relation* $\equiv$*.*

Proof: Let $(\Lambda,\, \Delta,\, f_0,\, u,\, \lambda,\, D)$, $(\Lambda',\, \Delta',\, f_0',\, u',\, \lambda',\, D') \in \mathcal{JF}\,(A)$ and $\Upsilon(A, V)$, respectively $\Upsilon'(A, V)$ be the corresponding Jacobi algebra extending structures. Since $\dim_k(V) = 1$, any linear map $r : V \to A$ is uniquely determined by an element $\alpha \in A$ such that $r(x) = \alpha$, where $\{x\}$ is a basis in V. We can easily see that the compatibility conditions from Proposition 5.2.5 applied to Jacobi flag datums take precisely the form given in the statement and hence the proof follows from Theorem 5.2.7 and Proposition 5.3.3. □

Remark 5.3.7 *In practice, in order to compute the quotient set* $\mathcal{JF}\,(A)/\equiv$ *constructed in Theorem 5.3.6 we shall use the decomposition (5.16) obtained in Corollary 5.2.9 by going through the following steps. First of all,* *we shall fix a pair* $(\Lambda, \lambda) \in A^* \times A^*$, *consisting of an associative algebra map* $\Lambda : A \to k$ *and a Lie algebra map* $\lambda : A \to k$ *satisfying the compatibility conditions (5.7). Secondly, for a given such pair* (Λ, λ) *we fix a scalar* $u \in k$ *and compute the set* $\mathcal{JF}^u_{(\Lambda,\lambda)}(A)$ *consisting of all triples* $(\Delta, f_0, D) \in \mathrm{End}_k(A) \times A \times \mathrm{End}_k(A)$ *such that* $(\Lambda, \Delta, f_0, u, \lambda, D)$ *is a Jacobi flag datum of* A. *Two triples* (Δ, f_0, D) *and* $(\Delta', f_0', D') \in \mathcal{JF}^u_{(\Lambda,\lambda)}(A)$ *are equivalent and we denote this by* $(\Delta, f_0, D) \equiv^u_{(\Lambda,\lambda)} (\Delta', f_0', D')$ *if and only if there exists* $\alpha \in A$ *such that (5.23)-(5.25) hold. Finally, we compute the quotient set* $\mathcal{JF}^u_{(\Lambda,\lambda)}(A)/\equiv^u_{(\Lambda,\lambda)}$ *—*$\mathcal{JH}^2(k, A)$ *will be the coproduct of these quotients sets over all triples* (Λ, λ, u)*—and then we list the isomorphism classes of the associated Jacobi algebras* $A_{(\Lambda, \Delta, f_0, u, \lambda, D)}$ *using (5.20)-(5.21). To conclude, using Theorem 5.3.6 and Corollary 5.2.9, we obtain:*

Corollary 5.3.8 *Let* A *be a Jacobi algebra. Then:*

$$\mathcal{JH}^2(k, A) \cong \sqcup_{(\Lambda,\lambda)} \Big(\sqcup_u \big(\mathcal{JF}^u_{(\Lambda,\lambda)}(A)/\equiv^u_{(\Lambda,\lambda)} \big) \Big) \tag{5.26}$$

where the coproducts in the right-hand side are made in the category of sets over all possible pairs (Λ, λ) *consisting of an associative algebra map* $\Lambda : A \to k$ *and a Lie algebra map* $\lambda : A \to k$ *satisfying (5.7) and over all scalars* $u \in k$.

Remark 5.3.9 *Let* A *be a Poisson algebra. The Poisson algebra version of Theorem 5.3.6 and Corollary 5.3.8 for computing* $\mathcal{PH}^2(k, A)$ *defined in Remark 5.2.8 are obtained as follows. First, we define the set* $\mathcal{PF}(A)$ *of all* Poisson flag datums *of* A*: it coincides with the set of all 6-tuples* $(\Lambda, \Delta, f_0, u, \lambda, D)$, *consisting of four* k*-linear maps* Λ, $\lambda : A \to k$, Δ, $D : A \to A$ *and two elements* $f_0 \in A$ *and* $u \in k$ *satisfying the following compatibilities for any* $a, b \in A$:

(PF0) $(\Lambda, \Delta, f_0, u)$ *is a flag datum of the associative algebra* A *and* (λ, D) *is a twisted derivation of the Lie algebra* A;

(PF1) $D(ab) = D(a)\, b + a\, D(b) + \lambda(a)\, \Delta(b) + \lambda(b)\, \Delta(a)$

(PF2) $\lambda(ab) = \lambda(a)\, \Lambda(b) + \lambda(b)\, \Lambda(a)$

(PF3) $\Delta([a, b]) = [\Delta(a), b] + \Lambda(a)\, D(b) - a\, D(b) - \lambda(b)\, \Delta(a)$

(PF4) $\Lambda([a, b]) = 0, \quad D(f_0) = 0, \quad \lambda(f_0) = 0$

(PF5) $D\big(\Delta(a)\big) - \Delta\big(D(a)\big) = \lambda(a)\, f_0, \quad \lambda\big(\Delta(a)\big) - \Lambda\big(D(a)\big) = \lambda(a)\, u$

(PF6) $2\,\Delta\big(D(a)\big) + 2\,\lambda(a)\, f_0 = u\, D(a) + [f_0, a], \quad 2\,\Lambda\big(D(a)\big) = -\lambda(a)\, u$

Then there exists a bijection $\mathcal{PH}^2(k, A) \cong \mathcal{PF}(A)/\equiv$, *where* $\equiv$ *is the equivalence relation defined exactly as in Theorem 5.3.6, but on the set* $\mathcal{PF}(A)$ *of all Poisson flag datums of the Poisson algebra* A.

Now, we will illustrate by examples the efficiency of Corollary 5.3.8 in classifying flag Jacobi algebras. The strategy followed will be that of Remark 5.3.7 imposed by the decomposition of $\mathcal{JH}^2(k, A)$ given in (5.26). Moreover, if A is a Poisson algebra we will also describe $\mathcal{PH}^2(k, A)$ in order to illustrate the difference between it and $\mathcal{JH}^2(k, A)$. The model is given below and we

Jacobi algebra	Algebra structure	Lie bracket
$J^{3,1}_{(\lambda_1, \lambda_2, u)}$ $(\lambda_1, \lambda_2, u) \in k \times k^* \times k$	$x^2 = 0$ $yx = xy = 2^{-1}\, u\, x$ $y^2 = -4^{-1}\, u^2 + u\, y$	$[y,\, 1] = -2^{-1}\lambda_1 u + \lambda_1 y$ $[y,\, x] = -2^{-1}\lambda_2 u + \lambda_2 y$
$J^{3,2}_{(\lambda_1, u, f)}$ $(\lambda_1, u, f) \in k^* \times k^2$	$x^2 = 0$ $yx = xy = 2^{-1}\, u\, x$ $y^2 = -4^{-1}\, u^2 + f\, x + u\, y$	$[y,\, 1] = -2^{-1}\, \lambda_1 u + \lambda_1\, y$
$J^{3,3}_{\delta}, \;\; \delta \in k^*$	$x^2 = 0, \; y^2 = \delta^2$ $yx = xy = \delta\, x$	abelian
$J^{3,4}_{u}, \;\; u \in k^*$	$x^2 = 0, \; y^2 = u\, y$ $yx = xy = 0$	abelian
$J^{3,5}_{(u,\, f, d_1, d_2)}$ $(u,\, f, d_1, d_2) \in k^4$	$x^2 = 0$ $yx = xy = 2^{-1} u\, x$ $y^2 = -4^{-1} u^2 + f\, x + u\, y$	$[y,\, 1] = d_1 x$ $[y,\, x] = d_2 x$

Table 5.2: Flag Jacobi algebras over $J^{2,\,1}$.

make the following convention: all undefined bracket or multiplication of two elements of a basis is zero. Let $J^{2,\,1}$ be the 2-dimensional Jacobi algebra of Example 5.1.2.

Proposition 5.3.10 *Let k be a field of characteristic $\neq 2$. Then:*

$$\mathcal{JH}^2\,(k,\, J^{2,\,1}) \;\; \cong \;\; (k \times k^* \times k) \sqcup (k^* \times k^2) \sqcup k^* \sqcup k^* \sqcup k^4 \quad (5.27)$$

and the equivalence classes of all 3-dimensional flag Jacobi algebras over $J^{2,\,1}$ are the Jacobi algebras $J^{3,1}_{(\lambda_1,\, \lambda_2,\, u)}$, $J^{3,2}_{(\lambda_1,\, u,\, f)}$, $J^{3,3}_{\delta}$, $J^{3,4}_{u}$, $J^{3,5}_{(u,\, f,\, d_1,\, d_2)}$ having $\{1,\, x,\, y\}$ as a basis and the multiplication and the bracket defined in Table 5.2.

Proof: To start with, we should notice that there is only one algebra map $\Lambda : J^{2,\,1} \to k$ namely the one given by $\Lambda(1) = 1$ and $\Lambda(x) = 0$. We denote $\lambda(1) = \lambda_1$ and $\lambda(x) = \lambda_2$, for some $(\lambda_1,\, \lambda_2) \in k^2$. Then $\lambda : J^{2,\,1} \to k$ is a Lie algebra map (the Lie brackets on $J^{2,\,1}$ and k are both abelian) and moreover (5.7) is fulfilled. Thus, the set of all pairs (Λ, λ) consisting of an algebra map $\Lambda : J^{2,\,1} \to k$ and a Lie algebra map $\lambda : J^{2,\,1} \to k$ satisfying (5.7) is in bijection with k^2. From now on we fix a pair $(\lambda_1,\, \lambda_2) \in k^2$ and a scalar $u \in k$. We will describe the set $\mathcal{JF}^{u}_{(\Lambda,\lambda)}\,(J^{2,\,1})$ consisting of all triples $(\Delta,\, f_0,\, D)$ such that $(\Lambda,\, \Delta,\, f_0,\, u,\, \lambda,\, D)$ is a Jacobi flag datum of $J^{2,\,1}$. We identify the k-linear maps Δ and $D : J^{2,\,1} \to J^{2,\,1}$ with their associated matrices with respect to the basis $\{1,\, x\}$:

$$\Delta = \begin{pmatrix} \Delta_{11} & \Delta_{12} \\ \Delta_{21} & \Delta_{22} \end{pmatrix} \qquad D = \begin{pmatrix} d_{11} & d_{12} \\ d_{21} & d_{22} \end{pmatrix} \qquad f_0 = f_1 + f_2\, x$$

for some $(\Delta_{ij}) \in k^4$, $(d_{ij}) \in k^4$, $(f_1, f_2) \in k^2$. We have to determine all these scalars such that axioms (JF0)-(JF10) are fulfilled for a fixed triple $(\lambda_1, \lambda_2, u) \in k^3$. Since this comes down to a laborious computation we will only provide a sketch of it. For instance, we can easily obtain that axiom (JF0) holds for a fixed triple $(\lambda_1, \lambda_2, u) \in k^3$ if and only if (Δ, f_0, D) is given by

$$\Delta = \begin{pmatrix} 0 & 0 \\ 0 & \delta \end{pmatrix} \qquad D = \begin{pmatrix} d_{11} & d_{12} \\ d_{21} & d_{22} \end{pmatrix} \qquad f_0 = \delta(\delta - u) + f\,x \tag{5.28}$$

for some δ, $f \in k$ and $(d_{ij}) \in k^4$ such that $\lambda_1 d_{12} = \lambda_2 d_{11}$ and $\lambda_1 d_{22} = \lambda_2 d_{21}$. We continue in a similar manner by testing the remaining compatibilities: for instance (JF2) and (JF4) hold automatically for such a triple (Δ, f_0, D) while we can easily see that (JF1) (resp. (JF3)) holds if and only if $d_{12} = -\lambda_2\delta$ (resp. $d_{11} = -\lambda_1\delta$). On the other hand we can prove that (JF5) holds if and only if $\lambda_2 f = 0$ and $\lambda_2\,\delta(2\delta - u) = 0$. Moreover, (JF8) holds if and only if $d_{11} = -2^{-1}\,\lambda_1 u$ and $d_{12} = -2^{-1}\,\lambda_2 u$). By eliminating the redundant compatibilities we can conclude that axioms (JF0)-(JF10) are fulfilled if and only if the triple (Δ, f_0, D) given by (5.28) satisfies the following equations:

$$d_{11} = -\lambda_1\delta = -2^{-1}\,\lambda_1 u, \quad d_{12} = -\lambda_2\delta = -2^{-1}\,\lambda_2 u, \quad \lambda_1 d_{22} = \lambda_2 d_{21} \tag{5.29}$$

$$\lambda_2 f = \lambda_2(2\,\delta - u) = \lambda_1\delta(2\,\delta - u) = f d_{22} = 0 \tag{5.30}$$

$$d_{21}(2\,\delta - u) = d_{22}(2\,\delta - u) = 0, \quad \lambda_1(2\,\delta^2 + \delta\,u - u^2) = 0 \tag{5.31}$$

Now, based on (5.30) we can decompose the set of all Jacobi flag datums of $J^{2,\,1}$ as follows

$$\mathcal{JF}\,(J^{2,\,1}) = J_1 \sqcup J_2 \sqcup J_3$$

where J_1 are those flag datums corresponding to $\lambda_2 \neq 0$, J_2 are associated to the case $\lambda_2 = 0$ and $\lambda_1 \neq 0$ while J_3 correspond to the case $\lambda_2 = \lambda_1 = 0$.

In what follows we only provide details for the first case, namely the one corresponding to $\lambda_2 \neq 0$; the other two cases are settled using an analogous treatment. If we denote $d_{22} := d \in k$, then equations (5.29)-(5.31) hold if and only if $f = 0$, $\delta = 2^{-1}\,u$ and $d_{21} = \lambda_2^{-1}\lambda_1 d$. Thus, $J_1 \cong k \times k^* \times k^2$ and the bijection is given such that the Jacobi flag datum $(\Lambda, \Delta, f_0, u, \lambda, D) \in \mathcal{JF}\,(J^{2,\,1})$ associated to $(\lambda_1, \lambda_2, u, d) \in k \times k^* \times k^2$ is given by

$$\Lambda(1) = 1, \;\; \Lambda(x) = 0, \quad \lambda(1) = \lambda_1, \;\; \lambda(x) = \lambda_2 \neq 0 \tag{5.32}$$

$$\Delta = \begin{pmatrix} 0 & 0 \\ 0 & 2^{-1}u \end{pmatrix} \quad D = \begin{pmatrix} -2^{-1}\,\lambda_1 u & -2^{-1}\,\lambda_2 u \\ \lambda_2^{-1}\lambda_1 d & d \end{pmatrix} \quad f_0 = -4^{-1}\,u^2 \tag{5.33}$$

Now, an elementary computation shows that the equivalence relation of Theorem 5.3.6 given by (5.23)-(5.25) on the Jacobi flag datums written for the set of triples $k \times k^* \times k^2$ becomes: $(\lambda_1, \lambda_2, u, d) \equiv (\lambda_1', \lambda_2', u', d')$ if and only if $\lambda_1' = \lambda_1$, $\lambda_2' = \lambda_2$, $u' = u$ and there exists $q \in k$ such that $d' = d + \lambda_2 q$. Since, $\lambda_2 \neq 0$, there exists such a q, namely $q := \lambda_2^{-1}(d' - d)$. This shows that $(\lambda_1, \lambda_2, u, d) \equiv (\lambda_1, \lambda_2, u, 0)$, for any $d \in k$ and the quotient set

$k\times k^*\times k^2/\equiv\;\cong k\times k^*\times k\times\{0\}\cong k\times k^*\times k$, which is the first component of $\mathcal{JH}^2(k,\,J^{2,1})$ in formula (5.27). The 3-parameter Jacobi algebra $J^{3,1}_{(\lambda_1,\,\lambda_2,\,u)}$ is precisely the unified product $J^{2,1}\ltimes k$ associated to the Jacobi flag datum corresponding to $(\lambda_1,\,\lambda_2,\,u,\,0)$ via the formulas (5.32)-(5.33), with the multiplication and the bracket as defined by (5.20)-(5.21). In a similar fashion we can prove that $J_2\cong k^*\times k^3$ and the bijection is given such that the Jacobi flag datum $(\Lambda,\,\Delta,\,f_0,\,u,\,\lambda,\,D)\in\mathcal{JF}(J^{2,1})$ associated to $(\lambda_1,\,u,\,f,\,d)\in k^*\times k^3$ is given by

$$\Lambda(1)=1,\;\Lambda(x)=0,\quad\lambda(1)=\lambda_1\neq 0,\;\lambda(x)=0 \tag{5.34}$$

$$\Delta=\begin{pmatrix}0&0\\0&2^{-1}u\end{pmatrix}\quad D=\begin{pmatrix}-2^{-1}\lambda_1 u&0\\d&0\end{pmatrix}\quad f_0=-4^{-1}u^2+f\,x \tag{5.35}$$

and $k^*\times k^3/\equiv\;\cong k^*\times k^2$ since $(\lambda_1,\,u,\,f,\,d)\cong(\lambda_1',\,u',\,f',\,d')$ if and only if $\lambda_1'=\lambda_1$, $u'=u$, $f'=f$ and there exists $q\in k$ such that $d'=d+\lambda_1 q$. The Jacobi algebra $J^{3,2}_{(\lambda_1,\,u,\,f)}$ is the unified product $J^{2,1}\ltimes k$ associated to $(\lambda_1,\,u,\,f,\,d:=0)$. Finally, one can show in an analogous manner that $J_3\cong(k^*\sqcup k^*)\sqcup k^4$ and the corresponding Jacobi algebras are the last three families of Table 5.2: $J^{3,3}_{\delta}$, $J^{3,4}_{u}$ and respectively $J^{3,5}_{(u,\,f,\,d_1,\,d_2)}$. □

Remarks 5.3.11 (1) *$J^{2,1}$ is a Poisson algebra and hence we can also compute the classifying object $\mathcal{PH}^2(k,\,J^{2,1})$. By testing which of the Jacobi algebras listed in Proposition 5.3.10 are Poisson algebras, we obtain that*

$$\mathcal{PH}^2(k,\,J^{2,1})\cong(k^*\times k)\sqcup k^*\sqcup k^*\sqcup k^3$$

and the equivalence classes of all 3-dimensional flag Poisson algebras over $J^{2,1}$ are the following four families of Poisson algebras $J^{3,1}_{(0,\,\lambda_2,\,u)}$, $J^{3,3}_{\delta}$, $J^{3,4}_{u}$ and $J^{3,5}_{(u,\,f,\,0,\,d_2)}$ of Table 5.2.

(2) *Similar to Proposition 5.3.10 we can describe all flag Jacobi algebras over any 2-dimensional Jacobi algebra listed in Example 5.1.2. In some cases the computations are straightforward: for instance we can immediately see that $\mathcal{JH}^2(k,\,J^{2,2})=\varnothing$ and, if $k\neq k^2$, then $\mathcal{JH}^2(k,\,J^2_d)=\varnothing$, for all $d\in S\subset k\backslash k^2$. Indeed, if $\Lambda: J^{2,2}\to k$ is an algebra map then, $\Lambda(x)=0$ since $x^2=0$ in $J^{2,2}$. By applying axiom (JF4) for $a:=x$ and $b:=1$, we obtain that $\Lambda(1)=0$ and we have reached a contradiction as Λ is a unitary algebra map. Thus, we obtained that there is no 3-dimensional Jacobi algebra containing $J^{2,2}$ or J^2_d as Jacobi subalgebras. The remaining two Jacobi algebras of Example 5.1.2 can be treated in a similar manner and are left to the reader.*

The recursive algorithm can be continued in order to describe all 4-dimensional flag Jacobi algebras. For instance, computations similar to those performed in Proposition 5.3.10 give the following:

Example 5.3.12 *Consider the Jacobi algebra J^3_{11} described in Table* 1. *Then:*

$$\mathcal{JH}^2(k,\,J^3_{11})\cong\{*\}\sqcup k\sqcup k^3\sqcup k^2\sqcup k^2\sqcup(k^*\backslash\{1\})\sqcup k\sqcup k\sqcup k\sqcup k\sqcup(k^*\backslash\{1\})$$

where $\{*\}$ *is the singleton set. The equivalence classes of all 4-dimensional flag Jacobi algebras over* J^3_{11} *are the Jacobi algebras with basis* $\{1, x, y, z\}$ *and the multiplication and the bracket defined below:*

$J^{4,1}$: $x^2 = x$, $y^2 = 0$, $z^2 = 0$, $xy = yx = 0$, $zy = yz = 0$, $zx = xz = z$, $[y, 1] = y$, $[z, 1] = 2z$, $[z, x] = 2$;

$J^{4,2}_{\gamma}$: $x^2 = x$, $y^2 = 0$, $z^2 = 0$, $xy = yx = 0$, $zy = yz = 0$, $zx = xz = \gamma y + z$, $[y, 1] = y$, $[z, 1] = z$, $[z, x] = -\gamma y + z$, where $\gamma \in k$;

$J^{4,3}_{\alpha, u, v}$: $x^2 = x$, $y^2 = 0$, $xy = yx = 0$, $zx = xz = \alpha - \alpha x + z$, $zy = yz = -\alpha y$, $z^2 = \alpha^2 + \alpha u + v x + uz$, $[y, 1] = y$, $[z, y] = \alpha y$, where $(\alpha, u, v) \in k^3$;

$J^{4,4}_{\alpha, b}$: $x^2 = x$, $y^2 = 0$, $xy = yx = 0$, $zx = xz = 0$, $zy = yz = \alpha y$, $z^2 = -\alpha^2 + \alpha^2 x + 2\alpha z$, $[y, 1] = y$, $[z, y] = by$, where $(\alpha, b) \in k^2$;

$J^{4,5}_{\alpha, u}$: $x^2 = x$, $y^2 = 0$, $xy = yx = 0$, $zx = xz = 0$, $zy = yz = \alpha y$, $z^2 = \alpha(\alpha - u) - \alpha(\alpha - u)x + uz$, $[y, 1] = y$, $[z, y] = -\alpha y$, where $(\alpha, u) \in k^2$;

$J^{4,6}_{\lambda}$: $x^2 = x$, $y^2 = 0$, $xy = yx = 0$, $zx = xz = x$, $zy = yz = 0$, $z^2 = x$, $[y, 1] = y$, $[z, 1] = -\lambda x + \lambda z$, where $\lambda \in k^* - \{1\}$;

$J^{4,7}_{a}$: $x^2 = x$, $y^2 = 0$, $xy = yx = 0$, $zx = xz = x$, $zy = yz = 0$, $z^2 = x$, $[y, 1] = y$, $[z, 1] = -x + ay + z$, where $a \in k$;

$J^{4,8}_{b}$: $x^2 = x$, $y^2 = 0$, $xy = yx = 0$, $zx = xz = x$, $zy = yz = 0$, $z^2 = x + by$, $[y, 1] = y$, $[z, 1] = -2^{-1}x + 2^{-1}z$, where $b \in k$;

$J^{4,9}_{a}$: $x^2 = x$, $y^2 = 0$, $xy = yx = 0$, $zx = xz = x$, $zy = yz = 2^{-1}y$, $z^2 = -4^{-1} + 4^{-1}x + ay + z$, $[y, 1] = y$, $[z, y] = -2^{-1}y$, $[z, 1] = -4^{-1} - 4^{-1}x + 2^{-1}z$, where $a \in k$;

$J^{4,10}_{a}$: $x^2 = x$, $y^2 = 0$, $xy = yx = 0$, $zx = xz = x$, $zy = yz = 2^{-1}y$, $z^2 = -4^{-1} + 4^{-1}x + z$, $[y, 1] = y$, $[z, y] = -2^{-1}y$ $[z, 1] = -2^{-1} - 2^{-1}x + ay + z$, where $a \in k$;

$J^{4,11}_{\lambda}$: $x^2 = x$, $y^2 = 0$, $xy = yx = 0$, $zx = xz = x$, $zy = yz = 2^{-1}y$, $z^2 = -4^{-1} + 4^{-1}x + z$, $[y, 1] = y$, $[z, y] = -2^{-1}y$ $[z, 1] = -2^{-1}\lambda - 2^{-1}\lambda x + \lambda z$, where $\lambda \in k^* - \{1\}$.

5.4 Classifying complements for Poisson algebras

In this section we deal with a special case of the unified product for Poisson algebras, namely the bicrossed product and its main applications. Throughout, all algebras are commutative, not necessarily unitary. Let P and Q be two given Poisson algebras. A Poisson algebra R *factorizes* through P and Q if P, Q are Poisson subalgebras of R such that $R = P + Q$ and $P \cap Q = \{0\}$. In this case Q is called a *Poisson complement* of P in R or a *P-complement of R*.

The concept of a matched pair of Poisson algebras was recently introduced in [189, Theorem 1]; we recall the definition following our notations and terminology since it will be a special case of the axioms (P1)-(P10) appearing in Corollary 5.2.3.

Definition 5.4.1 *A* matched pair *of Poisson algebras is a system* $(P,\, Q,\, \triangleleft,\, \triangleright,\, \leftharpoonup,\, \rightharpoonup)$ *consisting of two Poisson algebras P and Q and four bilinear maps*

$$\triangleleft : Q \times P \to Q, \quad \triangleright : Q \times P \to P, \quad \leftharpoonup : Q \times P \to Q, \quad \rightharpoonup : Q \times P \to P$$

such that $(P,\, Q,\, \triangleleft,\, \triangleright)$ *is a matched pair of commutative algebras,* $(P,\, Q,\, \leftharpoonup,\, \rightharpoonup)$ *is a matched pair of Lie algebras satisfying the following compatibility conditions for any a, $b \in P$ and x, $y \in Q$:*

$$x \rightharpoonup (ab) = (x \rightharpoonup a)\, b + (x \leftharpoonup a) \triangleright b + a\, (x \rightharpoonup b) + (x \leftharpoonup b) \triangleright a \tag{5.36}$$
$$x \leftharpoonup (ab) = (x \leftharpoonup a) \triangleleft b + (x \leftharpoonup b) \triangleleft a \tag{5.37}$$
$$x \triangleright [a,\, b]_P = [x \triangleright a,\, b]_P + (x \triangleleft a) \rightharpoonup b - a\, (x \rightharpoonup b) - (x \leftharpoonup b) \triangleright a \tag{5.38}$$
$$x \triangleleft [a,\, b]_P = (x \triangleleft a) \leftharpoonup b - (x \leftharpoonup b) \triangleleft a \tag{5.39}$$
$$\{x,\, y\}_Q \triangleright a = x \triangleright (y \rightharpoonup a) - y \rightharpoonup (x \triangleright a) \tag{5.40}$$
$$\{x,\, y\}_Q \triangleleft a = \{x \triangleleft a,\, y\}_Q - y \leftharpoonup (x \triangleright a) + x \triangleleft (y \rightharpoonup a) + (y \leftharpoonup a)x \tag{5.41}$$
$$(xy) \rightharpoonup a = x \triangleright (y \rightharpoonup a) + y \triangleright (x \rightharpoonup a) \tag{5.42}$$
$$(xy) \leftharpoonup a = x(y \leftharpoonup a) + (x \leftharpoonup a)y + x \triangleleft (y \rightharpoonup a) + y \triangleleft (x \rightharpoonup a) \tag{5.43}$$

The axioms defining a matched pair of Poisson algebras in Definition 5.4.1 are derived from axioms (P1)-(P10) of Corollary 5.2.3 if we ask for $\Upsilon(P, Q) = (\triangleleft,\, \triangleright,\, f := 0,\, \cdot,\, \leftharpoonup,\, \rightharpoonup,\, \theta := 0,\, \{-,\, -\})$ to be a Poisson extending structure of the Poisson algebra P through Q, where the cocycles f and θ are both trivial: $f(x, y) = \theta(x, y) := 0$, for all x, $y \in Q$.

Let $(P,\, Q,\, \triangleleft,\, \triangleright,\, \leftharpoonup,\, \rightharpoonup\,)$ be a matched pair of Poisson algebras. Then $P \bowtie Q := P \times Q$ is a Poisson algebra with the multiplication and the bracket defined for any a, $b \in P$ and x, $y \in Q$ by:

$$(a,\, x) \bullet (b,\, y) := \big(ab + x \triangleright b + y \triangleright a,\; x \triangleleft b + y \triangleleft a + xy\big) \tag{5.44}$$

$$[(a, x),\, (b, y)] := \big([a,\, b]_P + x \rightharpoonup b - y \rightharpoonup a,\; x \leftharpoonup b - y \leftharpoonup a + \{x,\, y\}_Q\big) \tag{5.45}$$

called the *bicrossed product* associated to the matched pair $\big(P,\, Q,\, \triangleleft,\, \triangleright,\, \leftharpoonup,\, \rightharpoonup\big)$. The bicrossed product of Poisson algebras is exactly the unified product $P \ltimes Q$ associated to a Poisson extending structure of the Poisson algebra P through Q having both cocycles trivial $f = \theta := 0$. The bicrossed product is the construction responsible for the so-called *factorization problem* which, formulated at the level of Poisson algebras, comes down to the following: for two given Poisson algebras P and Q, describe and classify all Poisson algebras which factorize through P and Q. The next result is [189, Theorem 1]. For the reader's convenience we also include a short and different proof of it based on Proposition 5.2.4.

Proposition 5.4.2 *Let P and Q be two given Poisson algebras. A Poisson algebra R factorizes through P and Q if and only if there exists $\big(P,\, Q,\, \triangleleft,\, \triangleright,\, \leftharpoonup,\, \rightharpoonup\big)$ a matched pair of Poisson algebras such that $R \cong P \bowtie Q$, an isomorphism of Poisson algebras.*

Proof: First we observe that any bicrossed product $P \bowtie Q$ factorizes through $P \cong P \times \{0\}$ and $Q \cong \{0\} \times Q$ and, via this identification, P and Q are Poisson subalgebras of $P \bowtie Q$ such that $P \bowtie Q = P + Q$ and $P \cap Q = \{0\}$. Conversely, assume that $R = (R, \cdot_R, [-,\, -]_R)$ is a Poisson algebra that factorizes through P and Q. Let $\iota : P \to R$ be the inclusion map and $\pi : R \to P$ the canonical k-linear retraction of ι, i.e., $\pi(p + q) := p$. Since $Q = \mathrm{Ker}(\pi)$ is a Poisson subalgebra of R we obtain that the cocycles f and $\theta : Q \times Q \to P$ constructed in the proof of Proposition 5.2.4 are both trivial: i.e., $f(x, y) := \pi(x \cdot_R y) = 0$ and $\theta(x, y) := \pi([x, y]_R) = 0$, for all x, $y \in Q$. Thus, the actions $(\triangleleft,\, \triangleright,\, \leftharpoonup,\, \rightharpoonup)$ defined for any $a \in P$ and $x \in Q$ by:

$$x \triangleright a \;:=\; \pi(x \cdot_R a), \qquad x \triangleleft a := x \cdot_R a - \pi(x \cdot_R a)$$

$$x \rightharpoonup a \;:=\; \pi\big([x,\, a]_R\big), \quad x \leftharpoonup a := [x,\, a]_R - \pi\big([x,\, a]_R\big)$$

make $\big(P,\, Q,\, \triangleleft,\, \triangleright,\, \leftharpoonup,\, \rightharpoonup\big)$ a matched pair of Poisson algebras while $\varphi : P \bowtie Q \to R$, $\varphi(a, x) := a + x$ becomes an isomorphism of Poisson algebras. □

As we have seen in the proof of Proposition 5.4.2, if a Poisson algebra R factorizes through P and Q, then we can construct a matched pair of Poisson algebras as follows:

$$x \triangleright a + x \triangleleft a = xa, \qquad x \rightharpoonup a + x \leftharpoonup a = [x,\, a] \tag{5.46}$$

for all $a \in P$ and $x \in Q$. Throughout, the above matched pair will be called *the canonical matched pair* associated with the factorization of R through P and Q. Proposition 5.4.2 allows for a computational reformulation of the factorization problem as follows: for two given Poisson algebras P and Q, describe and classify all bicrossed products $P \bowtie Q$ associated to all possible matched pairs $(P, Q, \triangleleft, \triangleright, \leftharpoonup, \rightharpoonup)$. The problem is far from being a trivial one.

Example 5.4.3 *Let k_0 be the trivial 1-dimensional Poisson algebra with basis $\{X\}$ and $\mathcal{H}$ the 3-dimensional Heisenberg Lie algebra with basis $\{H_1, H_2, H_3\}$ and the bracket defined by $[H_1, H_2] = H_3$. $\mathcal{H}$ admits a Poisson algebra structure with the associative multiplication given by $H_1^2 = H_3$. It can be easily seen by a straightforward computation that the bicrossed products corresponding to all matched pairs of Poisson algebras $(k_0, \mathcal{H}, \triangleleft, \triangleright, \leftharpoonup, \rightharpoonup)$ are the 4-dimensional Poisson algebras with basis $\{X, H_1, H_2, H_3\}$ listed below:*

$$\mathcal{H}^1_{(\beta, \alpha, \mu, \eta)} : X^2 = 0,\ H_1^2 = H_3,\ XH_1 = H_1X = \alpha\, H_3,\ XH_2 = H_2X = \beta\, H_3$$
$$[H_1, H_2] = H_3,\ [H_1, X] = \mu\, H_3,\ [H_2, X] = \eta H_3,\ \text{for } (\beta, \alpha, \mu, \eta) \in k^* \times k^3$$

$$\mathcal{H}^2_{(\xi, \mu, \eta)} : X^2 = 0,\ H_1^2 = H_3,\ [H_1, H_2] = H_3,\ [H_1, X] = \xi\, X + \mu\, H_2 + \eta\, H_3$$
$$\text{for } (\xi, \mu, \eta) \in k^3$$

$$\mathcal{H}^3_{(\eta, \mu, \tau)} : X^2 = 0,\ H_1^2 = H_3,\ [H_1, H_2] = H_3,\ [H_2, X] = \eta\, H_3$$
$$[H_1, X] = \mu\, H_2 + \tau\, H_3,\ \text{for } (\eta, \mu, \tau) \in k^* \times k^2$$

$$\mathcal{H}^4_{(\alpha, \mu, \tau, \eta)} : X^2 = 0,\ H_1^2 = H_3,\ XH_1 = H_1X = \alpha\, H_3,\ [H_1, H_2] = H_3$$
$$[H_1, X] = \mu\, H_2 + \tau\, H_3,\ [H_2, X] = \eta\, H_3,\ \text{for } (\alpha, \mu, \tau, \eta) \in k^* \times k^3$$

$$\mathcal{H}^5_{(\xi, \mu, \eta)} : X^2 = 0,\ H_1^2 = H_3,\ [H_1, H_2] = H_3$$
$$[H_1, X] = \mu\, H_3,\ [H_2, X] = \xi\, X + \eta\, H_3,\ \text{for } (\xi, \mu, \eta) \in k^* \times k^2$$

$$\mathcal{H}^6_{(\xi, \gamma, \mu, \eta)} : X^2 = 0,\ H_1^2 = H_3,\ [H_1, H_2] = H_3,\ [H_1, X] = \xi\, X + \xi\, \gamma^{-1}\, \eta\, H_2$$
$$+ \mu\, H_3 [H_2, X] = \gamma\, X + \eta\, H_2 + \xi^{-1}\, (\xi + 1)\, \eta\, H_3,$$
$$\text{for } (\xi, \gamma, \mu, \eta) \in (k^*)^2 \times k^2$$

In the rest of the section we deal with the converse of the factorization problem, called the bicrossed descent (or the classification of complements) problem, whose statement was given in the introduction. First we need to introduce the following concept:

Definition 5.4.4 *Let $(P, Q, \triangleleft, \triangleright, \leftharpoonup, \rightharpoonup)$ be a matched pair of Poisson algebras. A k-linear map $r : Q \to P$ is called a* deformation map *of the above* *matched pair if the following compatibilities hold for all* $p,\, q \in Q$*:*

$$r(p)r(q) - r(pq) = r\big(q \triangleleft r(p) + p \triangleleft r(q)\big) - q \triangleright r(p) - p \triangleright r(q) \quad (5.47)$$

$$r\big([p,\, q]\big) - \big[r(p),\, r(q)\big] = r\big(q \leftharpoondown r(p) - p \leftharpoondown r(q)\big) + p \rightharpoonup r(q) - q \rightharpoonup r(p) \quad (5.48)$$

We denote by $\mathcal{DM}\big(P,\, Q \mid (\triangleleft,\, \triangleright,\, \leftharpoondown,\, \rightharpoonup)\big)$ the set of all deformation maps of the matched pair $\big(P,\, Q,\, \triangleleft,\, \triangleright,\, \leftharpoondown,\, \rightharpoonup\big)$.

Example 5.4.5 *Consider the following matched pair between* k_0 *and* $\mathcal{H}$*:*

$$H_1 \triangleleft X = H_3, \;\; H_2 \triangleleft X = H_3, \;\; H_1 \leftharpoondown X = H_3, \;\; H_2 \leftharpoondown X = H_3$$

where the undefined actions are all equal to 0*. The corresponding bicrossed product is the Poisson algebra* $\mathcal{H}^1_{(1,\,1,\,1,\,1)}$ *from Example 5.4.3. Any deformation map associated to the above matched pair is given as follows:*

$$r_{(a_1,\, a_2)} : \mathcal{H} \to k_0, \;\; r_{(a_1,\, a_2)}(h_1) = a_1\, X, \;\; r_{(a_1,\, a_2)}(h_2) = a_2\, X, \;\; r_{(a_1,\, a_2)}(h_3) = 0$$

for some $a_1,\, a_2 \in k$*.*

The next result shows that to any deformation map r we can associate a new Poisson algebra called the r-deformation and, moreover, all complements of a given Poisson algebra extension $P \subset R$ can be described as r-deformations of a given complement.

Theorem 5.4.6 *Let* P *be a Poisson subalgebra of* R *and* Q *a given* P*-complement of* R *with the associated canonical matched pair* $\big(P,\, Q,\, \triangleleft,\, \triangleright,\, \leftharpoondown,\, \rightharpoonup\big)$*.*

(1) *Let* $r : Q \to P$ *be a deformation map of the above matched pair. Then* $Q_r := Q$*, as a vector space, with the new Poisson algebra structure defined for any* $q,\, t \in Q$ *by:*

$$q \cdot_r t = qt + t \triangleleft r(q) + q \triangleleft r(t) \quad (5.49)$$

$$[q,\, t]_r = [q,\, t] + q \leftharpoondown r(t) - t \leftharpoondown r(q) \quad (5.50)$$

is a Poisson algebra called the r*-deformation of* Q *and* Q_r *is a* P*-complement of* R*.*

(2) $\overline{Q}$ *is a* P*-complement of* R *if and only if there exists an isomorphism of Poisson algebras* $\overline{Q} \cong Q_r$*, for some deformation map* $r : Q \to P$ *of the above canonical matched pair.*

Proof: Let $P \bowtie Q$ be the bicrossed product associated to the canonical matched pair $\big(P,\, Q,\, \triangleleft,\, \triangleright,\, \leftharpoondown,\, \rightharpoonup\big)$. It follows from Proposition 5.4.2 that $R \cong P \bowtie Q$ as Poisson algebras.

(1) Although this claim can be proven by a very long but straightforward computation, we will provide a different and more natural approach. Given a

deformation map $r : Q \to P$, we consider $f_r : Q \to P \bowtie Q$ to be the k-linear map defined for all $q \in Q$ by:

$$f_r(q) = \big(r(q),\, q\big)$$

It turns out that $\widetilde{Q} := \mathrm{Im}(f_r)$ is a P complement of $R \cong P \bowtie Q$. We start by proving that $\widetilde{Q}$ is a Poisson subalgebra of $P \bowtie Q$. Indeed, for all $p,\, q \in Q$ we have:

$$\begin{aligned}
\big(r(p),\, p\big)\big(r(q),\, q\big) &= \big(\underline{r(p)r(q) + p \triangleright r(q) + q \triangleright r(p)},\\
&\quad p \triangleleft r(q) + q \triangleleft r(p) + pq\big)\\
&\overset{(5.49)}{=} \big(r(pq + p \triangleleft r(q) + q \triangleleft r(p)),\, p \triangleleft r(q)\\
&\quad + q \triangleleft r(p) + pq\big)\\
[(r(p),\, p),\, (r(q),\, q)] &= \big(\,\underline{[r(p),\, r(q)] + p \rightharpoonup r(q) - q \rightharpoonup r(p)},\\
&\quad p \leftharpoonup r(q) - q \leftharpoonup r(p) + [p,\, q]\,\big)\\
&\overset{(5.50)}{=} \big(r([p, q] + p \leftharpoonup r(q)\\
&\quad - q \leftharpoonup r(p)), p \leftharpoonup r(q) - q \leftharpoonup r(p) + [p, q]\big)
\end{aligned}$$

Therefore $\widetilde{Q}$ is a Poisson subalgebra of $P \bowtie Q$. Consider now $(p,\, q) \in P \cap \widetilde{Q}$. Since in particular we have $(p,\, q) \in \widetilde{Q}$ then $p = r(q)$. As we also have $(r(q),\, q) \in P$ we obtain $q = 0$ and thus $P \cap \widetilde{Q} = \{0\}$. Furthermore, for any $(p,\, q) \in R = P \bowtie Q$ we can write $(p,\, q) = (p - r(q),\, 0) + (r(q),\, q) \in P + \widetilde{Q}$. Hence, we have proved that $\widetilde{Q}$ is a P-complement of $P \bowtie Q$. We are left to prove that Q_r and $\widetilde{Q}$ are isomorphic as Poisson algebras. To this end, we denote by $\widetilde{f}_r$ the linear isomorphism from Q to $\widetilde{Q}$ induced by f_r. As we will see, $\widetilde{f}_r$ is a Poisson algebra map if we consider Q endowed with the Poisson structures given by (5.49) and (5.50). Indeed, for all $q,\, t \in Q$ we have:

$$\begin{aligned}
\overline{f}_r(q \cdot_r t) &\overset{(5.49)}{=} \overline{f}_r(qt + t \triangleleft r(q) + q \triangleleft r(t))\\
&= \big(\underline{r(qt + t \triangleleft r(q) + q \triangleleft r(t))},\, qt + t \triangleleft r(q) + q \triangleleft r(t)\big)\\
&\overset{(5.47)}{=} \big(r(q)r(t) + q \triangleright r(t) + t \triangleright r(q),\, qt + t \triangleleft r(q) + q \triangleleft r(t)\big)\\
&\overset{(5.44)}{=} \big(r(q),\, q\big)\big(r(t),\, t\big) = \overline{f}_r(q)\overline{f}_r(t)
\end{aligned}$$

and

$$\begin{aligned}\overline{f}_r([q,\,t]_r) &\stackrel{(5.49)}{=} \overline{f}_r([q,\,t] + q \leftharpoondown r(t) - t \leftharpoondown r(q))\\ &= \big(\underline{r([q,\,t] + q \leftharpoondown r(t) - t \leftharpoondown r(q))},\, [q,\,t] + q \leftharpoondown r(t)\\ &\quad -t \leftharpoondown r(q)\big)\\ &\stackrel{(5.48)}{=} \big([r(q),\,r(t)] + q \rightharpoondown r(t) - t \rightharpoondown r(q),\, [q,\,t] + q \leftharpoondown r(t)\\ &\quad -t \leftharpoondown r(q)\big)\\ &\stackrel{(5.45)}{=} \big[(r(q),\,q),\,(r(t),\,t)\big] = \big[\overline{f}_r(q),\,\overline{f}_r(t)\big]\end{aligned}$$

Hence we can conclude that Q_r is a Poisson algebra and this finishes the proof.

(2) Let $\overline{Q}$ be an arbitrary P-complement of R. As $R = P \oplus Q = P \oplus \overline{Q}$ we can find four k-linear maps:

$$a : Q \to P, \qquad b : Q \to \overline{Q}, \qquad c : \overline{Q} \to P, \qquad d : \overline{Q} \to Q$$

such that for all $q \in Q$ and $t \in \overline{Q}$ we have:

$$q = a(q) \oplus b(q), \qquad t = c(t) \oplus d(t) \tag{5.51}$$

It follows by an easy computation that $b : Q \to \overline{Q}$ is an isomorphism of vector spaces. We will denote by $\widetilde{b} : Q \to P \bowtie Q$ the composition $\widetilde{b} := i \circ b$, where $i : \overline{Q} \to R = P \bowtie Q$ is the canonical inclusion. Thus, by (5.51) we have $\widetilde{b}(t) = (-a(t),\, t)$ for all $t \in Q$. We will prove that $r := -a$ is a deformation map and $\overline{Q} \cong Q_r$. Indeed, $\overline{Q} = \mathrm{Im}(b) = \mathrm{Im}(\widetilde{b})$ is a Poisson subalgebra of $R = P \bowtie Q$ and we have:

$$\begin{aligned}\big(r(q),\,q\big)\big(r(t),\,t\big) = &\stackrel{(5.44)}{=} \big(r(q)r(t) + q \triangleright r(t) + t \triangleright r(q),\, q \triangleleft r(t)\\ &\quad +t \triangleleft r(q) + qt\big)\\ &= \big(r(t'), t'\big)\\ \big[\big(r(q),\,q\big),\,\big(r(t),\,t\big)\big] &\stackrel{(5.45)}{=} \big(\,[r(q),\,r(t)] + q \rightharpoondown r(t) - t \rightharpoondown r(q),\\ &\quad q \leftharpoondown r(t) - t \leftharpoondown r(q) + [q,\,t]\big) = \big(r(t''),\,t''\big)\end{aligned}$$

for some t', $t'' \in Q$. Hence, we have:

$$\begin{aligned}r(t') &= r(q)r(t) + q \triangleright r(t) + t \triangleright r(q),\\ t' &= q \triangleleft r(t) + t \triangleleft r(q) + qt \qquad (5.52)\\ r(t'') &= [r(q),\,r(t)] + q \rightharpoondown r(t) - t \rightharpoondown r(q),\\ t'' &= q \leftharpoondown r(t) - t \leftharpoondown r(q) + [q,\,t] \qquad (5.53)\end{aligned}$$

By applying r to the second part of (5.52), respectively (5.53), we obtain that r is a deformation map. Moreover, by a straightforward computation using (5.49), (5.50), (5.52) and respectively (5.53) it follows that $b : Q_r \to \overline{Q}$ is a Poisson algebra map and the proof is now finished. □

Examples 5.4.7 *Let k be an algebraically closed field of characteristic zero and let $\mathcal{H}^1_{1,1,1,1}$ be the bicrossed product described in Example 5.4.3. For any $(a_1, a_2) \in k^2$ consider $r_{(a_1, a_2)} : \mathcal{H} \to k_0$ the associated deformation map described in Example 5.4.5. Then, the corresponding $r_{(a_1, a_2)}$-deformation $\mathcal{H}_{r_{(a_1, a_2)}}$ of the Heisenberg Poisson algebra has the associative algebra structure and the Lie bracket given as follows:*

$$\mathcal{H}_{r_{(a_1, a_2)}}: \quad H_1^2 = (2a_1 + 1)\, H_3, \quad H_1\, H_2 = (a_1 + a_2)\, H_3,$$
$$[H_1,\, H_2] = (a_2 - a_1 + 1)\, H_3$$

If $a_1 = a_2 = 0$ then $\mathcal{H}_{r_{(0,0)}}$ coincides with $\mathcal{H}$. Moreover, for any $a \in k - \{2^{-1}\}$ the $r_{(a, -a)}$-deformation $\mathcal{H}_{r_{(a, -a)}}$ is isomorphic to the Heisenberg Poisson algebra $\mathcal{H}$, the isomorphism of Poisson algebras being given by:

$$\varphi : \mathcal{H}_{r_{(a, -a)}} \to \mathcal{H}, \quad \varphi(H_1) = \alpha\, H_1, \ \varphi(H_2) = (-2a + 1)\alpha^{-1}\, H_2, \ \varphi(H_3) = H_3$$

where α is a square root of $(2a + 1)$. However, if we consider $a_1 = 1$ and $a_2 = 0$ then we obtain the Poisson algebra $\mathcal{H}_{r_{(1,0)}}$ with the trivial Lie bracket and the multiplication given by: $H_1^2 = 3\, H_3$, $H_1\, H_2 = H_3$. Therefore, $\mathcal{H}$ and $\mathcal{H}_{r_{(1,0)}}$ are not isomorphic as Poisson algebras as a consequence of not being isomorphic as Lie algebras.

As the previous example shows, different deformation maps can give rise to isomorphic deformations. Therefore, in order to provide the classification of complements we need to introduce the following:

Definition 5.4.8 *Let $(P, Q, \triangleleft, \triangleright, \leftharpoonup, \rightharpoonup)$ be a matched pair of Poisson algebras. Two deformation maps $r, r' : Q \to P$ are called* equivalent *and we will denote this by $r \sim r'$ if there exists $\sigma : Q \to Q$ a linear automorphism of Q such that for any $q, t \in Q$ we have:*

$$\sigma(qt) - \sigma(q)\sigma(t) = \sigma(q) \triangleleft r'\big(\sigma(t)\big) + \sigma(t) \triangleleft r'\big(\sigma(q)\big) - \sigma\big(q \triangleleft r(t)\big) - \sigma\big(t \triangleleft r(q)\big)$$
$$\sigma\big([q, t]\big) - \big[\sigma(q), \sigma(t)\big] = \sigma(q) \leftharpoonup r'\big(\sigma(t)\big) - \sigma(t) \leftharpoonup r'\big(\sigma(q)\big) + \sigma\big(t \leftharpoonup r(q)\big) - \sigma\big(q \leftharpoonup r(t)\big)$$

The main result of this section which provides the answer to the bicrossed descent problem now follows:

Theorem 5.4.9 *Let P be a Poisson subalgebra of R, Q a P-complement of R and $(P, Q, \triangleleft, \triangleright, \leftharpoonup, \rightharpoonup)$ the associated canonical matched pair. Then $\sim$ is an equivalence relation on the set $\mathcal{DM}(P, Q \mid (\triangleleft, \triangleright, \leftharpoonup, \rightharpoonup))$ and the map*

$$\mathcal{HA}^2(P, Q \mid (\triangleleft, \triangleright, \leftharpoonup, \rightharpoonup)) := \mathcal{DM}(P, Q \mid (\triangleleft, \triangleright, \leftharpoonup, \rightharpoonup))/\sim \;\to \mathcal{F}(P, R),$$
$$\overline{r} \mapsto Q_r$$

is a bijection, where $\mathcal{F}(P, R)$ is the set of isomorphism classes of all P-complements of R. In particular, the factorization index of P in R is computed by the formula:

$$[R : P]^f = |\mathcal{HA}^2(P, Q \mid (\triangleleft, \triangleright, \leftharpoonup, \rightharpoonup))|$$

Proof: Two deformation maps r and r' are equivalent in the sense of Definition 5.4.8 if and only if the corresponding Poisson algebras Q_r and $Q_{r'}$ are isomorphic. The conclusion follows by Theorem 5.4.6. □

In what follows, for any $a \in k$, we denote by $\mathcal{H}_a$ the deformation of the Heisenberg Poisson algebra described in Example 5.4.7 associated to the pair $(0, a) \in k^2$. More precisely, the Poisson algebra structure on $\mathcal{H}_a$ is as follows:

$$\mathcal{H}_a: \quad H_1^2 = H_3, \; H_1 H_2 = a\, H_3, \; [H_1, H_2] = (a+1)\, H_3$$

Our next result provides a classification result for the k_0-complements of $\mathcal{H}^1_{1,1,1,1}$ from Example 5.4.3. In particular, the deformations $\mathcal{H}_a$ provide an infinite family of non-isomorphic 3-dimensional Poisson algebras.

Proposition 5.4.10 *Let k be an algebraically closed field of characteristic zero and $a, b \in k - \{-1, -2^{-1}, 0\}$. Then $\mathcal{H}_a$ and $\mathcal{H}_b$ are isomorphic as Poisson algebras if and only if $a = b$ or $a = -b(2b+1)^{-1}$. In particular, the factorization index $[\mathcal{H}^1_{1,1,1,1} : k_0]^f$ is infinite.*

Proof: Suppose $\varphi : \mathcal{H}_a \to \mathcal{H}_b$ is a Poisson algebra isomorphism, where $\varphi(H_1) = \Sigma_{i=1}^3 \alpha_i H_i$, $\varphi(H_2) = \Sigma_{i=1}^3 \beta_i H_i$, $\varphi(H_3) = \Sigma_{i=1}^3 \gamma_i H_i$, $\alpha_i, \beta_i, \gamma_i \in k$. Thus Theorem 5.4.9 implies $r_{(0,\,a)} \sim r_{(0,\,b)}$ and we obtain:

$$\gamma_1 = \gamma_2 = 0, \quad \alpha_1^2 + 2\,\alpha_1\,\alpha_2\, b = \gamma_3, \quad \beta_1^2 + 2\,\beta_1\,\beta_2\, b = 0 \tag{5.54}$$
$$\alpha_1\,\beta_1 + (\alpha_1\,\beta_2 + \alpha_2\,\beta_1)\, b = a\,\gamma_3, \quad (\alpha_1\,\beta_2 - \alpha_2\,\beta_1)\,(b+1) = (a+1)\,\gamma_3 \tag{5.55}$$

To start with, we point out that since φ is an isomorphism we must have $\gamma_3 \neq 0$. The last part of (5.54) implies $\beta_1 = 0$ or $\beta_1 + 2\,\beta_2\, b = 0$. Assume first that $\beta_1 = 0$. As φ is an isomorphism it follows that $\alpha_1 \neq 0$. Then the first part of (5.55) comes down to $\beta_2 = ab^{-1}\,\alpha_1^{-1}\,\gamma_3$. Using the second part of (5.54) we obtain

$$\beta_2 = ab^{-1}\,\alpha_1^{-1}\,\underline{\gamma_3} \overset{(5.54)}{=} ab^{-1}\,\alpha_1^{-1}\,(\alpha_1^2 + 2\,\alpha_1\,\alpha_2\, b) = ab^{-1}\,(\alpha_1 + 2\,\alpha_2\, b)$$

Using the second part of (5.54) and the above formulae for β_2, the last part of (5.55) becomes:

$$\begin{aligned}(a+1)\,\gamma_3 &= \alpha_1\, ab^{-1}\,(\alpha_1 + 2\,\alpha_2\, b)(b+1)\\ &= ab^{-1}\,(\underline{\alpha_1^2 + 2\,\alpha_1\,\alpha_2\, b})(b+1) \overset{(5.54)}{=} ab^{-1}\,\gamma_3\,(b+1)\end{aligned}$$

As $\gamma_3 \neq 0$ we obtain $ab^{-1}(b+1) = a+1$ which implies $a = b$. Assume now that $\beta_1 \neq 0$. Therefore, by the last part of (5.54) we get $\beta_1 = -2\,\beta_2\,b$. Now using the formulae for β_1, (5.55) becomes:

$$-\,\beta_2\,b(\alpha_1 + 2\,\alpha_2\,b) = a\,\gamma_3, \quad \beta_2\,(\alpha_1 + 2\,\alpha_2\,b) = (a+1)(b+1)^{-1}\,\gamma_3$$

As $\gamma_3 \neq 0$, we obtain $-a = b(a+1)(b+1)^{-1}$ which gives $a = -b(2b+1)^{-1}$. Therefore we proved that $\mathcal{H}_a$ and $\mathcal{H}_b$ are isomorphic Poisson algebras if and only if $a = b$ or $a = -b(2b+1)^{-1}$. Together with the fact that algebraically closed fields are infinite, we obtain that $[\mathcal{H}^1_{1,1,1} : k_0]^f$ is infinite. This finishes the proof. □

Bibliographical Notes

This chapter is based on the author's paper [24].

Bibliography

[1] Abrams, L. The quantum Euler class and the quantum cohomology of the Grassmannians, *Is. J. Math.* **117** (2000), 335–352.

[2] Adem, A., Milgram, R. J. *Cohomology of finite groups*, Berlin: Springer, 2nd Edition, 2004.

[3] Agaoka, Y. An algorithm to determine the isomorphism classes of 4-dimensional complex Lie algebras, *Linear Algebra Appl.* **345** (2002), 85–118.

[4] Agore, A.L. Categorical constructions, braidings on monoidal categories and bicrossed products of Hopf algebras, VUB Press, 199 + xx pages, ISBN 978-90-5718-212-9; PhD Thesis, Vrije Universiteit Brussel, 2012.

[5] Agore, A.L. Crossed product of Hopf algebras, *Comm. Algebra* **40** (2013), 2519–2542.

[6] Agore, A.L. Classifying complements for associative algebras, *Linear Algebra Appl.* **446** (2014), 345–355.

[7] Agore, A.L., Bontea, C.G., Militaru, G. Classifying coalgebra split extensions of Hopf algebras, *J. Algebra Appl.* **12** (2013), 1250227, 24 pages.

[8] Agore, A.L., Bontea, C.G., Militaru, G. Classifying bicrossed products of Hopf algebras, *Algebr. Represent. Theory* **17** (2014), 227–264.

[9] Agore, A.L., Bontea, C.G., Militaru, G. The classification of all crossed products $H_4\#k[C_n]$, *Symmetry Integrability Geom. Methods Appl.* **10** (2014), 049, 12 pages.

[10] Agore, A.L., Caenepeel, S., Militaru, G. Braidings on the category of bimodules, Azumaya algebras and epimorphisms of rings, *Appl. Cat. Structures* **22** (2014), 29–42.

[11] Agore, A.L., Chirvasitu, A., Ion, B., Militaru, G. Bicrossed products for finite groups, *Algebr. Represent. Theory* **12** (2009), 481–488.

[12] Agore, A.L., Militaru, G. Crossed product of groups. Applications, *Arabian J. for Science and Engineering*, **33** (2008), 1–17.

[13] Agore, A.L., Militaru, G. Extending structures II: The quantum version, *J. Algebra* **336** (2011), 321–341.

[14] Agore, A.L., Militaru, G. Schreier type theorems for bicrossed product, *Cent. Eur. J. Math.*, **10** (2012), 722–739.

[15] Agore, A.L., Militaru, G. Classifying complements for Hopf algebras and Lie algebras, *J. Algebra*, **391** (2013), 193–208.

[16] Agore, A.L., Militaru, G. Unified products for Leibniz algebras. Applications, *Linear Algebra Appl.* **439** (2013), 2609–2633.

[17] Agore, A.L., Militaru, G. Unified products and split extensions of Hopf algebras, *Contemporary Math. AMS* **585** (2013), 1–15.

[18] Agore, A.L., Militaru, G. Extending structures I: the level of groups, *Algebr. Represent. Theory* **17** (2014), 831–848.

[19] Agore, A.L., Militaru, G. Extending structures for Lie algebras, *Monatsh. für Mathematik* **174** (2014), 169–193.

[20] Agore, A.L., Militaru, G. Bicrossed products, matched pair deformations and the factorization index for Lie algebras, *Symmetry Integrability Geom. Methods Appl.* **10** (2014), 065, 16 pages.

[21] Agore, A.L., Militaru, G. The global extension problem, crossed products and co-flag non-commutative Poisson algebras, *J. Algebra* **426** (2015), 1–31.

[22] Agore, A.L., Militaru, G. Itô's theorem and metabelian Leibniz algebras, *Linear Multilinear Algebra* **63** (2015), 2187–2199.

[23] Agore, A.L., Militaru, G. Classifying complements for groups. Applications, *Ann. Inst. Fourier* **65** (2015), 1349–1365.

[24] Agore, A.L., Militaru, G. Jacobi and Poisson algebras, *J. Noncommutative Geom.* **9** (2015), 1295–1342.

[25] Agore, A.L., Militaru, G. Extending structures, Galois groups and supersolvable associative algebras, *Monatsh. für Mathematik* **181** (2016), 1–33.

[26] Agore, A.L., Militaru, G. Galois groups and group actions on Lie algebras, *J. Lie Theory* **28** (2018), 1165–1188.

[27] Aguiar, M., Andruskiewitsch, N. Representations of matched pairs of groupoids and applications to weak Hopf algebras. Algebraic structures and their representations, *Contemp. Math.* **376** (2005), 127–173, Amer. Math. Soc., Providence.

[28] Albeverio, S., Omirov, B.A, Rakhimov, I.S. Varieties of nilpotent complex Leibniz algebras of dimension less than five, *Comm. Algebra* **33** (2005) 1575–1585.

[29] Albeverio, S., Ayupov, Sh.A., Omirov, B.A., Khudoyberdiyev, A.Kh. n-Dimensional filiform Leibniz algebras of length $(n-1)$ and their derivations, *J. Algebra* **319** (2008) 2471–2488.

[30] Alekseevsky, D., Michor, P. W., Ruppert, W. Extensions of Lie algebras. Unpublished. ESI Preprint 881. arXiv:math.DG/0005042.

[31] Alekseevsky, D., Michor, P. W., Ruppert, W. Extensions of super Lie algebras, *J. Lie Theory* **15** (2005), 125–134.

[32] Alperin, J.L., Bell, R.R. *Groups and representations*, Springer-Verlag, New York, 1995.

[33] Amberg, B., Franciosi, S., de Giovanni, F. *Products of groups*, Oxford University Press 1992.

[34] Andrada, A., Salamon, S. Complex product structures on Lie algebras, *Forum Mathematicum.* **17** (2005), 261–295.

[35] Andrada, A., Barberis, M.L., Dotti, I.G., Ovando, G.P. Product structures on four dimensional solvable Lie algebras, *Homology, Homotopy Appl.* **7** (2005), 9–37.

[36] Andruskiewitsch, N. Notes on Extensions of Hopf algebras, *Can. J. Math.* **48** (1996), 3–42.

[37] Andruskiewitsch, N., Devoto, J. Extensions of Hopf algebras, *Algebra i Analiz* **7** (1995), 22–61.

[38] Andruskiewitsch, N., Natale, S. Examples of self-dual Hopf algebras, *J. Math. Sci. Univ. Tokyo* **6** (1999), 181–215.

[39] Andruskiewitsch, N., Schneider, H.J. Lifting of quantum linear spaces and pointed Hopf algebras of order p^3, *J. Algebra* **209** (1998), 658–691.

[40] Arad, Z., Fisman, E. On finite factorizable groups, *J. Algebra* **86**(1984), 522–548.

[41] Ardizzoni, A., Beattie, M., Menini, C. Cocycle deformations for Hopf algebras with a coalgebra projection, *J. Algebra* **324** (2010), 673–705.

[42] Ardizzoni, A., Menini, C., Stefan, D. A monoidal approach to splitting morphisms of bialgebras, *Trans. AMS* **359** (2007), 991–1044.

[43] Bahturin, Y. and Kegel, O. H. Universal sums of abelian subalgebras, *Comm. Algebra* **23** (1995), 2975–2990.

[44] Bai, C. Double constructions of Frobenius algebras, Connes cocycles and their duality, *J. Noncommu. Geom.* **4** (2010), 475–530.

[45] Barnes, D. On Levi's theorem for Leibniz algebras, *Bull. Australian Math. Soc.* **86** (2012), 184–185.

[46] Barnes, D.W., Newell, M.L. Some theorems on saturated homomorphs of soluble Lie algebras, *Math. Z.* **115** (1970), 179–187.

[47] Bavula V.V. The groups of automorphisms of the Witt W_n and Virasoro Lie algebras, *Czech. Math. J.* **66** (2016), 1129–1141.

[48] Benayadi, S. Structure of perfect Lie algebras without center and outer derivation, *Annales de la faculte de Science de Toulouse* **5** (1996), 203–231.

[49] Besche, H.U., Eick, B. and OBrien, E.A. A millennium project: Constructing small groups, *Internat. J. Algebra Comput.* **12** (2002), 623–644.

[50] Blattner, R.J., Montgomery, S. Crossed products and Galois extensions of Hopf algebras, *Pacific J. Math.* **137** (1989) 37–53.

[51] Bloh, A.M. On a generalization of the concept of Lie algebra, *Dokl. Akad. Nauk SSSR* **165** (1965), 471–473.

[52] Böhm, G., Nill, F., Szlachányi, K. Weak Hopf algebras. I. Integral theory and C*-structure, *J. Algebra* **221** (1999), 385–438.

[53] Borel, A., Mostow, G.D. On semi-simple automorphisms of Lie algebras, *Ann. of Math.* **61**(1955) 389–405.

[54] Božović, V. and Nicola, P. On group factorizations using free mappings, *J. Algebra Appl.* **7** (2008), 647–662.

[55] Bourbaki, N. *Lie groups and Lie algebras*, Chap. 1–3, Springer, Paris, 1989.

[56] Brandt, A. The free Lie ring and Lie representations of the full linear group, *Trans. AMS* **56** (1944) 528–536.

[57] Bryant, R.M., Stohr, R. On the module structure of free Lie algebras, *Trans. AMS* **352** (2000) 901–934.

[58] T. Brzeziński, T. Crossed products by a coalgebra, *Comm. Algebra* **25** (1997), 3551–3575.

[59] Brzezinski, T., Hajac, P.M. Coalgebra extensions and algebra coextensions of Galois type, *Comm. Algebra* **27** (1999), 1347–1367.

[60] Brzeziński, T., Wisbauer, R. *Corings and comodules*, Cambridge University Press, London Math. Soc. Lect. Note Ser. **309** 2003.

[61] Burde, D. Derived length and nildecomposable Lie algebras, *Scientific Buletin of the "Politehnica" University of Timisoara* (2013), 15–24.

[62] Burde, D., Ceballos, M. Abelian ideals of maximal dimension for solvable Lie algebras, *Journal of Lie Theory* **22** (2012), 741–756.

[63] Byott, N. P. Cleft extensions of Hopf algebras, *J. Algebra* **157** (1993), 405–429.

[64] Byott, N. P. Cleft extensions of Hopf algebras II, *Proc. London Math. Soc.* **67** (1993), 277–304.

[65] Caenepeel, S., Militaru G., Zhu, S. *Frobenius and separable functors for generalized module categories and nonlinear equations*, Lect. Notes Math. **1787** Springer Verlag, Berlin, 2002.

[66] Calderon Martin, A, J., Sanchez Delgado, J. M. On split Leibniz algebras, *Linear Algebra Appl.* **436** (2012), 1651–1663.

[67] Camacho, L.M., Cañetea, E.M., Gómez, J.R. and Redjepov, Sh.B. Leibniz algebras of nilindex $n-3$ with characteristic sequence $(n-3,2,1)$, *Linear Algebra Appl.* **438** (2013), 1832–1851.

[68] Canete, E.M, Khudoyberdiyev, A. Kh. The classification of 4-dimensional Leibniz algebras, arXiv:1301.7665.

[69] Cap, A., Schichl, H., Vanzura, J. On twisted tensor products of algebras, *Comm. Algebra*, **23**(1995), 4701–4735.

[70] Casas, J.M., Corral, N. On universal central extensions of Leibniz algebras, *Comm. Algebra* **37** (2009), 2104–2120.

[71] Casas, J.M., Ladra, M., Omirov, B. A., Karimjanov, I.A. Classification of solvable Leibniz algebras with null-filiform nilradical, *Linear Multilinear Algebra* **61** (2013), 758–774.

[72] Casas, J.M., Insua, M.A., Ladra, M., Ladra S. An algorithm for the classification of 3-dimensional complex Leibniz algebras, *Linear Algebra Appl.* **436** (2012), 3747–3756.

[73] Cayley, A. On the theory of groups, as depending on the symbolic equation $\theta^n = 1$, *Philosophical Magazine*, 4th series, **7** (1854), 40–47.

[74] Ceballos, M., Towers, D. A. On abelian subalgebras and ideals of maximal dimension in supersolvable Lie algebras, *J. Pure Appl. Algebra* **218** (2014), 497–503.

[75] Chekhov, L. and Mazzocco, M. Poisson algebras of block-upper-triangular bilinear forms and braid group action, *Comm. Math. Physics* **322** (2013), 49–71.

[76] Chen, Yongshan and Chen, Yuqun. Gröbner-Shirshov bases for metabelian Lie algebras, *J. Algebra* **358** (2012), 143–161.

[77] Chevalley, C., Eilenberg S. Cohomology theory of Lie groups and Lie algebras, *Trans. Amer. Math. Soc.* **63** (1948), 85–124.

[78] Christodoulakis, T., Papadopoulos, G.O., Dimakis, A. Automorphisms of real four-dimensional Lie algebras and the invariant characterization of homogeneous 4-spaces, *J. Phys. A-Math. Gen.* **36** (2003), 427–441.

[79] Christodoulakis, T., Terzis, P.A. The general solution of Bianchi type III vacuum cosmology, *J. Physics: Conference Series* **68** (2007), 012039.

[80] Cohn, P.M. A remark on the general product of two infinite cyclic groups, *Arch. Math. (Basel)* **7**(1956), 94–99.

[81] Conder, M. D. E., Isaacs, I. M. Derived subgroups of products of an abelian and a cyclic subgroup, *J. London Math. Soc.* **69** (2004), 333–348.

[82] Conway, J.H., Dietrich, H. and O'Brien, E. A. Counting groups: Gnus, moas, and other exotica, *Mathematical Intelligencer*, **30** (2008), 6–15.

[83] Covez, S. The local integration of Leibniz algebras, *Ann. Inst. Fourier* **63**(2013), 11–35.

[84] Cuvier, C. Algèbres de Leibniz: Définitions, propriétés, *Ann. Scient. Ec. Norm. Sup.* **27** (1994), 1–45.

[85] Dascalescu, S., Nastasescu, C., Raianu, S. *Hopf algebras. An introduction*, Marcel Dekker, New York, 2000.

[86] Dangovski, R., Drensky, V., Findik, Ş. Weitzenböck derivations of free metabelian Lie algebras, *Linear Algebra Appl.* **439** (2013), 3279–3296.

[87] Daniyarova, E., Kazachkov I., Remeslennikov, V. Algebraic geometry over free metabelian Lie algebra II: Finite field case, *J. Math. Sci.* **135**(2006), 3311–3326.

[88] de Graaf, W.A. Classification of solvable Lie algebras, *Experimental Math.* **14** (2005), 15–25.

[89] de Graaf, W.A. Classification of 6-dimensional nilpotent Lie algebras over fields of characteristic not 2, *J. Algebra* **309** (2007), 640–653.

[90] del Barcoa, V. J., Ovandob, G. P. Free nilpotent Lie algebras admitting ad-invariant metrics, *J. Algebra* **366** (2012), 205–216.

[91] DeMeyer, F., Ingraham, E. *Separable algebras over commutative rings*, Springer Verlag, Belin 1971.

[92] Demir, I., Misra, K. C., Stitzinger, E. On some structures of Leibniz algebras, arXiv:1307.7672.

[93] Doi, Y., Takeuchi, M. Cleft comodule algebras for a bialgebra, *Comm. Algebra* **14** (1986), 801–818.

[94] Donkin, S., Erdmann, K. Tilting modules, symmetric functions and the module structure of the free Lie algebra, *J. Algebra* **203** (1998) 69–90.

[95] Douglas, J. On finite groups with two independent generators. I,II, III, IV. *Proc. Nat. Acad. Sci. U. S. A.* **37** (1951), 604–610, 677–691, 749–760, 808–813.

[96] Drensky, V., Piacentini Cattaneo G. M. Varieties of metabelian Leibniz algebras, *J. Algebra Appl.* **01**, 31 (2002).

[97] Drinfeld, V. Quantum groups, *Proc. Intern. Congr. Math., Berkeley*, Vol. I (1987), 798–820.

[98] Elduque, A., Kochetov, M. *Gradings on simple Lie algebras Mathematical Surveys and Monographs AMS* **189** (2013).

[99] Erdmann, K., Wildon, M.J. *Introduction to Lie algebras*, Springer, 2006.

[100] Everett, C. J., Jr. An extension theory for rings, *Amer. J. Math.* **64** (1942), 363–370.

[101] Fernández Vilaboa, J. M., González Rodriguez, R and Rodriguez Raposo, A. B. Partial and unified crossed products are weak crossed products, *Contemporary Math. AMS* **585** (2013), 261–274.

[102] Farnsteiner, R. On the cohomology of associative algebras and Lie algebras, *Proc. Amer. Math. Soc.* **99** (1987), 415–420.

[103] Fialowski, A., Khudoyberdiyev, A. Kh., Omirov, B. A. A characterization of nilpotent Leibniz algebras, *Algebr. Represent. Theor.* **16** (2013), 1489–1505.

[104] Fialowski, M., Mandal, A. Leibniz algebra deformations of a Lie algebra, *J. Math. Physics* **49** (2008), 093512, 10 pages.

[105] Fialowski, A., Penkava, M. The moduli space of 3-dimensional associative algebras, *Comm. Algebra* **37** (2009), 3666–3685.

[106] Fialowski, A., Penkava, M., Phillipson, M. Deformations of complex 3-dimensional associative algebras, *J. Gen. Lie Theory Appl.* **5** (2011), 110102, 22 pages.

[107] Figueroa-O'Farrill, J. M., Stanciu, S. On the structure of symmetric self-dual Lie algebras, *J. Math. Phys.* **37** (1996), 4121–4134.

[108] Fisher, D.J., Gray, R. J., Hydon, P. E. Automorphisms of real Lie algebras of dimension five or less, *Journal of Physics A: Mathematical and General* **46** (2013), 225204.

[109] Fisman, E. On the product of two finite solvable groups. *J. Algebra* **80** (1983), 517-536.

[110] Fisman, E., Arad, Z. A proof of Szep's conjecture on nonsimplicity of certain finite groups, *J. Algebra* **108** (1987), 340–354.

[111] Flanigan, F. J. On the ideal and radical embedding of algebras, I. Extreme embeddings, *J. Algebra* **50** (1978), 153–174.

[112] Galitski, L. Yu., Timashev, D. A. On classification of metabelian Lie algebras, *J. Lie Theory* **9** (1999), 125–156.

[113] Garcia, G. A., Vay, C. Hopf algebras of dimension 16, *Algebr. Represent. Theory* **13** (2010), 383–405.

[114] Gauger, M. On the classification of metabelian Lie algebras, *Trans. Amer. Math. Soc.* **179** (1973), 293–329.

[115] Gavarini, F. The crystal duality principle: From Hopf algebras to geometrical symmetries, *J. Algebra* **285** (2005), 399–437.

[116] Gentchev, Ts. R. Factorizations of the sporadic simple groups, *Arch. Math. (Basel)* **47**(1986), 97–102.

[117] Giudici, M. Factorisations of sporadic simple groups, *J. Algebra* **304** (2006), 311–323.

[118] Gorbatsevich, V. V. On the level of some solvable Lie algebras, *Siberian Math. J.* **39** (1998), 872–883.

[119] Gorbatsevich, V. V. On some basic properties of Leibniz algebras, arXiv:1302.3345.

[120] Gorenstein, D., Herstein, I. N. On the structure of certain factorizable groups. II. *Proc. Amer. Math. Soc.* **11** (1960), 214–219.

[121] Grabowski, J. Abstract Jacobi and Poisson structures, *J. Geom. Phys.* **9** (1992), 45–73.

[122] Grabowski, J. Marmo, G., Perelomov, A. M. Poisson structures: Towards a classification, *Modern Phys. Lett. A* **8** (1993), 1719–1733.

[123] Grabowski, J., Marmo, G. The graded Jacobi algebras and (co)homology, *J. Phys. A* **36** (2003), 161–181.

[124] Grabowski, J. Brackets, *Int. J. Geom. Methods Mod. Phys.*, **10** (2013), 1360001.

[125] Gray, R.J. Automorphisms of Lie algebras, PhD thesis, Univ. of Surrey (2013). Available online at: https://www.surrey.ac.uk/maths/files/pgr/Theses/thesis_RGray.pdf

[126] Hajós, G. Ober einfache und mehrfache Bedeckung des n-dimensionalen Raumes mit einem Wurfelgitter, *Math Z.*, **47**(1941), 427–467.

[127] Hall, M. JR. The position of the radical in an algebra, *Trans. Amer. Math. Soc.* **48** (1940), 381–404.

[128] Han, J., Su, Y. Automorphism groups of Witt algebras, arXiv:1502.01441.

[129] Hochschild, G. Cohomology and representations of associative algebras, *Duke Math. J.* **14**(1947), 921–948.

[130] Hochster, M. Invariant theory of commutative rings, in: *Group actions on rings*, S. Montgomery (ed.), Contemporary Math. AMS **43** (1985), 161–179.

[131] Hofmann, K. H. Lie algebras with subalgebras of codimension one, *Illinois J. Math.* **9** (1965), 636–643.

[132] Hofstetter, I. Extensions of Hopf algebras and their cohomological description, *J. Algebra* **164** (1994), 264–298.

[133] Hölder, O. Bildung zusammengesetzter Gruppen, *Math. Ann.* **46** (1895), 321–422.

[134] Hong, Y. Extending structures for associative conformal algebras, *Linear Multilinear Algebra* **67** (2019), 196–212.

[135] Hong, Y. Extending structures and classifying complements for left-symmetric algebras, *Results Math.* **74** (2019), 32–74.

[136] Hong, Y., Su, Y. Extending structures for Lie conformal algebras, *Algebr. Represent. Theory* **20** (2017), 209–230.

[137] Hu, N., Pei, Y., Liu, D. A cohomological characterization of Leibniz central extensions of Lie algebras, *Proc. Amer. Math. Soc.* **136** (2008), 437–447.

[138] Humphreys, J.E. *Introduction to Lie algebras and representation theory*, Springer, 1972.

[139] Iovanov, M.C. Generalized Frobenius algebras and Hopf algebras, *Canad. J. Math.* **66** (2014), 205–240.

[140] Itô, N. Über das produkt von zwei abelschen gruppen, *Math. Z.* **62** (1955) 400–401.

[141] Jacobson, N. *Lie algebras*, Interscience Publ., New York, London, 1962.

[142] Jara Martinez, P., Lopez Pena, J., Panaite, F. and Van Oystaeyen, F. On iterated twisted tensor products of algebras, *Internat. J. Math.* **19** (2008), 1053–1101.

[143] Jiang, C.P., Meng, D.J., Zhang S.Q. Some complete Lie algebras, *J. Algebra* **186** (1996), 807–817.

[144] Kac, V. Graded Lie algebras and symmetric spaces (Russian), *Functional. Anal. i Prilozen* **2** (1968), 93–94.

[145] Kadison, L. *New examples of frobenius extensions*, Univ. Lect. Series **14** (1999), Amer. Math. Soc., Providence.

[146] Kassel, C. *Quantum groups*, Graduate Texts in Mathematics 155. Springer-Verlag, New York, 1995.

[147] Kath, I., Olbrich, M. Metric Lie algebras with maximal isotropic centre, *Math. Z.* **246** (2004), 23–53.

[148] Khudoyberdiyev., A.Kh., Omirov, B.A. The classification of algebras of level one, *Linear Algebra Appl.* **439** (2013), 3460–3463.

[149] Kinyon, M. K., Weinstein, A. Leibniz algebras, Courant algebroids, and multiplications on reductive homogeneous spaces, *Amer. J. Math.* **123** (2001), 525–550.

[150] Kirillov, A. Local Lie algebras, *Russian Math. Surveys* **31** (1976), 55–75.

[151] Knibbeler, V. Invariants of automorphic Lie algebras, arXiv:1504.03616.

[152] Kobayashi, S., Nomizu, K. *Foundations of differential geometry*, Wiley Classics Library, 1996.

[153] Kochetov, M. Gradings on finite-dimensional simple Lie algebras, *Acta Applicandae Mathematicae* **108** (2009), 101–127.

[154] Kock, J. *Frobenius algebras and 2D topological quantum field theories*, Cambridge University Press, London Math. Soc. **59**, 2003.

[155] Kraft, H., Procesi, C. *Classical invariant theory, a primer*, Available online at: http://jones.math.unibas.ch/ kraft/Papers/KP-Primer.pdf

[156] Kurdiani, R., Pirashvilli, T. A Leibniz algebra structure on the second tensor power, *J. of Lie Theory* **12** (2002), 583–596.

[157] Lambre, T., Zhou, G., Zimmermann, A. The Hochschild cohomology ring of a Frobenius algebra with semisimple Nakayama automorphism is a Batalin-Vilkovisky algebra, arXiv:1405.5325.

[158] Lang, S. *Algebra*, 3rd edition, *Graduate Texts in Mathematics*, Springer (2005).

[159] Laurent-Gengoux, C., Pichereau, A., Vanhaecke, P. *Poisson structures*, Vol. 347, Springer, 2013.

[160] Lecomte, P. Sur la suite exacte canonique asociée à un fibrè principal, *Bul. Soc. Math. France* **13** (1985), 259–271.

[161] Lichnerowicz, A. Le variétés de Jacobi et leurs algèbres de Lie associées, *J. Math. Pures Appl.* **57** (1978), 453–488.

[162] Lie, S. *Theorie der transformationsgruppen*, Vol. 1-3, Leipzig, 1888, 1890, 1893.

[163] Liebeck, M.W., Praeger, C. E., Saxl, J. The maximal factorizations of the finite simple groups and their automorphism groups, *Mem. Amer. Math. Soc.* **86** (1990), no. 432.

[164] Liebeck, M.W., Praeger, C. E., Saxl, J. On factorizations of almost simple groups, *J. Algebra* **185** (1996), 409–419.

[165] Liebeck, M.W., Praeger, C. E., Saxl, J. Regular subgroups of primitive permutation groups, *Mem. Amer. Math. Soc.* **203** (2010), no. 952.

[166] Loday, J.-L. Une version non commutative des algèbres de Lie: les algèbres de Leibniz, *L'Enseignement Math.* **39** (1993), 269–293.

[167] Loday, J.-L. and Pirashvili, T. Universal enveloping algebras of Leibniz algebras and (co)homology, *Math. Ann.* **296** (1993), 139–158.

[168] S. Lombardo, A.V. Mikhailov Reduction Groups and Automorphic Lie Algebras, *Comm. Math. Phys.* **258** (2005), 179–202.

[169] Lu, J., Wang, X., Zhuang, G. Universal enveloping algebras of Poisson Hopf algebras, *J. Algebra* **426** (2015), 92–136.

[170] Lu, J.H., Weinstein, A. Poisson Lie groups, dressing transformations and Bruhat decompositions, *J. Differential Geom.* **31** (1990), 501–526.

[171] Maillet, E. Sur les groupes échangeables et les groupes décomposables. *Bull. Soc. Math. France* **28** (1900), 7–16.

[172] Majid, S. Physics for algebraists: Non-commutative and non-cocommutative Hopf algebras by a bicrossproduct construction, *J. Algebra* **130** (1990), 17–64.

[173] Majid, S. Matched pairs of Lie groups and Hopf algebra bicrossproducts, *Nuclear Physics B, Proc. Supl.* **6** (1989), 422–424.

[174] Majid, S. *Foundations of quantum groups theory*, Cambridge University Press, 1995.

[175] Majid, S., Soibelman, Ya. S. Bicrossproduct structure of the quantum Weyl group, *J. Algebra* **163** (1994), 68–87.

[176] Majid, A. *Lectures on differential Galois theory*, University Lecture Series AMS **7** (1997).

[177] Marle, C. M. On Jacobi Manifolds and Jacobi bundles. In *Symplectic geometry, groupoids, and integrable systems*, MSRI **20** (1991), 227–246, Springer.

[178] Mason, G., Yamskulna, G. Leibniz algebras and Lie algebras, arXiv:1201.5071.

[179] Masuoka, A. *Extensions of Hopf algebras*, Trabajos Mat. 41/99 Fa.M.A.F. (1999).

[180] Masuoka, A. Abelian and non-abelian second cohomologies of quantized enveloping algebras, *J. Algebra* **320** (2008), 1–47.

[181] Mathew, A. The Galois group of a stable homotopy theory, *Adv. Math.* **291** (2016), 403–541.

[182] Mazzola, G. The algebraic and geometric classification of associative algebras of dimension five, *Manuscripta Math.* **27** (1979), 81–101.

[183] Medina, A., Revoy, Ph. Algébres de Lie et produit scalaire invariant, *Ann. Sci. Ècole Norm. Sup.* **18** (1985), 553–561.

[184] Michor, P. W. Knit products of graded Lie algebras and groups, Proceedings of the Winter School on Geometry and Physics (Srn, 1989). *Rend. Circ. Mat. Palermo* (2) Suppl. No. **22** (1990), 171–175.

[185] Militaru, G. The global extension problem, co-flag and metabelian Leibniz algebras, *Linear Multilinear Algebra* **63** (2015), 601–621.

[186] Molnar, R.K. Semi-direct products of Hopf algebras, *J. Algebra* **47** (1977), 29–51.

[187] Montgomery, S. *Hopf algebras and their actions on rings*, vol. 82 of CBMS Regional Conference Series in Mathematics, AMS, Providence, Rhode Island (1993).

[188] S. Montgomery. *Fixed rings of finite automorphism groups of associative rings*, Springer Lecture Notes in Mathematics **818** (1980).

[189] Ni, X., Bai, C. Poisson bialgebras, *J. Math. Phys.* **54** (2013), 023515.

[190] Ocneanu, A. Quantized groups, string algebras and Galois theory for algebras, in: *Operator algebras and applications*, London Math. Soc. Lecture Notes, **136** (1989), 119–172.

[191] Omirov, B.A., Rakhimov, I. S., Turdibaev, R. M. On description of Leibniz algebras corresponding to $\mathfrak{sl}_2$, *Algebr. Represent. Theor.* **16** (2013), 1507–1519.

[192] O. Ore. Structures and group theory. I. *Duke Math. J.* **3** (1937), 149–174.

[193] Ovando, G. Four dimensional symplectic Lie algebras, *Beiträge zur Algebra und Geometrie* **47** (2006), 419–434.

[194] Panaite, F. Equivalent crossed products and cross product bialgebras, *Comm. Algebra* **42** (2014), 1937–1952.

[195] Papistas, A.I. On the module structure of a group action on a Lie algebra, *J. Aust. Math. Soc.* **82** (2007), 237–248.

[196] Passman, D. *Infinite crossed products*, Academic Press, 1989.

[197] Pelc, O. A new family of solvable self-dual Lie algebras, *J. Math. Phys.* **38** (1997), 3832–3840.

[198] Petravchuk, A. P. On the sum of an almost abelian Lie algebra and a Lie algebra finite dimensional over its center, *Ukrain. Math. J.* **51** (1999), 707–715.

[199] Pierce, R.S. *Associative algebras*, Springer-Verlag, 1982.

[200] Peirce, B. Linear associative algebra, *Amer. J. Math.*, **4** (1881), 97–229.

[201] Popovych, R.O., Boyko, V.M., Nesterenko, M.O., Lutfullin, M. W. Realizations of real low-dimensional Lie algebras, *J. Phys. A: Mathematical and General* **36** (2003), 73–37.

[202] Poroshenko, E. N., Timoshenko, E. I. Universal equivalence of partially commutative metabelian Lie algebras, *J. Algebra* **384** (2013), 143–168.

[203] Praeger, C. E., Schneider, C. Factorisations of characteristically simple groups, *J. Algebra* **255** (2002), 198–220.

[204] Radford, D. E. *Hopf algebras*, Series on Knots and Everything **49**, World Scientific Publishing 2012.

[205] Rakhimov, I.S., Hassan, M.A. On isomorphism criteria for Leibniz central extensions of a linear deformation of μ_n, *Internat. J. Algebra Comput.* **21** (2011) 715–729.

[206] Rakhimov, I.S., Husain, S.K. Classification of a subclass of nilpotent Leibniz algebras, *Linear and Multilinear Algebra* **59** (2011) 339–354.

[207] Rakhimov, I.S., Rikhsiboev, I.M., Basri, W. Complete lists of low dimensional complex associative algebras, arXiv:0910.0932v2.

[208] Rédei, L. Zur Theorie der faktorisierbaren Gruppen I, *Acta Math. Acad. Sci. Hung.* **1** (1950), 74–98.

[209] Rognes, J. Galois Extensions of Structured Ring Spectra/Stably Dualizable Groups, *Memoirs AMS* **192** (2008), 137 pages.

[210] Rotman, J.J. *An introduction to the theory of groups*. Fourth edition. Graduate Texts in Mathematics 148, Springer-Verlag, New York, 1995.

[211] Rovi, A. Hopf algebroids associated to Jacobi algebras, *Int. J. Geom. Methods Mod. Phys.* **11** (2014), 1450092.

[212] Schauenburg, P. The structure of Hopf algebras with a weak projection, *Algebr. Represent. Theory* **3** (1999), 187–211.

[213] Schneider, H.-J. Some remarks on exact sequences of quantum groups, *Comm. Algebra* **21** (1993), 3337–3358.

[214] Singer, W. Extension theory for connected Hopf algebras, *J. Algebra* **21** (1972), 1–16.

[215] Study, E. Über Systeme complexer Zahlen und ihre Anwendungen in der Theorie der Transformationsgruppen, *Monatsh. für Mathematik* **1** (1890), 283–354.

[216] Su, Y., Zhub, L. Derivation algebras of centerless perfect Lie algebras are complete, *J. Algebra* **285** (2005), 508–515.

[217] Svobodova, M. Fine gradings of low-rank complex Lie algebras and of their real, *Symmetry Integr. Geom.* **4** (2008), 039, 13 pages.

[218] Sweedler, M.E. Cohomology of algebras over Hopf algebras, *Trans. Amer. Math. Soc.* **133** (1968), 205–239.

[219] Sweedler, M.E. *Hopf algebras*, Benjamin, New York, 1969.

[220] Szechtman, F. Equivalence and normal forms of bilinear forms, *Linear Algebra Appl.* **443** (2014), 245–259.

[221] Szép, J. Über die als Produkt zweier Untergruppen darstellbaren endlichen Gruppen. *Comment. Math. Helv.* **22**(1949), 31–33.

[222] Szép, J., Rédei, L. On factorisable groups, *Acta Univ. Szeged. Sect. Sci. Math.* **13**(1950), 235–238.

[223] Szép, J. Zur Theorie der endlichen einfachen Gruppen, *Acta Sci. Math. Szeged* **14**(1951), 111–112.

[224] Takeuchi, M. Matched pairs of groups and bismash products of Hopf algebras, *Comm. Algebra* **9** (1981), 841–882.

[225] Thrall, R.M. On symmetrized Kronecker powers and the structure of the free Lie ring, *Amer. J. Math.* **64** (1942), 371–388.

[226] Turkowski, P. Solvable Lie algebras of dimension six, *J. Math. Phys.* **31** (1990), 1344–1350.

[227] Vaes, S., Vainerman, L. Extensions of locally compact quantum groups and the bicrossed product construction, *Adv. Math.* **175** (2003), 1–101.

[228] Walls, G. L. Groups which are products of finite simple groups. *Arch. Math. (Basel)* **50** (1988), 1–4.

[229] Wever, F. Über invarianten von lieschen ringen, *Math. Annalen* **120** (1949), 563–580.

[230] Wiegold, J., Williamson, A. G. The factorisation of the alternating and symmetric groups., *Math. Z.* **175** (1980), 171–179.

[231] Zappa, G. Sulla costruzione dei gruppi prodotto di due dati sottogruppi permutabili tra loro, *Atti Secondo Congresso Un. Mat. Ital., Bologna, 1940*, pp. 119–125. Edizioni Cremonense, Rome, 1942.

[232] Zhu, C., Van Oystaeyen, F., Zhang Y. On the (co) homology of Frobenius Poisson algebras, *Journal of K-theory: K-theory and its Applications to Algebra, Geometry, and Topology* **14** (2014), 371–386.

[233] Zusmanovich, P. Central extensions of current algebras, *Trans. Amer. Math. Soc.* **334** (1992), 143–152.

Index